Praise

T0028559

"Why did Americans nearly ... ...n? In *The Bald Eagle*, [Jack E.] Davis, who won a Pulitzer Prize for *The Gulf*, a clever history of 'America's Sea,' has written a double biography: a history of the species and a history of the symbol. . . . *The Bald Eagle* is the rare natural history that plays as a comedy. It's a dark comedy, however, because its lessons are not easily transferable to our broader, ongoing ecological catastrophe. . . . [A] moving portrait of a species victimized for its own evolutionary successes."

—Nathaniel Rich, *Atlantic*

"*The Bald Eagle* is compelling and paints a dignified portrait of the famous bird, within and outside of American culture. The author's occasional playful tone lightens the mood during its darker moments and even helps to underline the hypocrisy of the treatment of this bird of prey, simultaneously esteemed and maligned. . . . This is a history that turns the tables on Americans; the creature that embodied the scrappiness of the early nation is now a model of resilience we can only hope to emulate."   —Olive Fellows, *Christian Science Monitor*

"An impressive work of scholarship . . . [I]f you have any questions about [this] bird, Mr. Davis's *The Bald Eagle* is a great place to look for answers."   —Bill Heavey, *Wall Street Journal*

"As Davis recounts in his engaging and highly detailed cultural and natural history of the unofficial national bird (Davis points out that no president or Congress has ever signed a proclamation or law making it official), Americans, intentionally and unintentionally, devised all sorts of ways to kill off these majestic raptors that some lauded as 'the monarch of the air.' . . . Davis deftly brings alive the bald eagle as a real animal, separate from both the myths of its rapaciousness and the symbolic majesty that at times has made the birds emblems for organizations ranging from the National Rifle Association to the National Wildlife Foundation."   —Matt Jaffe, *San Francisco Chronicle*

THE
BALD
EAGLE

*The Improbable Journey
of America's Bird*

JACK E. DAVIS

Liveright Publishing Corporation

*A Division of W. W. Norton & Company
Celebrating a Century of Independent Publishing*

For information about permission to reproduce selections from this book,
write to Permissions, Liveright Publishing Corporation,
a division of W. W. Norton & Company, Inc.,
500 Fifth Avenue, New York, NY 10110

For information about special discounts for bulk purchases,
please contact W. W. Norton Special Sales at
specialsales@wwnorton.com or 800-233-4830

Manufacturing by Lakeside Book Company
Book design by Lovedog Studio
Production manager: Anna Oler

Library of Congress Cataloging-in-Publication Data

Names: Davis, Jack E., author.
Title: The bald eagle : the improbable journey of America's bird / Jack
E. Davis.
Description: First edition. | New York, NY : Liveright Publishing
Company, [2022] | Includes bibliographical references and index.
Identifiers: LCCN 2021054960 | ISBN 9781631495250 (hardcover) |
ISBN 9781631495267 (epub)
Subjects: LCSH: Bald eagle—United States. | Bald eagle—
Conservation—United States.
Classification: LCC QL696.F32 D375 2022 | DDC 598.9/43—dc23/
eng/20211117
LC record available at https://lccn.loc.gov/2021054960

ISBN 978-1-324-09410-4 pbk.

Liveright Publishing Corporation,
500 Fifth Avenue, New York, N.Y. 10110
www.wwnorton.com

W. W. Norton & Company Ltd.,
15 Carlisle Street, London W1D 3BS

1 2 3 4 5 6 7 8 9 0

*To the memory of Marina Powdermaker
and Geoff Sutton*

In an eagle there is all the wisdom of the world.

—John Fire (Lame Deer),
   in *Lame Deer, Seeker of Visions*

CONTENTS

Part Three
NEW SCIENCE AND NEW ATTITUDES

Part Four
RESTORATION

THE BALD EAGLE

PROLOGUE

Standing at the rocky edge of a stream, a bald eagle scans its surroundings with luminous eyes. Satisfied, it unfurls its wings straight up, drops into a crouch, and, in one fluid motion, pushes skyward, sweeping its seven-foot wingspan out and down. The eagle rises, slowly at first, legs and feet hanging like pendants. Its powerful wings beat hard against the air, and it continues to rise. Limbs draw up for the journey ahead.

Nesting season has come to its inevitable end, and migration has begun. As fall slips toward winter, bald eagles across western Alaska have been called to a distant place, a place they pursue annually throughout their lifetimes—as did their ancestors in their lifetimes. Their autumnal exodus from seasonal territories proceeds not in a mass or in crowded waves. There are no herding calls sounded to a group, no flocks gathering for takeoff, no wedges of flight piercing the sky. Instead, each bird—even the young ones not long from the nest—follows its own impulse to leave, and over a period of weeks, thousands of eagles set out individually on a solitary journey duplicated by multitudes.

Seeking a southwesterly course, wind in their faces, sun on their backs, they eventually roll out over the Aleutian Islands. Each island in the trackless sea below is an ancient link in a migration route mapped in the eagles' evolutionary memory. Tracing it from several hundred to several thousand feet up, they continue to fly solo, loosely spaced a half mile or more apart in streams twenty to thirty miles long, soaring on favorable winds where they exist and pumping wings providently where they don't.

Each passing day grows stingier with light, granting barely six hours between the dim interludes of dawn and dusk. Well before sunset, the journeying birds descend to some remembered resting place en route. They fish for renourishment and then settle down in trees or on rock ledges for the night. The next morning, they fish again and then lift above the dewy haze one by one to push on. If the weather is clear, these daylight fliers will travel a hundred or more miles before another night of rest.

A favored stopover is the island of Amaknak, a four-day journey from their mainland territories, if they don't linger. Amaknak is also the final destination and winter residence for many of the eagles. As it comes into sight, they wheel toward its green and rocky hills.

On descent, primary flight feathers splay and twist; tail feathers pitch upward and downward. Horizontal wings dance on fickle air currents. Heads dip forward, and keen eyes pick out landing spots as each nimble bird floats in. Legs reach down and toes spread to meet the upward-surging ground in a near balletic landing. Wings close as a final bow.

⌒

MIGRATING FROM ONE PART of Alaska to another doesn't seem like much of a winter retreat, but bald eagles can tolerate cold weather. What they cannot abide is thick ice that prevents them from catching fish. The coastal waters around Amaknak don't freeze solid—a fact well known to them. Not only are fish accessible, but so, too, are tens of thousands of wintering waterfowl, easily pluckable fare. If the winter prospects are plentiful a few hundred miles away, why should an eagle from the hyperborean climes of Alaska bother to fly as far south as Washington or Oregon or, as some mystifyingly do, all the way to New Mexico?

The migrators that arrive on Amaknak join hundreds of balds that reside there year-round. Food that is abundant in the winter is even more so in warmer months. On a map, the Aleutian chain looks like a trail of breadcrumbs that falls away from the mainland in a southwestward arc. The islands create an archipelago divide between the

Bering Sea and the North Pacific Ocean. The convergence of these two waters stimulates a gushing wellspring of marine life that draws to Amaknak some forty million spring- and summertime nesting seabirds—shearwaters, kittiwakes, fulmars, petrels, cormorants, and albatrosses. Balds nest alongside them.

At 3.3 square miles, Amaknak is one of the larger of the breadcrumbs. On its north side looking out to the Bering Sea is Iliuliuk Bay, which was once an undisturbed natural habitat of seals and sea otters and the pre-European home of Unangan, the indigenous people of the region. By the late eighteenth century, the island had become an outpost of Russian fur traders, who called the Natives "Aleut." They also named the bay "Dutch Harbor," yet there is no reliable evidence that the Dutch ever settled their namesake water. Perhaps the Russians first heard about Amaknak Island from Netherlander whalers or seal and sea lion hunters who would slip into the protected bay to trade with the Native peoples and take on fresh provisions or escape storm-ravaged seas.

Dutch Harbor is wrapped in the protective embrace of conical hills that range from a few hundred to more than a thousand feet in height. They are the kind of hills you might imagine on a volcanic island, which Amaknak is. Unpopulated by humans, they are treeless yet green with vegetation when not white with snow. In the summer the green spreads like a lush quilt embroidered with red salmonberries, purple orchids, yellow-green honeysuckles, and chocolate-brown chocolate lilies. Exposed rock in the hills adds shades of gray, brown, blackish brown, and rust.

There is plenty of rust of the iron oxide kind around the island too, where the air is perpetually wet and salty. Up in the hills, rust is found on the metal hardware of old gun batteries from World War II (the Japanese bombed Dutch Harbor twice), but most of it is down around the harbor. It's on trucks, cars, chain-link fences, and manhole covers. It's rubbed to a burnish on handrails, door handles, and steel stair treads. It's on loading cranes, shipping containers, and crab traps. It's on the winches, booms, anchors, and bulkheads of fishing trawlers. It bleeds down the sides of their deckhouses and hulls.

For as long as Dutch Harbor has existed in name, it has been a commercial fishing port, and a phenomenally productive one at that. In 2005 the Discovery Channel began filming the reality television series *Deadliest Catch* at Dutch Harbor, following snow- and king-crab fishing boats and crews out into the cold and often turbulent Bering Sea. In 2020 the network's most popular reality show filmed its sixteenth season, and Dutch Harbor finished the year as the top commercial fishing port in the United States for the twenty-fourth consecutive year.*

The bald eagles may be blind to the film cameras, but Dutch Harbor's commercial success doesn't elude them. The 700–800 million pounds of fish and shellfish hauled into port are another call to the eagles' winter migration. Like pelicans and gulls, balds will eagerly take a free offering coming their way.

The green and rocky hills overlooking the bay make eminent perches for these fishing raptors, although not necessarily preferred ones. Wild though they may be, the eagles come down from the hills like upland villagers to the market, situating themselves amid the clutter of the island's human population. When *Deadliest Catch* first arrived, their number hovered around six hundred to the human population's roughly forty-five hundred. Today, hardly a rooftop, streetlamp, fence rail, or satellite dish has not been visited by one of these feathered fishers. Eagles even congregate at the historic Russian Orthodox church and wrap their toes and talons around the crosses atop its two steeples, positioning themselves closer to the higher power than the worshippers inside do. A most coveted high seat is a towering lattice-boom cargo crane used for loading and unloading the trawlers' giant fishing nets. When the boom is idle, it is rare that the top end isn't hosting a white-headed lookout. From this vertiginous pinnacle, an eagle can survey the activity on the fishing docks from one end of the island to the other. Like crows on a wire, they line up on the gunwales of trawlers and the gables of fish houses, baldly eyeing every catch. Dozens will wait and watch as boats offload cornucopias of the sea, looking

---

* The ranking is based on poundage and value of catch combined.

for fish to spill or, perchance, a friendly crew member to toss one their way. Gangs of industrious eagles can be seen cleaning out remnants in drying fishing nets, undisturbed by boat crews. The net cleaning is the least of their brazen acts.

In the frost-heaved parking lot of the Safeway grocery, a dozen eagles once descended on a box of freshly caught pollock in the bed of a Nissan pickup. A local woman posted a video online of the impromptu jubilee. "What did you guys find in that truck?" she asks playfully as her smartphone camera zooms in, wholly ignored by the partakers. The video went viral. That one and others the woman posted began attracting nearly a million viewers a month on YouTube. Some of the most popular of these videos feature an eagle pair that regularly visited her front porch and competed with her territorial cat Gizmo for prime perching spots on the railing.

Bald eagles are so numerous and common that direct encounters with locals became inevitable. One once descended on a sixteen-year-old to pluck a slice of cheese pizza out of his hand. A pair with a nest at the post office regularly dive-bombed customers. If they were just about any other bird, they would likely have been banished for their mischief. But calling the eagles "Dutch Harbor pigeons" is about as far as locals come to decrying their presence.

Dutch Harborites more or less acknowledge the right of the wild birds to live beside them. Before the US Fish and Wildlife Service relocated the nest from the post office, customers cheerfully donned hardhats. The sixteen-year-old carried his scars from the pizza raptor proudly, and the owner of the Nissan pickup attributed his loss to carelessness. Fishers appreciated the net cleaning, and the cantankerous Gizmo sometimes surrendered his spot on the porch railing after trading stares with the feathered intruders.

Fifty and more years earlier, these ubiquitous avian residents in Dutch Harbor would have received a hostile reception. Commercial fishers viewed them not with a grin and a chuckle but with a grunt and a sneer. Proclaiming eagles as nuisance wildlife that were stealing fish and biting into profits, they shot, netted, and clubbed their supposed competition. Down in the lower forty-eight states, eagles endured the same punishments for similar purported crimes. It didn't matter that this so-called nuisance bird was a symbol of America.

⌒

DESPITE THE HOSTILITY, THE bald eagle has been associated with higher principles and better attributes since 1782, when Congress made it the central figure on the Great Seal of the United States. Representing fidelity, self-reliance, strength, and courage, the founding bird quickly attained a vaunted perch in America's iconography. Its visage appeared on the US capitol dome and pediments, hard and paper currency, business and sports-team logos, coat buttons and cuff links, and the 101st Airborne's sleeve patch. It was a hood ornament, door knocker, money clip, and chest tattoo—the stuff of trinkets and gewgaws.

Yet no animal in American history, certainly no avian one, has to the same extreme been the simultaneous object of reverence and recrimination. For centuries, eagles risked their lives flying across American skies.

People familiar with the history of the bald eagle generally know that in the mid-twentieth century, DDT nearly eliminated the species in the contiguous US. Fewer know that before then the bald eagle's existence was gravely threatened by something as equally toxic as that lethal chemical pesticide: myths, lies, and insensitivity. False accusations fostered a cross-eyed vision of a morally depraved predator and thief, and degenerate scavenger—a bird more corrupt than the pilfering garden crow. In the American mind, the species served no real benefit beyond presiding as the embodiment of the national emblem and other insignia.

We can attribute the deterioration of conditions in the late twentieth century to collateral damage from the reckless use of a pesticide. Earlier, however, depredation had been deliberate. For more than a century before DDT's commercial introduction, Americans prosecuted a fierce coast-to-coast assault directly against their flagship bird. The aggression was nothing short of premeditated and acceptable—legally and socially. Anyone anywhere in the position to do so squeezed the trigger and carried out the bird's execution—respectable, hardworking, churchgoing people who thought they were doing no more harm than pulling up an annoying dandelion.

Yet the bald eagle's prominence as a cultural and commercial symbol was never greater. No fine calculus seemed to exist between myth and reality. This invented dichotomy was the wildlife counterpart to the culturally contrived noble and savage Indian and the happy darkie and black-beast rapist. Americans treasured the bald eagle's image even as they detested its living complement. Reverence kept company with scorn. Respect with disrespect. Idolization with vilification. Immortality coincided with murder. Before Euramericans, when North America was still a land of indigenous peoples, the bird of freedom that would lose its freedom encountered no such stormy contradictions. It flew as freely as the wind.

Cresting in the second half of the nineteenth century, the violence perpetrated against bald eagles compared with the onslaught that the American bison and passenger pigeon suffered in the same era. The sagas of the latter two animals are well known, recounted in a passel of books written since the eleventh-hour rescue of the prairie-storming bovine and the extinction of the sky-darkening migrator. The account of the destructive malice that tormented the founding bird, by contrast, lies among the missing pages of a prescribed history.

The recorded memory of nations tends to be less comprehensive than selective, clotted with romanticism that obscures truth. Yet allowing the truth to die rather than preserving it can also silence the heroism in laudable acts. The good can be lost with the bad. In the case of bald eagles, history has failed to give the successful effort to stop the slaughter its fair due.[1]

Saving bald eagles from oppressive indignities and ultimate destruction is a story worth expanding on and celebrating. It's a story with two parts that elevates shining moments in the American experience.

All this is to say that twice, not once, the United States nearly lost its flagship bird from the wild, and twice people aided its return. The first time, the Bald Eagle Protection Act of 1940 sought to stop the century-long bloodshed and establish a peaceful accord between

Americans and the persecuted species. Unfortunately, reconciliation quickly coincided with the emergence of another troubling conflict, brought on by the introduction of DDT for everyday use. The chemical pesticide was supposed to improve the quality of human life. Yet whatever social benefits DDT delivered, they were rapidly undermined by the damage to human and wildlife habitation, most noticeable in the genocidal blitz against wild bird populations.

For their own part in the rescue efforts, bald eagles managed to extricate themselves from past assaults and exhibit a will to carry on, a will that we might view in heroic terms. While policymakers and wildlife experts advanced an agenda for restoring the population of this imperiled species, the birds pursued an evolutionary instinct for self-preservation. Humans did not alone determine their destiny; bald eagles plotted out their own story as natural beings.

THE PAGES OF HISTORY you have in hand offer a comprehensive account of a singular avian species and its relationship with a nation and its people. The book's nine chapters with vignettes—discrete introductory stories that precede each chapter and pivot the larger narrative—in essence constitute a biography of both the living bird and the symbol. Looking at the two as one reveals that the place of these hallmark birds in the American pantheon runs deeper than mere national icon and opens out to more than conflict and reconciliation. At the nation's founding, bald eagles were only one defining element of the rarefied natural environment of North America, an environment that constituted the wellspring of an original national identity. Within a century after adoption of the Great Seal, restless Americans and new immigrants broke out across the continent and into the modern age of manufacturing and consumption. Inevitably, nature as resource eclipsed nature as cultural identity. America's attachment to its organic foundation, a source of its exceptionalism among other nations, would soon be forgotten.

In the 1960s, a shift in consciousness rescued the country from the onset of unsparing ecological disaster. The incentive for change, manifested in the modern environmental movement and a spate of farsighted environmental laws, grew out of many factors, not least of which was the near complete disappearance of what was by then a beloved species. Its decimation and inspiring revival helped raise awareness of early connections made between American exceptionalism and nature. The charismatic bird became more than a patriotic totem. It evolved into a measure of how well Americans were advancing a healthier intercourse with nature, a place where people, not just wildlife, lived.

The Bald Eagle ultimately concerns itself with connections—humans to eagles specifically, and to nature generally—and how such connections weakened and strengthened throughout our history. The book is as much about values too, seemingly opposing ones—patriotism and environmentalism—yet in the American historical context they are complementary at their core. The bald eagle's magnificence reflected values that Americans associated with themselves and their country. Letting the hallmark bird die out, ending the migration flights along the Aleutian Islands and across the lower forty-eight states, risked betrayal to the country's national and natural heritage. Losing the species would have been tantamount to cowardice. Setting up the conditions that facilitated its revival was an act of courage, one that turned failure into triumph, and self-recrimination into virtue.

In few places, then, are patriotism and environmentalism so clearly complementary and connected as in the life history of *Haliaeetus leucocephalus*.

INTRODUCTION
Haliaeetus leucocephalus:
The Species

IN SOUTHERN TEXAS, A SHOUT AWAY FROM MEXICO, DOWN
near the echoey bottom of Seminole Canyon, some unknown person of an earlier age painted an image of an eagle on a limestone wall.
The Paleo artist likely used a brush made from animal hair and paint
from pulverized red ochre, a natural pigment of iron oxide and clay
extracted from the area's siltstone. Hundreds of other paintings, which
may be as old as eight thousand years, appear on the canyon's limestone as craggy shadows of a civilization hidden in the muted depths
of time, a civilization that passed briefly through the canyon's solemn
geological existence.

This hard and harsh hollow in the earth has existed through the
ages frequently devoid of the spoken word and human travail. Well
before creative hands met limestone walls, the canyon was habitat
to reptiles, insects, and mammals. Avian species, including circling
birds of prey, possessed the canyon too. Twenty miles away and still
in this hill country of "rock rubble and thin soil," as a Texas geologist
described it, is another canyon. On a cliff above, a pair of eagles nested
and fledged their young over many seasons in the late nineteenth century. Their domestic ritual inspired a name, Eagle's Nest Canyon,* cre-

---

* Today, the canyon's official name is Mile Canyon, but many still call it Eagle's
Nest Canyon or Canyon of the Eagles.

ating a link between the ancient and the modern, a Paleo artist and white settlers of a more recent time. The latter raised a cliffside town that survived for a short period, and named it Eagle's Nest. A nearby stream became known as Eagle's Nest Creek, and a land bridge, Eagle's Nest Crossing.[1]

The high ridges, deep gorges, and streams of South Texas make suitable habitat for the two eagle species that live in North America: the golden and the bald. Goldens favor mountains and rock ledges for nest locations. They feed mostly on birds and terrestrial vertebrates. The high-top outcroppings around the two canyons put goldens in strategic positions to look down at prey slithering and skittering around the rock rubble and to take flying prey out of the sky.

While golden eagles may have lent their identity to the name of the town, canyon, and crossing, bald eagles more closely fit Eagle's Nest Creek. The environment around the two canyons is dry, but it is not without come-hither water. Both canyons spill rock to the edge of the Rio Grande, and lying between them and flowing into the grander water is the Pecos River. Rivers equal fish equal bald eagles. Balds eat what goldens eat, but their tastes run to fish—saltwater and fresh. Seizing prey from land, water, and sky, balds are the rare airborne species that feeds at all three tables. They are birds of prey, meaning that they hunt live animals for food, and they are raptors, which are birds of prey with hooked beaks and large talons bearing fatally sharp ends. Hawks, falcons, vultures, owls, kites, caracaras, and ospreys are all raptors. More so than the others, except vultures and perhaps caracaras, balds also scavenge.

Either species, the golden or the bald, could have been the model for the eagle painted on Seminole Canyon's limestone. Maybe both were. The Paleo artist portrayed the eagle in flight overhead with outstretched wings. Its tail feathers are long, its head is turned to the left. Goldens and balds are similar in body size and among the largest of the world's sixty-eight species of eagles. The adult bald weighs about the same as the golden but stands slightly taller and has a longer wingspan, up to eight feet. It has white head and tail feathers and a yellow beak, to the golden's dark tail and head and grayish-black beak. Both have yellow toes and dagger talons. None of these highlights are evident in the red-ochre eagle of Seminole Canyon. But one important difference between golden and bald eagles makes a case for the painted one being a bald: the legs. Those in the painting are devoid of feathers close to the toes. The golden's legs and feet are covered down to the toes. "Booted," this feathering is called, whereas the bald's legs and feet are "unbooted," making them more practical for dipping into the water for fish.

This anatomical trait is common with the *Haliaeetus* genus, known as the sea eagle, to which the bald belongs. In Greek, *hals* means "salt" or "sea," and *aetos* means "eagle." The bald's species name, *leucocephalus*, distinguishes the bird from other *Haliaeetus* members and means "white-headed." So, the two words together, *Haliaeetus leucocephalus*, the bald's scientific name, translate to "white-headed sea eagle."

Nine other species join balds in the *Haliaeetus* genus: the African, Madagascan, Pallas's, Steller's, Sanford's, white-bellied, white-tailed, gray-headed, and lesser sea eagles. The others are scattered across the globe and live everywhere except in South America and Antarctica. On occasion, sea eagles from Asia summer in the Aleutian Islands. Goldens, which belong to the *Aquila* genus, range across the Northern Hemisphere into Europe and Asia.

Balds are endemic to North America; they live nowhere else in the wild. Outside of a few renegades, they are continental homebodies. The nineteenth-century naturalist John Burroughs, an essayist in the Thoreauvian tradition, spoke eloquently of the bald eagle's range. He wrote, "[The] continent is his home. I never look upon one without emotion; I follow him with my eye as long as I can. I think of Canada, of the Great Lakes, of the Rocky Mountains, of the wild and sounding sea-coast. The waters are his, and the woods and the inaccessible cliffs. He pierces behind the veil of the storm, and his joy is height and depth and vast spaces."[2]

Burroughs also called the bald a "bird of large ideas, he embraces long distances." Now and then, one or two of these North Americans will stray to Central America or the Virgin Islands. In 1973, a most hearty one, perhaps blown by a storm, flew to southern Ireland, only to be shot down like enemy aircraft by an unscrupulous marksman. Fourteen years later, another truant, a juvenile, made the same inconceivable journey. No one shot this one. Wildlife rangers netted the famished stray and replenished it with venison. Then, as if to remind it that it was native to another continent, they accompanied it back across the Atlantic, in a plane. If the two soloist birds followed the historic 1927 route of Charles Lindbergh in the *Spirit of St. Louis* from Newfoundland to Ireland, which is likely, they would have winged nearly two thousand miles.[3]

Descendants of kites, eagles have been around for more than thirty million years. Fossil evidence of bald eagles dates to at least a million years ago. They are represented by two existing subspecies, distinguished by their regional preferences and physical size. *H. l. washingtoniensis* generally abides in Canada, Alaska, and the northern continental states, and is larger than its southern cousin, *H. l. leucocephalus*, which nests in the southern US, the Baja California peninsula, and northern Mexico. Between nesting seasons, each cousin migrates into the region of the other—a behavior that has prompted many avian authorities to disregard the subspecies divisions altogether.

Current prevailing estimates put the continent-wide bald eagle population before America became America at between 250,000 and 500,000. The higher number is likely more representative of the actual population. When we consider Native Americans and the native bird together before European contact, it is clear that the bald eagle's population was not only extremely abundant but widespread.

As the people of those timeless canyonlands on the Rio Grande knew bald eagles, so too did the various mound builders around the Gulf of Mexico, as well as the Yamasee and Guale of the Southeast coast; the Powhatan of the Potomac and Chesapeake; the Wampanoag and Mohegan of the Northeast; the Naskapi Innu farther north; the Ojibwa of the Upper Peninsula; the Lakota and Dakota of the Black Hills region; the Kiowa of the High Plains and Rocky Mountains; the Inuit of the farthest North; the Aleut of the Aleutian Islands; the Salish around Puget Sound; the Chumash of the Southwest Coast; the Akwa'ala of the Baja Peninsula; the Pueblo peoples of the Southwest; and innumerable others living up and down long and short rivers, around big and small lakes, and everywhere seaside.

The Spanish, who in the 1560s settled northeastern *La Florida*, where the indigenous Timucua had lived for thousands of years, knew bald eagles too.* But the first non-Native artist known to compose an image of *Haliaeetus leucocephalus* on American soil was an Englishman. The image exists as an etching and is part of the collection of John White, a naturalist and artist who in 1585 sailed with a group of colonists to Roanoke Island, in present-day North Carolina. Roanoke was supposed to be England's first North American colony, and White its governor. The soil, claimed a contemporary observer, was the "most plentifull, sweete, fruitfull and wholesome of all the worlde." The colony nevertheless failed, and the settlers mysteriously disappeared. All that survived was the haunt of unhappy memories, an abbreviated scientific report by one of White's colleagues, and seventy-seven drawings.[4]

The artwork is now housed at the British Museum in London and depicts birds, fish, reptiles, and indigenous people. Twenty-seven images of avifauna have survived with the collection. Some art experts speculate that whereas White painted the fish and the reptiles, a servant named Christopher Kellett, a limning painter (one who illustrates for manuscripts), composed the birds. Someone, likely one of the two, wrote at the top of the bald eagle image, "Nahyapuw. The Grype, almost as bigg as an Eagle." *Nahyapuw* is apparently the name for the eagle in Carolina Algonquin, the extinct language of the indigenous people of the area. *Grype* would be the English word "gripe," which is the white-tailed *Haliaeetus* of the British Isles.[5]

The Roanoke eagle has a white tail and head, and is unmistakably *Haliaeetus leucocephalus*. It has full possession of its perch, the stubby remains of a tree that, in its bumps and contortions and reticulated bark, suggests a subdued reptilian form. Oversize yellow feet and curling black talons heighten that allusion. The Roanoke eagle is no carica-

---

* The French settlement in northeastern Florida, at Fort Caroline, was short-lived after the Spanish forced out their rivals.

ture though. Its carriage is erect, and its features sharp. It is handsome, dignified, unyielding, unassailable, and eloquent.

With their population sweeping from one end of the continent to the other, bald eagles had the power to mesmerize and inspire. Their animating qualities pierced the worldviews of Native peoples and Euramericans alike, and both fell under the raptors' charismatic spell. As they had Native cultures, bald eagles lifted the gaze of the Europeans who came and settled after the Roanoke colonists and began calling themselves Americans. When they formed an independent republic, they put the captivating bird's image not on canyon rock but on just about everything else, and they made the bald eagle their national bird.

Or did they?

Don't-tread-on-me stare. (Preston Cook Collection)

Part One

A BIRD FOR A NEW NATION

ONE

◆ ───── *Two Myths* ───── ◆

THE UNITED STATES ADOPTED THE BISON AS ITS NATIONAL mammal in 2016. Before that, it made the oak the national tree and the rose the national flower. But the United States has never done what every state of the union has done: adopt a designated feathered representative. The national bird is, *officially*, nonexistent. Every reader right now is thinking or saying out loud—shouting even—*What?! Of course the US has a national bird. It's the bald eagle!* Even US government websites make that claim. But the government is, well, wrong.

The position for national bird remains vacant. No president or Congress has ever signed a proclamation or passed a law to fill it. Ever since Congress affixed the bald eagle on the Great Seal of the United States in 1782, the white-headed raptor has falsely basked in the position of national bird. Yet tomorrow, the day after, or next week the president could, with a small group of vested onlookers gathered around, put an executive pen to an executive order to grant the honor of national bird to the cardinal, oriole, jay, or, yes, sidewalk pigeon.

Benjamin Franklin wanted the turkey displayed on the Great Seal and as a national symbol. Everybody knows that too. Or did he? Any American familiar with the least bit of bald eagle trivia seems to know about Franklin's legendary displeasure over the bald eagle's reigning status and his desire to appoint the turkey in its place. The basis for this alleged preference lies in his taking exception to the obnoxious penchant of bald eagles for stealing fish from ospreys. He believed that the turkey, which would never commit such a degenerate crime, was, by comparison, a model citizen of the avian community. The truth is that

Franklin's desire for the turkey ranks with the best of fictional fable, alongside young George Washington's cherry-tree honesty and John Henry's steel-driving tenacity.

The tale of Franklin and the birds dates to the early nineteenth century, after he shared his views of the eagle and turkey in a 2,129-word letter to his daughter, Sarah Bache. He devoted the majority of those words to expressing his disapproval of the Society of the Cincinnati. The society was founded in 1783, a year after the adoption of the Great Seal, to preserve the ideals, values, and memory of the French and Continental army officers who served in the Revolutionary War. Only those officers—not common soldiers—and their firstborn male descendants could join the society. Washington was a founding member, as was Alexander Hamilton, Aaron Burr, James Monroe, Henry Knox, John Paul Jones, Light Horse Harry Lee, even King Louis XVI. Critics, who included John Adams and Thomas Jefferson, accused the society of validating a hereditary nobility common to Europe but foreign to the democratic ideals of the US. Franklin derided Cincinnati as an "Order of Knights" formed in "direct Opposition to the solemnly declared Sense of their Country." Induction came with a medal, a consolation for members who, Franklin maintained, were envious of the "ribbands and crosses they have seen hanging to the button holes of foreign officers."[1]

The centerpiece of the medal, and suspended from a blue ribbon edged in white, was a bronze bald eagle. Initially, many of the American veterans invited to join the organization rejected the society. Then, Major Pierre Charles L'Enfant, a French military engineer who would go on to design Washington, DC, arrived from France with a "bundle of eagles." As Thomas Jefferson told the story, the eagle medals "at once, produced an entire revolution of sentiment" in favor of joining the society.[2]

Franklin was on a tear when he mentioned the organization's insignia in the letter to his daughter. His ranting about democratic values and elitism pivoted him into a summary assessment of the bird on the Great Seal. "For my own part I wish the bald eagle had not been chosen as the representative of our country. He is a bird of bad moral character" who "does not get his living honestly." The species that purportedly suffered the bald's crimes the most was the industrious osprey, a.k.a. the "fish hawk." "Perched on some dead tree," the eagle "watches the labour of the fishing hawk; and when that diligent bird has at length taken a fish, and is bearing it to his nest for the support of his mate and young ones, the bald eagle pursues him, and takes it from him." Franklin did not mince words. In his estimation, the bald was a lazy, thieving breed, a false noble, and "rank coward." To drive home his point, he drew a comparison with the turkey, a "little vain and silly tis true" but "much more respectable" and full of "courage." He also appreciated the turkey as "peculiar to" the United States, while eagles "have been found in all countries."[3]

He was, of course, wrong about the bald eagle's range, and that of the turkey too, which lives throughout much of the Western Hemisphere. Whether the turkey is a plucky bird, as Franklin maintained, is arguable. It is certainly shrewd, having endured among Americans, who consider it most suitable for the serving plate. But it isn't an exceptionally dignified-looking animal, as Franklin admitted: a "little vain and silly tis true." From skinny neck up, it's all jiggle and shrivel framing round-eyed befuddlement. The gobbling doesn't help. Its pear-shaped body suggests docility, the opposite of the bald eagle's broad-shoulder mettle (some heraldic depictions in Europe showed eagles posing with wings open in the manner of a strong man flexing his biceps). The turkey is a decent sprinter, yet anything but a flying ace. It's a close-to-ground aviator with an aerial range of about a quarter mile, although its usual is from ground to tree branch—where it perches at night in fear of predators. The bald fears none, and it is capable of divine heights, in line with the ambitions of a rising nation.

No one knows whether Franklin's commentary was even sincere. You could never tell with him. The lover of a good joke—his especially—he confessed elsewhere that he was characteristically given to "Prattling, Punning and Joking." One can add counterfeit. He disguised his hugely popular *Poor Richard's Almanac* as the musings of an unprosperous farmer named Richard Saunders. Saunders was a fiction, and the almanac was parody and fable from Franklin's pen and printing press. The "best satires" that came off that press, says historian Jill Lepore, were "relentlessly scathing social and political commentary attacking tyranny, injustice, ignorance," the same sort of writings with which he flailed the Society of the Cincinnati. His views of less serious matters, such as the temperament of a bird, could change for literary convenience. *Poor Richard's* once described the bald eagle not as cowardly but as "fierce." Franklin was America's Mark Twain before Mark Twain. Even in letters, writes Lepore, "he couldn't always govern his wit."[4]

Nor could he control his duplicity. On July 4, 1789, the year in which the US Constitution went into force, the Society of the Cincinnati elected Franklin an honorary member, an honor he accepted. Nor did he object when the French referred to Franklin the foreign diplomat as the "Eagle of the West." In his experiments generating electricity using Leyden jars, he once wired up a turkey, "about ten pounds weight," to see whether electrocution would kill it. It did. He was later quoted as saying, "Birds killed in this manner eat uncommonly tender." Was he joking?[5]

As for the letter, Franklin more than likely composed his complaint as an opinion piece masquerading as a private missive that he intended to leak to the public. Politics was not a subject he discussed with his daughter—at least, not in correspondence—and he never sent the letter in question to her. He never apparently leaked it either. The public got wind of it after his grandson William Temple Franklin published a collection of his late elder's writings. That was twenty-seven years after Franklin's death. By then, Franklin's opinion would not have swayed anybody. The bald eagle as a symbol was immensely popular within the government and among the people.[6]

It is no surprise, then, that the potshots Franklin took at the bald eagle created something of a chuckling sensation. His criticisms and caprice, however, are only a small part of the story in establishing—discovering, actually—a founding bird. It is true that he objected to the bald eagle's exalted cultural status. It is equally true that he never expressly said in a letter or anywhere else that he preferred the turkey as a national symbol. As it happens, he did not want the turkey; he wanted Moses. Yes, the biblical Moses.

SEARCHING FOR
A SEAL

A T THE FAR SIDE OF SUMMER AND THE REVOLUTIONARY War, in 1782, a messenger headed north out of Philadelphia on horseback. In his satchel, inked on rolled-up parchment, was an authorization for a prisoner exchange with the British. Eight years of war was reaching its end, with the messenger riding toward it carrying something more than a pro forma document. For several days, his horse took him along rutted cart roads and narrow Indian traces, through green and shade clinging to the waning season, before reaching Newburgh, New York, where he dismounted at a local militiaman's farmhouse. Built of granite fieldstone as solid and tight as a fortress, and affording a favorable lookout over the Hudson River, the stalwart dwelling served as headquarters for the commander of the Continental army. It was an authentic George-Washington-slept-here place. On the river below, bald eagles nested from winter through spring and fished year-round.

Unrolled, the authorization at first glance had the look of any of the hundreds of official documents that the Continental Congress had sent to Washington over the years. Toward the upper left corner of the buff parchment were the familiar signatures of the secretary and president of Congress, Charles Thomson and John Hanson. Embossed above their names was an image that no one outside of Congress had ever seen on an official document. It was the new Great Seal of the United States. At its center was the raised profile of a handsome bald eagle, with a shield on its chest, arrows and olive branch in its talons, and an "*E pluribus unum*" scroll unfurled and clenched in its beak. The impulse to run his fingers over the embossed form would likely have struck even someone as reserved as "our trusty and well beloved George Washington," as the document addressed him.[7]

A nation's seal, in the simplest of terms, is its government ID badge and credentials card, verifying an official authentication of state. In English law and European tradition, no agreement, charter, or grant was *factum* until sealed. A country partaking in international affairs without a seal was like a graduate without a diploma, a practicing lawyer without a license, or John Hancock without his signature. It was like Ben Franklin reading in his rocking chair, as was his wont, fully disrobed. There was a certain degree of undress to it all.

Congress had issued continental currency; authorized an army, navy, and post office; executed articles of war; and fought that war nearly to completion—all without a national signature in the form of a seal. Legitimizing the burgeoning republic's standing in global affairs alongside patronizing, powder-wigged nations called for, at minimum, a national flag and seal. The Americans flew their first flag, the Continental Colours, on New Year's Day 1775 and put up a new rendition, with thirteen red and white horizontal stripes and a blue canton with thirteen white stars, in September 1777. An official seal was still five years away.

The long-awaited inaugural appearance of the seal marked a rite of passage for the young sovereign republic and the sovereign creature on the seal's front. Placing a native bird there had merit. Indigenous peoples and Euramericans regarded the bald eagle as the king of avian species. In both size and bearing, it conveys intrepidness and integrity. Next to the California condor, it is the second-largest American raptor, weighing as much as fourteen pounds and measuring up to thirty-seven inches from head to talon. Add to that a white corona and a dignified, somewhat aloof countenance that favors a lord-of-the-manor aspect, and you see nobility, a bird you're not inclined to argue with.

These traits—sovereignty and nobility—are, of course, human projections. As did Franklin, we anthropomorphize the bald eagle because doing so helps us relate to the winged breed and its world, and to decide where both fit with ours. Yet birds do not organize their world as we do ours, from the lowly to the kingly. Contrary to what storybooks and

Disney animators might have us believe, beyond humans the animal kingdom has no curtseying or genuflecting, no tributes paid or jesters afoot. Still, similarities between humans and birds exist.

The typical day in the life of most wild animals revolves around two elemental preoccupations: finding food and avoiding becoming someone else's. In that context, wild animals recognize who has power over whom, and the preyed-upon react how we might in the presence of a potential assailant: by choosing between fight or flight.

When the dingo is on the prowl, for example, the pugilist kangaroo does not stop to put up its dukes. It flees. In the face of danger, opossums play possum, while chickens often do the same when raptors like bald eagles are circling. The adult bald is an apex predator perched on the uppermost link of the food chain, a senior associate of the highest trophic level, a gourmand of others and *plat du jour* of none. But its power does not go uncontested. Jays will swoop in to mob an eagle that enters their breeding territory, just as they do house cats snooping about bird feeders. Crows can be similarly unrelenting, and it is tantalizing to think that they would welcome living up to the animal-group name we ascribe to them: "murder." Small birds don't always back off either. Franklin once observed that the "little king bird no bigger than a sparrow attacks [the eagle] boldly and drives him out of the district." Kingbird challenging the king of birds.[8]

In the human realm, angry mobs are an intrinsic testimony to the supremacy of rulers, carnivorous raptors in their own way. Pharaohs, caesars, sultans, shahs, tsars, rajas, caciques, kings, and queens have, throughout their existence, dealt with uprisings. Sovereigns have also regarded various species of eagles as fellow monarchs of another domain. Many adopted them as an official bird and used their impressive image on coats of arms and national seals—in heraldry, as it is called.

The founders of a great democratic republic, the United States, did the same. They had been disgruntled subjects of a monarchy, men who abhorred crowns and thrones and imperial power. Yet they also recognized that their infant nation emphasizing rule by the people (specifically, property-owning white men) required some level of investment in conventional officialdom.

The bald eagle did not at first stand out as an obvious choice for a national symbol. Nothing did. The endeavor to design a seal that appealed to congressional members was not as arduous as winning the war, but it was trying in its own way and went on for nearly as long. Like battlefield defeats preceding ultimate victory, several failed attempts to draft a suitable design came first, involving three successive congressional committees, as well as nine congressional delegates, one congressional official, three artists, and one consultant—fourteen men in all. Of all those men and committees, no one envisioned a winsome eagle at the seal's physical and metaphoric center until two relatively obscure historical figures—William Barton, a twenty-eight-year-old underemployed lawyer moonlighting as an artist, and Charles Thomson, the fifty-two-year-old secretary of the Continental Congress—finally saw traits in the avian native that were worthy of emulation. Important foundational inspiration had passed down to them too, from one Pierre Eugène du Simitière, an immigrant who was eager to apply his creative talents to shaping the new nation.

CONGRESS SET IN MOTION the quest for a seal after endorsing what would become the most lauded document in US history, the Declaration of Independence, on the most celebrated day, July 4, 1776, and in the country's most historic city, Philadelphia.

At the time, more than thirty thousand people lived in Philadelphia, most of them along a gridiron of streets that English Quakers had laid out nearly a hundred years earlier, nearly a hundred miles upriver from the Atlantic Ocean. Despite its distance from the sea, Philadelphia emerged as America's leading maritime city. With timbered hardwood floated down from northern forests, its shipbuilding eclipsed even that of Boston, and its shipment of goods exceeded that of both Boston and New York. Philadelphia had the first successful lending library and first medical school, which, along with the College of Philadelphia and the American Philosophical Society, nourished a surplus of scholarly and scientific minds. As beneficiaries of Pennsylvania's commitment to religious freedom, Anglicans, Catholics, Quakers, Mennonites, Pietists, and Jews delivered spiritual guidance throughout the city. Philadelphia also had efficient postal routes, cobbled streets, whale-oil streetlamps, and a big bell cast in tin and copper—not yet permanently cracked and named the Liberty Bell.

Throughout the war with England, the Continental Congress met in Pennsylvania's statehouse, redbrick and twenty-three years sturdy except for the rotting wooden steeple above its clock tower, where that big copper-and-tin bell hung. The statehouse was showing its age but wore a new name. It had been called the capitol building of the Pennsylvania colony until two days earlier, July 2. Delegates had voted for independence that day, formally shedding the label of "colony" for that of "state," indeed declaring thirteen principal colonies to be states united.

With some fifty delegates going about their work, the statehouse day and night was awash with breeches, white stockings, doublets, a cassock or two, and tricorne hats. Hats removed and placed to the side revealed the bare scalp of John Adams and the wiry auburn mane of Thomas Jefferson. Powdered wigs were few.* A common accessory of the British elite, they were expensive, but they had also fallen out of favor with many of the patriots. Refusing to wear one was a political statement, a symbolic rallying cry for union.

Two days after the vote for independence, on the calendar date that the nation would proclaim as its day of birth, Thomas Jefferson presented his draft of the "declaration on independence" for Congress's scrutiny. The delegates picked up quills and undertook some considerable editing. They deleted phrases, changed words, and streamlined leggy sentences. Jefferson's "a people who mean to be free" tightened into "a free people." His "unremitting injuries" became "repeated injuries." His "sacred & undeniable truths" resolved into "self-evident truths." Franklin, unwigged, a drapery of hair falling from his speckled dome, was responsible for that one. Jefferson cringed. He called the changes "mutilations."[9]

---

* There were exceptions. Give-me-liberty-or-give-me-death Patrick Henry was known to wear a wig, at least in the early days of Congress.

He nevertheless voted with the rest of the membership in favor of the Declaration's adoption. Congress then ordered its preservation, with editorial changes, on parchment. The new government was without a national seal, but Congress intended to have one by the time the Declaration was ready for the delegates' signatures within the next weeks. To accomplish the adoption of a seal, they followed convention: they created a committee. As the day's last order of business, likely completed after dinner, they confirmed the appointments of the seal committee's members: Franklin, Jefferson, and Adams. Charles Thomson recorded their names in his secretary's ledger. In a letter to Maryland delegate Samuel Chase, Adams implied that independence itself hung on the adoption of a seal. "As soon as an American Seal is prepared," he wrote, "I conjecture the Declaration will be subscribed by all the Members" of Congress.[10]

On July 8, the day before Adams wrote Chase, bells, including the one atop the statehouse, tolled across the city. On a platform outside, John Nixon, a colonel and future general in the Continental army, gave the first public reading of the Declaration of Independence. United by rebellious passion, several from the gathered crowd stormed the statehouse, pulled down King George's coat of arms from a wall in the courtroom, and took it away several blocks to the city common, where they put it to flames.

The Franklin-Jefferson-Adams seal committee was seven weeks away from proposing a replacement to hang over the ghost mark left by the king's coat of arms.

OFTEN CALLED THE "GREAT seal" and sometimes assigned as a national coat of arms, a country's seal professes allegiance, authority, and membership among nation-states. Armorial "bearings"—the general term for seals and coats of arms—date to the ancient civilizations of Mesopotamia, Egypt, Greece, and Rome. The earliest devices were cut in stone and other hard materials and formed in a cylindrical shape so that the imprint could be rolled into clay. Eventually, the devices made their way into ivory, gemstones, and metal and led to the development of the signet ring, which a person of position wore. An earl, let's say, when signing an important personal or business document, would seal the deal by pressing his ring in wax or rolling its inked face on paper.

Before nation-states had seals, monarchs and lords had them. Popes, cardinals, and bishops had them. So, too, did barristers, merchants, and prominent families, as well as surveyors, mapmakers, and engineers. Rarely without were corporations, guilds, fraternal organizations, sporting clubs, schools, and castles. Cities, towns, counties, parishes, and states still have them, and every US state and territory hangs its seal on government walls. Americans know their seals as round, but seals can be square, rectangular, triangular, elliptical, or shield-shaped. The seal of the territory of Guam is the outline of an eye turned vertical. Some seals are the shape of the featured object itself, such as Japan's imperial seal of a very nonimperialistic yellow chrysanthemum.

Seals are, indeed, works of art. They are an illustrative profile of the possessor, intended to impart a smart, positive, and memorable impression that celebrates greatness or outstanding attributes. The classic model is found in the seals of older European nations and families. At their center, the devices display a patterned shield with a supporter on each side, and often include a crest, helmet, or crown. A must is a motto conveying strength, achievements, honors, aspirations, and, if a personal seal, noble bloodlines. Some designs are allegorical, proclaiming distinctive geographic features. In Guam's vertical eye, for instance, a coconut palm lingers on the shore of a peaceful blue bay. Or a design might interpret a vernacular history. The seal of Kansas looks the way a state history textbook could read. It's a visual narrative, a diorama in miniature, extolling conquest, expansion, pioneer life, and commerce—to wit, propounding a story of progress.

Animals are common figures on national coats of arms—much more so than human figures. The shield on the royal coat of arms of the United Kingdom has a crowned lion supporting the left side and a white unicorn the right, both standing on their hind legs—a position in heraldic language known as "rampant." Lions are quite popular in vintage coats of arms, which also favor hounds, stags, and dragons. Newer nations that have broken free from the tentacles of empire have been inclined to look to homegrown species. Australia has a lion and a shrike (a bird), neither of which lives in Australia, yet also a kangaroo and emu, both of which do. Jamaica has a crocodile. The Solomon Islands have a crocodile too, along with a shark. Guyana took up with a pair of jaguars, Botswana a couple of zebras, the Congo a duo of elephants, Gabon a twosome of panthers, and Namibia a brace of oryxes (a type of antelope).

Birds turn up frequently. They range from the dove (Cyprus) to the flamingo (the Bahamas) to the condor (Chile, Bolivia, and Ecuador). Eagles outnumber all other birds by far. They've made the roster of political and military emblems since Mesopotamia. In early Greek mythology the eagle was the messenger of Zeus. Arguably, the most recognizable and historic standard (a flagless staff) is that of the Roman legion. It and its finial eagle (*aquila* in Latin) date to the late tenth century and the Holy Roman emperor Otto III, king of Italy and Germany. In excited public assemblies, citizens expressed their sentiments by thrusting *aquila* standards toward heaven. In Rome, the Romans did as Natives of North America did: they regarded the eagle as a divine messenger. The *aquila*, writes the literary scholar Janine Rogers, "is the ancestor of eagle emblems" throughout much of the Old World. Its image migrated with conquest across the land and seas. Charlemagne had an eagle. Napoleon had one. Sometimes eagleless conquerors encountered preexisting eagle emblems—in the Middle East and North Africa, for example. In the twelfth century, the personal emblem of Saladin, sultan of Egypt and Syria, was likely an eagle. Today the Saladin eagle is a symbol of Pan-Arabism, appearing on the coats of arms of Egypt, Palestine, Jordan, Yemen, and Iraq.[11]

There was no need for the *aquila* or Saladin eagle to emigrate to the Americas; Native peoples had long embraced the symbolism of native animals. The great aboriginal city-state of Tenochtitlán was built on an island in a lake where, according to Aztec legend, the founders spotted an eagle atop a flowering cactus, eating a smaller bird. That eagle of legend, a golden, is now on the Mexican national flag.

In some cases, the eagle depicted on official insignia is ornithologically correct, such as the African fish eagle on Zambia's national flag and South Sudan's seal. Panama's coat of arms bears a harpy eagle, Indonesia's a Javan hawk-eagle. More often, though, emblem eagles are a stylized nonspecies. Perhaps the best recognized is Germany's eagle, first introduced by Charlemagne and given a complex and, as a fascist symbol also adopted by Italy, controversial history. Germany has featured its bird as both single- and double-headed. The wings have almost always been displayed spread and the talons open, accompanied by a menacing expression. With a rigid pose and uniform straight lines depicting open wings, standing atop a wreathed swastika, the eagle of the Third Reich conveyed a defiantly intractable nature. On the current coat of arms, adopted in 1950, the German eagle is black with red beak, tongue, and talons against a gold shield—a descendant of the Roman *aquila*.

Some have argued that the American heraldic eagle falls within Roman lineage. Influences of ancient Rome existed at the time of the nation's founding in architecture, political ideals, and governmental structure. They were important to the design of the United States seal too, but the eagle on its front did not come down from the ancients. It is all American.

WITH FRANKLIN CHAIRING, THE seal committee began by searching for ideas and inspiration. Essential to developing the badge of a nation was a familiarity with heraldry, the art and science of seals and coats of arms, which was prominent in Europe and had been around since the ancients. Of the fifty-six signatories of the Declaration, some thirty-four had coats of arms. Heraldry was around and about the rebellious land, yet it occupied an ambiguous place in Anglo-American society. It suggested distinguished heritage but was not necessary to conferring social status. Although it did not raise doubt about one's patriotism, it could, like powdered wigs, come across as too British. The wigless Adams, Jefferson, and Franklin were keen visionaries of the new nation, well-read in the classics, and culturally sophisticated. Presumably, they could bring the needed knowledge to creating a proper seal.

In truth, the committee members were working with little more than a basic familiarity with heraldry. Each had a family coat of arms, although Adams was the only one who had come by his legitimately, which is to say, passed down from ancestors. Jefferson and, apparently, Franklin had purchased their coats of arms, just as, Jefferson once said, one might purchase "any other coat."[12]

Virtually everyone who has written about the committee has said that its smartest move was to retain an artist who knew something about heraldry. The one chosen for the task was Pierre Eugène du Simitière, a Swiss-born naturalized citizen who had come to Philadelphia by way of the West Indies and New York. During his professional life, Du Simitière would complete countless portraits of famous men, leaving valuable images for history, yet history has not privileged the artist with the same of himself. One nineteenth-century chronicler of the seal's journey referred to him as a "little French West Indian." In actuality, we don't know whether Du Simitière was of mixed lineage or pure European stock, whether he was dark- or light-skinned or some shade between, whether he was tall or short, bushy or bald. He was likely slender. Thirty-eight years old when he answered the committee's call, he was the archetypal starving artist, which was a somewhat bewildering condition when placed beside the social and professional connections he cultivated.[13]

Adams knew Du Simitière before the committee hired him. A bachelor and something of a shipwrecked soul, he had assembled a sizable collection in natural history that he kept at his rented quarters and occasionally opened to private viewing. The collection made him known to many of the intellectual lights of the city. Adams was a sometime visitor of the collection who regarded its curator as a "very curious Man." In March 1776, after Continental forces secured Boston from the British, Congress wanted to honor Washington—powder-haired but not powder-wigged—in some way. Adams asked Du Simitière to develop an image of Washington for a gold medal to award the general. Although Congress rejected the design, Adams considered Du Simitière's work to be "well executed" and "very ingenious."[14]

Du Simitière also possessed a professional's familiarity with heraldry. While living in the West Indies, he had developed a fascination with the art of flags, coinage, and coats of arms. He filled notebooks with drawings and watercolors of various designs, honing skills in preparation of a specialty. Even before settling in Philadelphia in 1774, he took an entrepreneurial interest in the growing discord between Britain and the colonies. Each new state, he reasoned, would need a coat of arms, as would the budding nation. To be invited to join the committee was surely a summit he had dreamed of scaling.[15]

Du Simitière and the venerable three congressional delegates each worked up their own ideas for a seal—the artist with sketches and the others in writing. Evident in their conceptions is the influence of the most common literature of the day, classical works and the Bible. Adams suggested duplicating *The Judgement of Hercules*, by Simon Gribelin, a French-born engraver who spent his adult life and career in England. A young Hercules stands at the center of a fruitful, paradisiacal setting flanked by female personifications of Virtue and Vice. Scantily clad and coquettish, Vice beckons Hercules on a flowery path of worldly pleasures, while Virtue points knowingly to the narrow and winding uphill road to duty and honor, the presumed path for America.

Allegory was as well the notion of "Dr. F," as Adams referred to Franklin in a letter home to his wife, Abigail. Seventy years old, round in the middle, Adams's senior by three decades, Jefferson's by nearly four, Franklin proposed Moses confronting Pharaoh (wearing a crown, an apparent allusion to King George III) charioting across the Red Sea. A religious parable had been advanced by this "thorough Deist," as Dr. F called himself. Although he persistently rebelled against his Puritan upbringing, Franklin was a complicated man of complex philosophy. He "jettisoned virtually all theological beliefs," writes one of his biographers, while embracing faith as a guide to "disciplined, benevolent success in this life." In Franklin's rendition for the Great Seal, Moses stands beneath a pillar of fire in the sky and closes around Pharaoh and his army the waters that Moses had parted for the Israelites. Franklin added a motto: "Rebellion to Tyrants is Obedience to God."[16]

Jefferson, deist consort, turned to the same pages in the Bible. He envisioned the children of Israel in the wilderness, their journey guided by a cloud overhead during the day and a pillar of fire in the sky at night. He also recommended devising a medallion-style seal with two sides. On the reverse, he conjured Hengist and Horsa, the brothers of fifth-century legend who led the Germanic invasion of Britain.

If Jefferson's latter image had been realized, this symbol of Anglo-Saxonism would most certainly have proved controversial, later if not sooner. For one thing, King George employed German missionaries to fight the American people. For another, the American people were a confluence of bloodlines from around the world. Cities, the epicenters of revolt, were robustly pluralistic, with Protestants and Catholics and Jews and Muslims from northern, southern, and eastern Europe, and all around the Mediterranean. Manhattan was a Babel of nearly two dozen languages sounding on the streets and in the shops, to say nothing of the mother tongues of the city's twenty thousand Africans and West Indians, representing one-third of its population, most of them enslaved, all of them subjected to oppressive black codes. Nationwide, enslaved Blacks, along with a minuscule number of free Blacks, accounted for nearly twenty percent of the total population.

Du Simitière, the "little French West Indian," pulled out drawing paper and pencil to try to capture something of this diversity. At the center of his concept for a seal, he placed a shield with a supporter on each side, imitating the concept that was common to European heraldry. To the left of the shield, he drew the Goddess of Liberty and to the right a buckskin-clad, musket-bearing American soldier. Within the shield, Du Simitière drew a smaller shield and around it even smaller ones of the thirteen states—a design known as a "roll of arms." Within the inner shield he penciled in rough renditions of the seals of England, Scotland, Ireland, Germany, France, and Holland—the "six principal nations of Europe," as Adams put it, "from whom Americans have originated." If Congress had adopted the design, the US seal would have also included a rose, a lion, and fleur-de-lis. There would have been a tiny pitch-black eagle too, borrowed from Hapsburg Germany's seal, on which it had been a standard since the thirteenth century.[17]

The committee reported to Congress on August 20. Nearly all the delegates had signed the Declaration of Independence eighteen days earlier, quashing Adams's speculation, shared with Samuel Chase, that the adoption of a seal would give stimulus to the delegates' pens. The committee brought forward Du Simitière's concept for the seal's face, calling it the "Arms of the United States of America." One principal change came with it: the buckskin soldier had been discharged from his duties and replaced by the Goddess of Justice, bearing a sword and the scales of justice. Franklin's encounter between Moses and Pharaoh made the reverse side.

The would-be artisans of American symbolism had merged traditional heraldic symbols with biblical allegory, acknowledging European lineage while extolling the prevailing belief among Americans that they, like the Israelites, were a chosen people. The concept had some logic to it, the execution not so much. The visual was too busy, crowded with imagery, which was evidently one reason for its failure to excite fellow delegates. Congress, Secretary Thomson noted for the record, ordered the proposal "to lie on the table." And there it died.[18]

Du Simitière would eventually fare little better than the committee's proposal, despite a series of fruitful possibilities. For a while, he gave drawing lessons to Jefferson's oldest daughter, Martha. Using his private collection, he opened the first natural history museum in the Northeast, and he became an elected member of the American Philosophical Society and recipient of an honorary degree from the College of New Jersey (later renamed Princeton). The states of Georgia, Delaware, and New Jersey commissioned seals from him, none of which, incidentally, he gave an eagle, although New Jersey got a horse and Delaware a cow.

Which brings to mind creamware. In 1779, at the urging of congressional delegate John Jay, George Washington visited Du Simitière in his home to sit for a portrait. Du Simitière's is the first-known bust portrait of the founder, yet virtually unknown. The portrait was popular in England, particularly as a motif on creamware—lead-glazed bowls, cups, and pitchers, cream colored and cream filled. But in the States, the portrait passed with little notice. There was a moment of recognition in 1791 when Du Simitière's Washington appeared on the face of a privately struck one-cent piece. But Washington never assented to his leonine profile appearing on the coin, which had a short life in circulation. On its reverse side was the Great Seal of the United States that Congress had authorized nine years earlier, and at its center was the bald eagle, although Du Simitière was not the one who conceived the eagle for the seal.

So many qualifiers—"yet," "but," and "although"—marked Du Simitière's life, along with one big "might have": the lucky break that might have delivered him from poverty never materialized. He lived not ten more years after serving his commission with the committee—another victim of the short life expectancy of that triumphant but disease-ridden age. He died a month past his forty-seventh birthday, destitute.

Yet, he lived long enough to see three of his ideas immortalized on the seal that finally made the grade, and to see the exaltation of that regal bird.

THE REGAL EAGLE IS of our imagination. "A bird," observed the writer Polly Redford in a perceptive book on raccoons and eagles published in 1963, "is not noble any more than a dog is faithful, a pig greedy, a donkey stubborn, or a fox sly." And a bald eagle is not bald either, not in the way of a turkey, wood stork, vulture, or emu, and not like Franklin or Adams.[19]

The person who first called North America's sea eagle a bald eagle was more than likely a man or woman in seventeenth-century British America, when people were surrounded by an array of wild and domesticated animals. Etymologists believe that this person took "bald" from "piebald." The descriptive "piebald" peaked in common parlance in the early twentieth century and dates to the sixteenth. It originated from the name of a black-and-white songbird common across the British Isles and Europe, the magpie. "Piebald" refers to dapples of white against a dark color, combining usually black and white or brown and white, as in the magpie or a piebald horse. The adult bald eagle's contrasts are blackish brown and white. The immature eagle is nearly chocolate in color, including its head, and patchy all over, defiantly patchy, with no pure-white outer feathers.

It is possible that piebald was not the etymon for this American eagle that looked like no eagle back in the British Isles. Perhaps the naming person was reaching not to the "pie" in piebald but to its root, "bald." Bald is a more bare-bones description that commonly denotes a spot, splotch, or blaze of white. *Cheffyl bal*, a Welsh phrase, refers to a horse with a white mark on its face. Well into the nineteenth century, many people, including John James Audubon, called the bald eagle the "white-headed eagle."

"Bald" also means, of course, being destitute of feathers or hair; and, yes, his contemporaries called John Adams bald, for his slick dome. Some observers have speculated that people took to using the name "bald eagle" because the bird's white head was mistaken for a bare one. Audubon wrote, "It is only necessary for me to add, that the name by which the bird is universally known in America is that of *Bald Eagle*, an erroneous denomination, as its head is as densely feathered as that of any other species, although its whiteness may have suggested the idea of its being bare." But the scalps of birds without head feathers are usually a shade of red, gray, or brown, never brilliant white. To see the eagle as bald in a John Adams way, you would have to be pretty much as blind as a bat.[20]

Maybe the birth of the name had nothing to do with the eagle's handsome head. Maybe, instead, it was inspired by an ugly habit, the one Franklin famously ranted about. "Bald" is synonymous with "flagrant," "blatant," and "brazen." Early on, British settlers recognized that this uniquely American bird had a devilish appetite for stealing food from other birds, often in midair. *Look! Did you see that eagle, friend Charles? I'll be go to hell. It did rob a fish from that other bird that caught it. That is one bald eagle.* The person who named the indigenous eagle may have been thinking of its brazenness rather than its whiteness. Or maybe alleged bad behavior and a piebald appearance were together the default, and "bald" served as a double entendre.

Seven thousand feathers clad the bald bird *Haliaeetus leucocephalus*. North America has no other avian species with a white head and white tail and the eagle's dark body coloring. The most prominent beyond the white feathers are the blackish-brown ones of the back and breast and the dark-gray and gray-brown ones of the wings. Balds develop their all-white feathers between ages four and six, a signal of reaching sexual maturity. Their beaks turn too, slightly earlier, from black to yellow, as do their eyes, from brown to pale yellow.

The eyes of the bald eagle are among the most striking in nature. They are "clear and round," wrote an anonymous observer of American history's most celebrated bald eagle, named Old Abe, who went onto the battlefield with Union forces during the Civil War. The "iris is a brilliant straw color, and appears like the sky, changing in luster just as his moods change; the pupil is large, intensely black and piercing, contracting and expanding with microscopic and telescopic action at every light and shade."[21]

The late eighteenth-century and early nineteenth-century artist Charles Willson Peale, who kept a live eagle for many years, called the strength of the eagle's eyes "really astonishing." A ten-pound adult's eyes are nearly the size of a full-grown man's and, according to some sources, four or more times stronger. Using a good pair of field glasses, that man can match, even exceed, the bird's distance vision, but the eagle will still outsee him. With tens of millions more photoreceptor cells—rods and cones—the eagle's eye has much sharper depth and color perception, and a vaster field of vision. One of its most impressive ocular assets is the ability to focus on two objects simultaneously and distinguish between object and background. From more than a mile out, a bald eagle might spot a squirrel in a tree in one direction and a rabbit in a field in another, both furred in natural camouflage yet neither hidden from the monarch in its realm. Able to discern the finest of differences in colors, shadows, and textures, birds of prey have the most acute eyesight of living vertebrates. When you call someone an "eagle eye," you're drawing on a metaphor that dates to sixteenth-century England, and one that is true to the source.[22]

What surely caught the eye of the Founders was the bird's fierce-looking gaze. It is permanently fixed by a bony upper ridge—known as the supraorbital ridge—that angles slightly down and over its forward-facing eyes. For the winged hunter, this anatomical adaptation works much like a sun visor; for us, the aggressive brow reinforces the image of the eagle's formidable nature. Add to that the jet-black pupils and luminous yellow irises, with the absorptive eyes surveilling over a lethal raptor's beak, and you have the ultimate don't-tread-on-me stare.

⌒

THE AMERICAN BIRD WAS an ideal metaphor for the patriotic fervor of the time and a young nation staring hard into its future viability. By contrast, the first seal committee's conceptual metaphor was paradoxical. The ordinarily forward-looking minds of Franklin, Jefferson, and Adams looked backward and conceived something derivative. They had consulted Old World tradition, yet these were men who were also turning tradition on its head with self-evident truths, unalienable rights, liberty, equality, and happiness vigorously pursued. Congress's rejection should not have surprised anyone, except that the decision left the United States a country without a seal.

Secretary Thomson still had nothing to press into laws of the nation and orders to Washington. Diplomats resorted to using their personal seals on official state papers for want of a national one. The perpetual heraldic undress, wrote Polly Redford, "was a shocking state of affairs." Yet three and a half years crept by before Congress appointed a new committee.[23]

Straightaway, the second committee sought the aid of a consultant, Francis Hopkinson. Another Philadelphian and a Declaration signer, his powdered hair long and fixed in a braided queue at the back, Hopkinson was a former member of Congress, a lawyer, writer, musician, virtuoso on the harpsichord, and the first American composer to commit notes to paper. His compositions included such numbers as "The Battle of the Kegs" and "The New Roof, a Song for Federal Mechanics." Hopkinson's creative mind was also versed in heraldry. He had a proven track record. The first official American flag, adopted by congressional resolution in June 1777, materialized in part from his artistic vision. Complementing the lines of the thirteen stripes, he arranged the thirteen stars in horizontal rows. This is the flag we don't hear about, the one that flew before the flag with the stars in a circle, which we have always been told Betsy Ross hand-stitched in her sunlit parlor at the request of George Washington, between sewing his buttons, which she never did.*

Stars and stripes went on the front of the next concept seal. Starting with the shield in the middle, Hopkinson eliminated Du Simitière's crazy quilt of nineteen escutcheons—a clean sweep that surely drew few if any protests—and filled in thirteen diagonal stripes. The same number of stars float above like stardust from heaven. For supporters, Hopkinson chose Liberty and a uniformed soldier. On the reverse, he ditched Moses and Pharaoh for a second rendition of Liberty. Hopkinson had recently designed the Continental fifty-dollar bill, and in a couple of years he would participate in mocking up the first US coin. Both satisfied Congress. But not Hopkinson's seal. A congressional resolution, bearing no seal, dispassionately concluded that the "fancy-work of F. Hopkinson ought not to be acted on."[24]

THERE WAS STILL NO bald eagle on American blazonry—not on Continental flag, coin, paper bill, or tabled seal proposal. That indigenous rapturous figure would later prove to be what was missing for a proposal to measure up to expectations—that exceptional addition that no one had been able to put a finger on, although Thomson would do so in another year, while the eagle waited in the wings.

Thomson had occupied his executive position since Congress's first days, and arguably no one was more frustrated by the shocking state of *sine insignia non* that the country found itself in. Pressing matters of war and nation-birthing were still playing out. There were several more battles over which British and American forces would trade wins. There was the treasonous plot to surrender West Point concocted by Benedict

---

* The 1777 flag resolution did not specify the shapes or arrangement of the stars or the dimensions of the stripes or even the flag itself. Beyond thirteen stripes and a number of stars indicating the number of states, creativity ruled until President William Howard Taft issued an executive order in 1912 establishing specifics for the arrangement and design of the stars (five points, one up) and proportions of the flag.

Arnold, whose apprehension remained imperative (he evaded capture and would die of natural causes in England in 1801). There was the no small matter of implementing the recently ratified Articles of Confederation, the original and ill-fated blueprint for a federal government. The Americans were anticipating victory, and peace talks were imminent. That meant a treaty. A treaty required the seals of participating nations.

Most likely at Thomson's urging, Congress organized a third seal committee, which followed previous practice in turning to an outside consultant. William Barton was the choice of Thomson and the committee. Yet another Philadelphian, Barton came from an accomplished family and as a respected member of the bar. All the same, for much of his adult life he petitioned tirelessly for positions in government, including the chief clerkship of the State Department under Jefferson. He humbled himself in letters, flattered recipients, dropped names, and rarely found his petitions answered favorably. He had a wife and eventually nine children to support, and when you read his solicitations—so many of them—it's hard not to wonder whether plaguing financial struggles were driving him. Or was there some internal raging pressure, conferred by his kin and class, to land in the beam of recognition shone by the elite of his day? His younger brother, Benjamin Smith Barton, a professor of natural history, regularly exchanged letters with Jefferson; one of his uncles, David Rittenhouse, a famous astronomer, was admired by those whom Barton admired.

Barton thought of himself as a man of the sciences of numismatics (currency) and heraldry, and indeed he would be invited into the membership of the American Philosophical Society. "When very young, I made myself acquainted with this science," he wrote, referring to heraldry in a letter to George Washington. Barton expressed his fear that those who unfavorably regarded heraldry as a social prop for an emerging American elite would ensure its demise in democratic society. He therefore thought to write "Your Excellency," as he duteously addressed Washington, to ask his "favor" to lend his estimable name to heraldry's compatibility with a nation of equals. Washington responded at length, opting politely to "proceed on other grounds" and favor whatever sentiments arose among the people.[25]

After Thomson commissioned Barton's consulting and artistic services for the third seal committee, Barton contributed more to the American experiment than egalitarian heraldry ever would. A couple of tries at realizing a national seal were required, though. In Barton's first attempt, he stationed to one side of Hopkinson's shield a crowned Genius of America, a "Maiden with loose Auburn tresses," and to the other an armored warrior with a plumed helmet—a vision out of the storied past of knights and castles. Another helmet, this one disembodied, rested on top of the shield, and on the helmet a gamecock perched. Below, in the center of the shield, Barton envisioned a Doric column topped by an eagle.[26]

Finally, a bird—not just one but two—one of which was an eagle. The eagle's position on the shield, however, was subordinate to the cock, which, Barton believed, was "distinguished for two most excellent Qualities, Necessary in a free Country, viz *Vigilance* & *Fortitude*."[27] The committee rejected Barton's first try. Was it the knight? America had no such species. Was it the avian ranking, the eagle positioned below the cock? The eagle, after all, was the king of birds.

In his second effort, Barton replaced the armored foreigner with a uniformed Patriot. Better. He also expelled the cock and promoted the eagle to the cock's supreme perch on top of the disembodied helmet. Better again. But the eagle was not a bald. It was all white, a nonspecies, a feathered image without ornithological credentials, without existing character traits to associate with the country and people it was supposed to represent.[28]

Whether it was the persistent cluttered imagery or the ornithological distortion, Congress didn't care for the latest design or have confidence in the committee's continuing its work. Dissolving this third committee, Congress created another: a committee of one, Charles Thomson. It referred the task of getting the job done to its perpetual secretary, the one person who had been on the front lines of each heraldic effort.

Thomson himself appears on American blazonry. Look at the back of the two-dollar bill. The image there includes all but a few of the delegates gathered in the Assembly Room of the Pennsylvania State House on July 4. John Hancock, president of the Continental Congress, sits at the table in the foreground as Jefferson hands him the draft of the Declaration. A tall man stands to Hancock's right. That's Thomson, overseeing the assembly with the fingers of his right hand resting on his giant ledger book, as if gesturing to the grand moment's place in the historical record. The portrait is a facsimile of John Trumbull's famous 1818 oil-on-canvas *Declaration of Independence.*

The painting has hung in the US Capitol rotunda since 1826, the year of the nation's fiftieth birthday. Thomson had passed away two years earlier, at the remarkable age of ninety-four. An Irish-born Philadelphian, teacher of Greek and Latin, cofounder of the American Philosophical Society, and distiller of rum, Thomson was an early leader among the Sons of Liberty. He earned the respect of John Adams as the "Sam. Adams of Philadelphia." Like many of his station, Thomson sat for an etched portrait executed by Pierre du Simitière, who gave his subject a heavy brow and an abundant nose. Yet missing from the interpretation are Thomson's facial scars, souvenirs of his tenure as secretary of the Continental Congress. He served throughout the body's existence, possessing such powers that some referred to him as America's prime minister.[29]

This is where the scars come in. Thomson's authority spared him no contempt. South Carolina delegate and noted slave trader Henry Laurens couldn't stand Thomson, an outspoken opponent of slavery.* Others accused him of all sorts of intolerable offenses. Pennsylvania's James Searle claimed that Thomson misquoted him in the record, and the two crossed swords on the statehouse floor, using walking sticks. "Yesterday Mr. Searle cained the secy. of Congress," Massachusetts delegate Samuel Holten noted in his diary, "& the secy returned the same salute." Each lanced the other's face.[30]

Thomson was a man of his convictions, and of his country. Think-

---

* Thomson adopted the attitude many Pennsylvania Quakers took toward the mistreatment of Indians and the immorality of slavery. In a letter to Jefferson, he described slavery as a "cancer that we must get rid of." Jefferson hedged on the issue, but as patriots, the two men were one. Keith Thomson, *Jefferson's Shadow: The Story of his Science* (Yale University Press, 2012), 120.

ing more like an American than a progeny of Britain, he brought allegory and imagery to home shores in a seal sketched by his own hand.
With the newest version, he jettisoned everything except the shield
and its red and white stripes, which he aligned as an inverted chevron. He offered the shield no supporters other than an eagle, signify-
ing self-reliance. A constellation of thirteen stars, the only ones in the
image, float in a radiant sky above the eagle's head. The bird's left talon
grips thirteen arrows and its right an Assyrian olive branch, deriving
from antiquity. From the eagle's beak, a banner streams with the words
E Pluribus Unum—"out of many, one"—a common Latin acclamation
for independence. Du Simitière had included the motto in the proposal
of the first committee, and Thomson liked it.

He was adamant about the eagle too. He made the eagle big and
bold, and he was unequivocal about which kind it should be: an "American bald eagle." The "Head & tail," he wrote in his instructions, "are
white[,] the body & wings of lead or dove colour."[31]

On the seal's reverse side was another reference to antiquity: an
unfinished pyramid. Replacing the pyramid's top was a triangle with
the eye of providence displayed at the center of a glory—indicating
celestial rays of the sun. The eye and glory were another of Du Simitière's influences, executed over the shield in his drawing for the
first committee. Above and complementing the glory, Thomson
included the words *Annuit Cœptis*, which translates to "favors our
undertakings," which is to say, "God favors." Across the pyramid's
foundational course of stone, Thomson penned in "MDCCLXXVI,"
1776. That, too, came from Du Simitière. Underneath all was the
line *Novus Ordo Seclorum* (meaning "new order of the ages"), which
Thomson borrowed from the Roman poet Virgil's *Fourth Eclogue*,
composed around 42 BC. A single supporter of the shield, expressions of peace and firmness, and the identifiable indigenous raptor
spoke to Virgil's sentiment.

Before a final image with written description went to Congress,
Barton fortunately made a few adjustments to Thomson's sketch.
Thomson had drawn the eagle with a crest at the back of its head.
The crest was an anatomical flaw; worse, it gave the eagle the look of

a chicken wearing a fright wig. Barton tamed the crest slightly and made improvements to a mortifying pair of legs that looked like turkey drumsticks. Thomson had the tips of the bird's open wings inverted, pointing downward. Barton turned them up, allowing the bird, as he explained in a letter to Thomson, an exalted appearance of "supreme Power & Authority." While the seal's canvas was no longer overtaxed with illustrations, the expressed and implied values were many: "Union & Strength"; "Vigilance, Perseverance & Justice"; "Virtue and preservation"; and "Hardiness & Valour."[32]

On June 20, 1782, Congress finally approved a proposal for a seal, and with that the bald eagle became an American symbol.

⌒

WHERE THOMSON ACQUIRED THE inspiration for enlisting the services of the American raptor is anybody's guess. Of the thirty-four coats of arms of the Declaration's signatories, one—that of Caesar Rodney, delegate of Delaware—displayed three nondescript eagles. Otherwise, lions—on nine coats of arms—ruled. Benjamin Randolph, onetime Philadelphia landlord of Jefferson and maker of the portable desk on which Jefferson wrote the Declaration of Independence, displayed an eagle on his shop sign and business card. Thomson surely walked past the shop on Chestnut Street countless times and may have bought furniture from Randolph. With open wings and threatening talons, Randolph's eagle was dramatic yet an amorphous species with a feather crest behind its head. Although several species of eagles with naturally crested feathers exist, including one called a crested eagle, none are native to Europe or North America.

Still, crested eagles in imagery were standard. In that regard, the eagle on Randolph's sign was similar to another eagle that appeared on a series of coins from the reign of King Charles V in the 1500s. Notably, the Charles V eagle is also grasping thunderbolts and an olive branch in its talons. Did those talons and feather crest catch Thomson's eye? The two-hundred-year-old Charles V coins would have been scarce in the States, yet prized among collectors. Du Simitière featured a rare-coin collection at his museum. One or more of these minted eagles may have been part of it, and perhaps Thomson saw, even admired, them.*

Again, so many qualified guesses for so many questions. Here's another theory: around the time the Charles V coins were used as a medium of exchange, books on emblems and symbols had become popular across Europe. A few of these books survived the years and made the crossing to America. Thomson doesn't seem to have had one in his library, but Barton did. So, too, did Franklin—a 1597 volume published by the Bavarian scholar Joachim Camerarius. The first emblem featured in its pages is of an eagle posturing with wings inverted very much like the one Thomson drafted for the Great Seal, although without a shield. This particular bird has raised feathers behind its head and neck, and a flailing tongue. In similarity with the Charles V coins, there is positioned next to one talon an olive branch and next to the other a thunderbolt—symbols, as the caption reads, of peace and war. A few more pages in, another eagle carries an arrow across its back. Was Thomson working off these images when he came up with his feather-crested bald eagle with olive branch and arrows? Perhaps.[33]

Conceivably, the secretary had found his muse by simply looking out his window or strolling along the waterfront. Bald eagles were prevalent and conspicuous in the East, especially around lakes and rivers and bays, beside which cities had grown into their Americanness. Philadelphians shared a skein of fish-frothing creeks and two significant rivers, the Schuylkill and the Delaware, with assorted terrestrial and aerial wildlife. From its dual rise in the Catskill Mountains, sliding past Philadelphia and on to Delaware Bay, the Delaware River was a four-hundred-mile thoroughfare of fish and piscatorial birds. By

---

* The current seal of New York has a crested eagle perched atop a globe atop a shield, an image that many assume was part of the state's original 1778 seal and maintain was Thomson's model. But the provenance of this image can be affirmed only as far back as 1882, when the legislature devised a new seal while making a number of unsuccessful attempts to find a written description of the original seal. See, for example, George Earlie Shankle, *State Names, Flags, Seals, Songs, Birds, Flowers and Other Symbols* (Greenwood, 1970), 45.

some accounts, a bald eagle's nest occupied every mile or two along the banks of rivers on the Eastern Seaboard.

A river is a place where, like the water in it, life never stops moving, no matter the day, the season, the heat, the ice. What moved more than anything in the Delaware was spawning American shad. This is the legendary fish that allegedly saved Washington's starving men at Valley Forge, although the truth, as always, is more complicated.[34]

The Delaware's estuary was the largest breeding ground of the American shad. A sound nearly as common in Philadelphia as horse hooves on cobbles was pushcart vendors yodeling, "Shad-e-o! Shad-e-o!" The city's most popular fish was a native of the East Coast, ranging geographically from Newfoundland to Florida, and physically from two to eight pounds, a nice size at the lower end for an eagle to catch and carry. The abundant and appetizing food source attracted the settlement of Native peoples eight to ten thousand years before the founding of Philadelphia. The last of them, the Lenape, tolerated Swedish arrivals of the seventeenth century before the Swedes gave way to English Quakers, Germans, Scots, Dutch, and French. The Europeans, as did enslaved Africans, fished the river for food and market and named one of their settlements Fishtown, which became a part of Philadelphia. Unlike the Lenape, the eagles never left—that is, until DDT. As with any commercial fishing port with a fleet of boats and huddle of smokehouses and processing houses, Fishtown served up splendid opportunities for opportunistic eagles.[35]

So, too, did sportfishing. William the Conqueror introduced fly-fishing, along with castles and feudalism, to the British Isles, and British piscators took their fly rods to the Americas. Around Philadelphia, the Schuylkill River (pronounced "*SKOOL*-kil" or, vernacularly, "SKOO-kul") and its tributaries with braided and babbling water were favored spots for sportfishers, typically casting for perch, salmon, and striped bass rather than shad. Margaret Penn, daughter of the colony's founder William Penn, wrote her brother back in England, requesting that he send her a "four jointed strong fishing Rod and Reel with strong good Lines." Wherever an angler took up a position, an eagle or two often swerved in to do the same just feet away—standing waterside, wings folded, leering sidelong, in a moment of biological symbiosis between bird and human. An uncased fly rod arcing a horsehair line across the air signaled possibilities to the bird: the thrashing hooked fish that got away from the fisher; the live one that sprang the creel basket; the successfully caught and unhooked ones set up on the bank, twitching and awaiting the cleaning knife; to say nothing of the kindly angler tendering a gift to the feathered bystander. The point is that nesting, surveilling, fishing, and prospecting eagles were frequently on display for Philadelphians.[36]

A BALD EAGLE FISHING is quite a sight. It starts from a far-off perch or from passes made overhead. When it sees flashing in the water, it will mark its prey and surrender to gravity, using it to its advantage. Wings open, feet back, it drops toward the river or "fishy sea," as the poet Wallace Stevens described this "sovereign sight" of this sovereign bird. Turning into a long approach, it puts legs and feet out front and then eases them down for a nimble yet deadly touch-and-go. Open talons hit the water smartly and lift out a fish. Momentum then urges the eagle, often looking down at its wriggling bounty, into an upward arc on pulsing, "dazzling wings."[37]

A bald taking its catch from the water while in flight was a show-stopping performance for early Americans. Seeing one steal from another bird in midair was an even more stunning execution, and a not unfamiliar one around Philadelphia. Quite often, an eagle will race in to force another eagle to release a fish it has caught. Such assaults turn into aerial duels, with the contestants flying in somersaults and twirls, going talons against talons. If the fish drops and the aggressor catches it in the air, the assailed becomes the aggressor, and the fish might go back and forth between the contestants multiple times.

An equally common target is a fish-wielding osprey, the bird that Franklin judged to be the chief victim of the larcenous bald eagle. An eagle will follow the osprey's activities as intently as it does prey. For its part, the osprey knows instinctively that, after snagging a fish from the water, its airborne food may end up in the talons of another. In eagle country, everything is up for grabs. The eagle's own instinct is to pursue, from behind, above, or the side, to conduct a harassing flyby until the osprey lets loose its catch. When the fish drops, the attacker dives, hitting speeds of up to one hundred miles per hour, to make a catch before the plunder hits the water—a jaw-dropping performance. If the eagle misses, it will retrieve the fish, too stunned to swim for it, at the water's surface.

The osprey is a favored mark because its diet is nearly one hundred percent fish, and few birds are more industrious and skilled at catching them. It is anatomically built for the job. With a body weight about one-third the eagle's and a wingspan nearly as broad, the osprey has a power-to-mass ratio that seemingly defies physics and physiology. An agile cargo carrier, it is capable of ferrying up to ninety percent of its weight, which at the uppermost is four and a half pounds. Even then, with fish in tow, the osprey is lighter than an adult eagle without cargo. At very best, an eagle can carry off not quite half its weight—less if pulling that weight out of water.

The talon of the rear, or central, toe of the eagle and osprey pierces into its prey; the three front toes grip it. The central toe of the osprey is shorter than the bald's by half an inch, but its talons are more curved and longer than the eagle's two-inch talons. As is common with raptors, both birds have tendons in their toes and feet with a serrated side that serves as a series of adjustable stops, as in a zip tie (except release is possible anytime)*, that enables them to lock their talons around prey without exerting their muscles.

The osprey hunts from the air, looks straight down, and sees to several feet below the surface. An eagle's line of sight extends more forward, penetrating not far below the surface. When dropping in for a catch, the eagle angles toward the water off a horizontal plane and reaches its feet and toes eight inches at most into the water, intending to continue in flight. The osprey fishes similarly yet hesitates for a moment at the water before rising skyward. It will also at times turn

---

* It's a myth that eagles drown because they cannot unlock their talons from an oversize fish.

into a hard vertical dive, transforming into an avian missile locked onto a target. Indeed, it looks like a missile at this point, or a militant yoga master, with legs thrust straight forward, tight against its body, feet out in front of head, speed pulling feathers sharply back in the effect of action lines in a drawing. Piercing the water, the osprey shoots three to four feet below. Employing both techniques, the osprey has a kill rate as high as seventy percent. A fishing eagle's is somewhere around thirty. So, the eagle steals.[38]

For eating the hard-won lunch of another, exploiting its labors, the bald eagle reaped its share of derision from around Philadelphia and around the country, not least of all from "Dr. F." But the Founding Father's lively charges against the founding bird's scruples, we now know, were unfair. Behavior that Franklin attributed to sloth and moral depravity, scientists today attribute to intelligence. They have a name for larceny of the bald eagle kind: "kleptoparasitism." Nearly two hundred bird species are party to it. Gulls seize food from terns. Crows from starlings. Juvenile oystercatchers from adult oystercatchers. Frigate birds, known as pirates of the sea, from tropic birds. Blackbirds pilfer. They snatch from thrushes. Cormorants from cormorants. In a karmic twist, balds are sometimes the mark, the patsy, the chump. Golden eagles do to them what they do to ospreys. So do red-tailed hawks and even, occasionally, the aggrieved osprey. Muggings in the wild aren't perpetrated solely by birds, of course. Brown bears liberate foxes and coyotes of their spoils. Lions do the same to hyenas, and hyenas get back at lions, cackling as they flee with the eatable goods. But it's not just the big victimizing the small. Gulls swipe from pelicans, sea otters, and beach tourists.

Scientists say stealing is more efficient, that it conserves energy. All animals give priority to budgeting stored energy. Efficiency often makes the difference between life and death. Most birds operate at the margin of depletion. Outside nest-building, mating, and migration, virtually all their movements are related to hunting and gathering. Often, instead of conducting surveillance maneuvers as do terns and pelicans, bald eagles will hunt from a perch if one is handy and situated advantageously. To many observers before and long after Franklin, the "sitting about" was evidence of the work ethic of a malingerer. But science interprets this method of hunting as adaptive and expedient behavior, enabled by the eagle's keen eyesight and swift actions. And from that energy-saving perch, eagles look out for the better-fishing osprey with bounty in its talons. Kleptoparasitism is smart, innate foraging, and the behavior has less to do with brutishness and body size than with brain size. Not among the artful pilferers is Franklin's respectable turkey—supersize body, bantam-size brain locker.

The osprey might be a more successful fisher, but the bald eagle is still a master fisher, and a vigorous one. Once chicks show up, one or both of the parents will catch an average of four to five fish a day to feed their hungry brood, adding variety now and then with a rodent or rabbit. If balds depended on ospreys to supply their nourishment half as much as some of their detractors claimed, neither osprey nor eagle would survive.

LIVING BY THEIR WITS, bald eagles flourished on the Eastern Seaboard and, despite false judgements about their moral character, gained the honor of representing the nation on the Great Seal of the United States. Yet the raptor with striking profile and unflinching gaze had not been without faunal competitors.

Before Charles Thomson came to Congress's rescue with the bald eagle, before this sleeper candidate for national emblem arose to prophetic fame, there were a number of other hopeful animal representatives. To emphasize diversity and unity, some Americans envisioned a flock of birds, a drove of sheep, and, depending on one's opinion about American independence, a kettle of fish. To honor industry, pluck, and shrewdness, respectively, others suggested the beaver, bison, and white-tailed deer. Preferable among citizens living in the backcountry and southern states were rattlesnakes, wildcats, hornets, and alligators—extolling aggression and territoriality, qualities considered necessary to survival (these animals would one day find a calling as sports-team mascots). In the end, venom, claws, stingers, and teeth lacked, to borrow words from the historian David Hackett Fischer, the desired "gravitas of a national emblem."[39]

The British, with their cultural preference for four-legged mammals, had their own opinions, of course. One embittered Tory proposed a zebra bearing thirteen stripes, with the hindmost standing in for Massachusetts, that antitax, Boston Tea Party instigator of revolution. Others envisioned a jackass or mongrel dog, while some believed that the bald eagle was just about right, that its thieving and brutish ways and alleged moral laxity said everything that could be said about the American character.

François-Jean de Beauvoir, Marquis de Chastellux of France, a man of letters and military bearing partial to good Madeira, who fought beside Washington during the war, thought the British should partake in a little self-examination before finding fault with others. To Chastellux, there was no better symbol than the bald eagle for the nation that he expected would elevate the moral and political order of Western society to new heights. The Great Seal was six months old when, in a letter to an American correspondent, Chastellux looked to winged creatures to make his point: "It seems as if the English, in every thing, wish only for a *half liberty*. Leave owls and bats to flutter in the doubtful perspicuity of a feeble twilight; the American eagle should fix her eyes upon the world."[40]

The six years required to devise a proper seal for the United States proved worth the wait. Upon its adoption, Jefferson was reputedly quite pleased, and he extolled the bald eagle as a "free spirit, high soaring and courageous." Over the years, countless newspaper articles and popular books have attributed these words to the Founding Father and failed seal designer. In 1999, a White House speechwriter included them in an address that President William Jefferson Clinton gave at an Independence Day ceremony while standing next to a bald eagle named Challenger. That Jefferson actually uttered these words is probably no truer than, say, a turkey having been in the running for national symbol. Still, they reflect the appropriateness of Thomson's ultimate permutation. According to Fischer, the "dignity and grace" and "virtue and respectability" (Franklin's complaint notwithstanding) associated with the bald eagle and its physical traits "were exactly the symbols that Congress desired in the last years of the War for Independence."[41]

The bald eagle seemed fated for its position as founding bird. It was the picture of the nation's full-fledged independence and sovereignty. As the Great Seal's central figure, *Haliaeetus leucocephalus* gave the national emblem standing in heraldry and among the seals of other nations. America's bird was inherently stalwart and handsome and would grow even more handsome as engravers over the years fine-tuned its features. Although cousin eagles had long appeared on European coats of arms, they were at best secondary figures—ornithologically abstruse, one-dimensional, improvisational, more type and logo than faithful representation. None of those other birds was the bald eagle—itself an original like the United States, a native of North America, claimed by no other land. It was contemporary, authentic, enduring.

TWO

✦ —— *Buttons and Coins* —— ✦

THE GREAT SEAL OF THE UNITED STATES WAS NOT A MONTH old, and George Washington had not yet received that first embossed congressional directive at his Hudson River headquarters, when the bald eagle as national symbol went on full ceremonial display at the residence of the French ambassador in Philadelphia. The occasion was the birthday celebration of the dauphin, the heir apparent to the French throne. In the courtyard of the ambassador's residence, reported the *Pennsylvania Packet*, was a "hall of the most excellent architecture" designed specifically for the gala. At the "farthest extremity of the hall" hung the French coat of arms. At the "other extremity" hung the shield of the United States "supported by the American bald eagle."[1]

The French ambassador had preempted an American unveiling of the founding bird on its new patriotic perch. Those who were not among his fifteen hundred privileged guests would have to wait for the August 1 issue of the *Columbian Magazine*, which ran an illustrated spread on the Great Seal. The splendidly poised raptor's elevation to America's bird represented a significant expansion of its emblematic horizons. Before 1782, eagle imagery was present but not prevalent in daily life around the former colonies. Engravers and metalsmiths occasionally included renditions on lockets, tokens, tankards, and hilts of swords. But the feathered figure embodied no patriotic glow or collective point of view. As the late art critic Philip Isaacson wrote, the "emotional precedent" that might have elevated the bald eagle to cultural icon did not yet exist.[2]

There was no question of the white-headed bird's popularity by the time of Washington's presidential inauguration in April 1789 in New York City, where it made fashion and architectural statements. The US Constitution replaced the Articles of Confederation that year, and under the Constitution's auspices the new Congress once again approved the Great Seal, with the bald eagle, as the national insignia. Congress also commissioned Pierre Charles L'Enfant, a French-American military engineer, to convert New York's old city hall into the new nation's Federal Hall, which was to be Congress's first home under the US Constitution. L'Enfant had served on Washington's staff at Valley Forge and subsequently designed the eagle on the Society of the Cincinnati's insignia. For his latest commission, he installed a bold relief on the pediment of the Federal Hall. The *Pennsylvania Packet* described what might catch the eye:

> The attic story is composed of ornamental figures, festoons, and trophies crowned with a pediment, on which a large eagle, surrounded with a glory, appears bursting from a cloud, and carrying thirteen arrows and the arms of the United States.[3]

Directly below during the inaugural ceremony on the upstairs portico, Washington raised his right hand to be sworn in. He wore an unassuming, dark, broadcloth suit, except his coat had been dressed up with metal buttons manufactured in England and stamped with the likeness of the American eagle.

Five months later, when he set out on an official tour of the states, the bird of his buttons seemed everywhere. On the road to Boston, Washington's travels on horseback and by carriage took him through southern New England, treating him to the leafy yellows and oranges of the season and to the peoples' fervent patriotism and new way of expressing it. Residents and business owners along his route had drawn silhouettes of the bald eagle in soap or whitewash on windowpanes—on every pane in some cases—and backlit them with candles.

Reaching Boston, Washington was greeted by a raw northwesterly wind and fall colors blown asunder. The chilly conditions, he wrote in his diary, were offset by the warm welcome of a "vast concourse of people." The president and impresario of independence was truly the man of the hour—not to mention the century. In doorways, at open windows, and on rooftops, "well dressed Ladies and Gentlemen" cheered as he rode past, sitting presidentially tall on his white horse. On Market Street, an untold number of others were "thronged" about a "handsomely ornamented" arch under which the procession passed. Wrapped in bunting, as was much of the city, the arch had been custom-built for the occasion. At the top center, a cloth-canopy spire sheered up eighteen feet above the arch. Presiding on the canopy's peak, where a cross would rise from a church's steeple, was a large wooden eagle. It was likely fabricated by the master shipcarver Simeon Skillin Jr. Not long after, Skillin and his brother would cut and chisel most of the decorative woodwork for the USS *Constitution*, which included a spread eagle at the stern.[4]

After leaving Boston, Washington was twice made the namesake there: the city rededicated part of the Boston Post Road in his honor, and locals unceremoniously denominated a citywide illness, believed to have spread from his admirers standing in the cold, as the "Washington influenza."

Washington had become more than a name and national hero. He was now a moral force, a virtue, a motif. The man's apotheosis was the country's, and the founding bird had landed as his avian counterpart, all but perched on his forearm. As emblem and unyoked species, the bald eagle certified the freedom for which Washington had fought. Some years later, the editors of *Harper's Magazine* would reflect on this "symbol of daring and defiance" that had "ever been foremost." During the war, on the day American forces crossed the Saint Lawrence River in preparation for taking Quebec City from the British, South Carolina congressional delegate Thomas Lynch wrote to Washington, "I mean not to anticipate your Determination, but only to approve your Design to hover like an Eagle over your Prey, always ready to Pounce [on] it when the proper Time comes." The painter Edward Savage, who was thirteen when fighting broke out between the British and

Americans, had Washington sit for at least three portraits. In one, the Society of the Cincinnati's bald eagle medal is pinned to the left lapel of the victorious general's uniform.[5]

Clear-eyed and self-possessed—embodying strength, grace, and resolve—man and bird stood for the professed case-hardened traits of an endeavoring people. Symbolically, metaphorically, they were two sides of the same coin.

Indeed, the two were popular candidates for the real thing. In 1791, when Washington was in the third year of his presidency, a private minting firm in Birmingham, England, shipped a cask of copper one-cent pieces to the States for consideration by congressmen and cabinet officers—and the president himself. The prototypes were based on Pierre Eugène du Simitière's bust profile of Washington. Washington was on the front, and the bald eagle on the back. Using images of the most popular man and, arguably, most popular indigenous bird in the young republic was smart branding for a business looking to expand its opportunities in a new market. The next year, another outfit submitted to the US Mint a gold coin bearing the twin icons.

Privately, Washington the person wished for honor and admiration and took a shine to the gold coin, valuing one in his pocket as a keep-sake (in 2018 it sold at auction for $1.7 million). Publicly, however, he opposed pomp, ceremony, and iconography that suggested monarchy, and none of the prototypes entered officially into circulation. Not until 1932, the bicentennial of Washington's birth, did the nation's first president appear on a government-issued coin, the twenty-five-cent piece, with America's bird on the reverse, perched with wings spread on a bundle of arrows and framed by two olive branches. Founding Father and founding bird have remained in circulation together ever since.*

---

* In 1999 the eagle came off the back of the quarter to make room for designs commemorating the fifty states, national territories and jurisdictions, and National Park Service sites. In 2021 the mint replaced the commemorative design with one of Washington crossing the Delaware. Quarters with bald eagles are no longer issued, but they remain in circulation.

Not comfortably at first, though. The quarter's debut touched off a minor panic. Critics greeted the new coin with the claim that the bird on the back was not ornithologically correct. Because the eagle was cast on a monometallic surface unsuitable for revealing the coloration of a bald, purists scrutinized the legs, maintaining that they were fully booted with feathers, and thus those of a golden. The George Washington Bicentennial Committee quickly stepped in and turned to Robert Bruce Horsfall, a respected wildlife illustrator, for help in verifying the species.

Taking a magnifying glass to the coin, Horsfall focused on the bird's legs between the heel and toes, where the offending feathers were said to exist. When looking at a bird's leg, people are apt to notice its backward-bending knee—except that's not a knee. It's the heel, which was of great interest to Horsfall. Of equal importance to him was the bone extending from the heel down to the toes, which unschooled observers might take to be a shinbone. It's called the metatarsus. On some birds, such as eagles, the metatarsus is fairly short; on others, such as wading birds, it is quite long. The metatarsus, remarkably, is the bird's foot, with the toes (usually four) at the end. Like a ballet dancer, birds tend to walk on their toes with their heels in the air (as do dogs, cats, and horses). The true knee is up close to the body and connected to the heel by the long tibiotarsus bone. Avian knees, in fact, bend forward, in the same direction as human knees.

The zigzagging configuration of bird legs is an evolutionary design that came down from theropod dinosaurs. It's a masterwork of anatomy that lends itself to the bird's ability to run, walk, perch, hop, paddle, and spring into flight. Once in the air, a bird can fold its legs up like an aluminum beach chair and pull them close to the body. For landing, the complex of bones and joints makes them good shock absorbers. When a bird sits on a branch and bends its legs, the leg muscles retract to pull the toes tight around the branch or twig. The bird can now sleep without having to exert muscles to stay on its perch.

When Horsfall held the glass before his eye to examine the engraved bird's legs, he saw that its feet were featherless and not those of a golden. The critics and the committee alike, according to the *New York*

Times, had been horrified by the possibility that an engraver might have coupled the native-born father of their country with an "international eagle instead of the bald or American variety." What was important here was that the minted bird match both Charles Thomson's chosen species for the seal and the founding natural environment of the country. Like George Washington and the white-headed indigenous bird, the natural environment had for 150 years stood out as conspicuously American.[6]

NATIONAL IDENTITY, NATIONAL EXPANSION, AND EVERYWHERE THE "MONARCH OF THE AIR"

H ERE'S A MYTH THAT IS NOT QUITE AS FAMOUS AS Franklin's eagle impeachment but was once popular enough to be repeated across generations. It starts with a battle early in the American War of Independence. Which battle, no one knows for sure. The Patriots were taking on the most powerful military in the world—well trained, highly disciplined, lavishly equipped, and vastly experienced. Untested in combat for the most part, the Americans were determined yet unsure of themselves. As the fighting intensified, the din of armed conflict attracted the attention of a pair of bald eagles. The two took to the sky and circled overhead, screeching. Hearing them, the Patriots looked up. The "raucous cries" of the sovereign birds excited "raucous cries" from the Patriots, who shouted, "They are shrieking for freedom." Taking example from the eagles' conviction, the Americans went on to defeat the British.[7]

The story was popularized by the writer Maude M. Grant. Born in the year of the nation's centennial in Monroe, Michigan—home of George Armstrong Custer and Mary "Mother" Jones—Grant published storybooks for young people. Some were patriotic and some were nature studies—themes not diametrically opposed in her day. Grant's eagle tale appeared in *The Junior Instructor*, a home reader for children and their parents. "The Eagle, Our National Emblem" suggests that the nation and its symbol were meant to be. It's an origin story of nature's giving providential validation to the allegiant and honorable struggle of a new republic.

Grant's tale is a charming one but not without problems. The event it depicts is not true, for one. The other problem has to do with the eagles. If the fighting that Grant wrote about had been the least bit loud—and armed engagement tends to be loud—no one would have heard the eagles overhead. Balds do not put out decibel levels that pierce ears or cut through gunfire. Thrusting their head back as far as possible, opening their mouth wide to the sky, fluttering their tongue, belting out their best, eagles release a call that is a lively and rapid *chee-chee-chee-chee-chee*, followed by a *tchu-tchu-tchu*. The sound brings to mind a squealing gull or chirping teakettle. Eagles also wield a guard call, or peal, when humans or trespassing birds approach. Males usually sound the warning, exerting a *kark-kark-kark*. Yet the intensity of this alarm is short of an exclamatory clarion and seems unbefitting an imposing raptor and cultural icon.

That's why, when Hollywood filmmakers first thought to incorporate a screaming bald eagle into a scene, they realized they could not let the chittering bird speak its own lines. Sound technicians were compelled to search for a voice-over performer. The golden eagle would not do. Its call is only slightly bolder than the bald's. The bird they ultimately cast—a bird one-quarter the weight with one-quarter the wingspan—was the red-tailed hawk. Sharp and intrepid, the hawk's cry could both inspire fighting patriots and raise the hairs of moviegoers and TV watchers. Its commanding voice is so favorable to the on-screen American bird that audiences have not been the wiser. For example, throughout the eight-year run of the popular television show *The Colbert Report*, a show that incidentally traded on fakery, a bald eagle streaked through the opening title sequence while a dubbed-in redtail pricked the ears of viewers.

In defense of this otherwise charismatic screen star, we might consider that the way our ears and minds perceive its voice is not necessarily how other living species do. Scientists don't know much about the sound and vocal behavior—bioacoustics, they call it—of bald eagles beyond what we hear. They don't know whether the balds' voice and cadence are learned or innate. They don't know why these birds sound the way they do. We know that most birds vocalize to beckon a potential mate—the sleepless mockingbird that sings bittersweetly through the night, for one. After two pair up, the female's phonetic signal of sexual willingness follows at some point. An egg-sitting, or brooding, bird, including an eagle, will call when it wants its mate to bring food or assume its shift in the nest. Birds also, of course, vocalize to mark their territory. This is what the songsters are doing when their melodies rise with the sun in the morning. They break—with some exceptions, such as field sparrows—from late morning to late afternoon and then reprise their chorus as the light turns golden toward sunset. Birds sound out to maintain conversational contact with a mate, to indicate their hunger to parents, and to communicate fear to their group. Many flocking birds announce their readiness for flight departure—a "saddle up" call. Sandhill cranes do this, and they have some twenty more kinds of vocalizations for other purposes.

Only animals fully grasp the meaning of the sounds they make, and they likely don't characterize them as we do. Even if, say, rabbits regard the eagle's voice as piteous, they know they are hearing the sound of a dangerous predator that likes to eat rabbits. For its part, the eagle must know that the recognition of its voice, meek or stout, is enough to keep away intruders. And those erstwhile intruders are not fooled by what we humans consider a meek voice. Lewis and Clark seemed to have had the correct insight when they trekked through the country's newest territorial acquisition and observed that prey and "carnivorous competitors" alike feared the sound of an eagle.[8]

Starting a century and more before Hollywood began cinematizing myth, people typically gave the bald eagle a voice it didn't possess. Journalists, poets, essayists, book writers, story writers, outdoor writers, and even naturalists said the bald eagle screamed, screeched, and shrieked. In their eyes and ears, it had to; it was America's bird. The paragon of courage and vigor, grasping thirteen arrows in its left talon, could not be a squeaker.

The bald eagle was also the exemplar of North American nature, the basis of a distinct national identity, one that bespoke exceptionalism. Linking nature with national identity was one of its most important roles as a symbol. When the country was young, Americans looked to indigenous sources to give expression to communal pride in their country's rise to international legitimacy. The indigenous bald eagle was the embodiment of nature and nation together, corresponding with the genetic forging of a republic that, not unlike Indian cultures, extracted its selfhood from the prodigious physical geography and all that conspicuously rose from, grew across, lived on, fed upon, ran along, flowed through, gouged into, nested atop, and flew over that geography. Biologically, the bald eagle resided at the top of the domestic animal world that the North American environment had engendered; emblematically, it alluded to national identity willed by nature. Maude Grant could not have chosen a more ideal animal for telling her patriotic tale.

Still, the founding bird was not the lone testament to special indigenous attributes of the republic. Some articles of the wild were buried like time capsules that, when unearthed, as if on cue, helped advance the notion of American exceptionalism.

IN 1705, THE STORY goes, near the rural village of Claverack in New York's Hudson River valley, a rock rolled down a hill and stopped at the feet of a Dutch farmer. The farmer bent over and picked up the rock—except it wasn't a rock. It was a five-pound fossilized tooth, the origin of which the farmer could only guess. It was odd and oddly fascinating, enough so for him to trade it to a man in town for a bumper of rum. No one, a Boston newspaper reported, knew "whither the Tooth be of Man or Beast." Reading Genesis 6:4 in their Bibles, many people believed the tooth was that of a gargantuan people who lived before the Great Flood. The Puritan minister and Salem witch trial conspirator Cotton Mather took the tooth as irrefutable evidence that biblical giants had walked the Earth. Eventually, the fossil made its way to the colonial governor of New York, Edward Hyde, Lord Cornbury, a cousin of the queen, who shipped it off to London with a label that read, "tooth of a Giant."[9]

Other rock-size teeth turned up in South Carolina. Enslaved Africans there proposed that these finds were not from giants of Genesis but were like the giants of their distant homelands: elephants. Teeth and bones that continued to surface at Claverack resembled teeth and bones that were being hacked out of ice in Siberia. Long tusks were found too. European anatomists named the giant creatures *mamants*. English speakers Anglicized *mamant* into "mammoth." Natural scientists then applied "mammoth" to the discoveries in New York and subsequent finds in the Ohio River valley. The early mammoths of North America were really mastodons, a distant and extinct relative of elephants. Like the bald eagle, the mastodon is strictly an American species.

Big-bone fever had swept the country by then, much like bald eagle iconography. The mastodon reinforced the belief in the existence of many more unknown natural wonders that made America a special place. Nature endowed the nation and its people with something enviable: the North American continent's land and landmass, agreeable climate, profuse wildlife, ripening fruits, and unparalleled physical beauty.

Culturally, the US lacked a definable aesthetic and remained an outpost of Europe. Living in London with her ambassador husband, Abigail Adams wrote home to a friend, "I will not dispute what every person must assent to; that the fine arts, manufactures, and agriculture have arrived at a greater degree of maturity and perfection" in Europe. Art, architecture, literature, and music—and the early proposals for the Great Seal—had come in on the tide from English, French, Greek, and Italian shores.[10]

While cultural independence awaited its day, nothing set America apart from its rivals more than did nature. The English philosopher and Enlightenment thinker John Locke wrote in 1689, "In the beginning all the World was *America*." Locke was referring to the Edenic fruits of the land that a man could parlay, as fruits of his labor, into being "Lord of his own Person and Possessions, equal to the greatest" (by "Person," the philosopher did not mean enslaved Africans, displaced Indians, or women). Locke knew of no such opportunities advanced by earthly offerings that were available to the same degree elsewhere.[11]

The homeland of Anglo newcomers to America was largely barren and unalluring by comparison. Nature in America was bigger, more diverse, and accessible. Nature was the new nation's backyard and front yard. In the introduction of a scenery album he published in 1840, Nathaniel P. Willis put it succinctly: "Nature has wrought with a bolder hand in America." The word "mammoth" matched America's physicality, and the bones of Claverack were tangible evidence of its organic singularity—of its superiority, argued many, including Franklin, Jefferson, and both John Adamses. In her letter from postwar London conceding that America trailed in the arts, Abigail Adams also wrote of the new nation's superiority in the natural world: "Do you know that European birds have not half the melody of ours? Nor is their fruit half so sweet, nor the flowers half so fragrant, nor their manners half so pure, nor their people half so virtuous." The esteem that the American Founders held for the country's natural heritage affirmed Charles Thomson's genius in putting on the front of the national seal the bald eagle—that native bird, representative of natural America and a nation of free people (the preceding exceptions notwithstanding). Thomson's selection was also an early step toward an original aesthetic based on an organic American palette rather than European duplications.[12]

No ONE WORKED HARDER than the artist Charles Willson Peale to stimulate the public's interest in nature as a national attribute. Peale was one of the most celebrated, and certainly one of the most colorful, painters of early America. Tinkerer, patriot, soldier, scientist, he was born in Maryland, where, as a young man, he exhibited a talent for painting. Tall and slim, at least in his self-portraits, he studied painting abroad, as did most noted American artists of his generation, before returning to Maryland on the eve of rebellion. In 1776, Peale moved to Philadelphia to construct a history on canvas of the Patriots and their cause. He joined the Sons of Liberty, signed up with the militia, rose to the rank of captain, and painted on some of the battlefields where he also fought. He was a multitasker and polymath bursting with ideas and enthusiasm. He developed a device he called a polygraph that duplicated letters, another that pared apples; he also invented a smokeless fireplace and a portable steam bath. Despite Peale's manic energy, people found him engaging. Washington liked him, Franklin respected him, and Jefferson admired him.[13]

After the war Peale returned to painting full-time, specializing in portraiture. Beginning when Washington was a colonel in the Virginia militia, before his hair turned gray and sideburns fell to his earlobes, before he had to put in false teeth (made not of carved wood but of ivory, brass, gold, and human pearly whites) to present the proper image, Peale captured his likeness with pen, pencil, and brush some sixty times, seven from life. Paying attention to bone structure was integral to Peale's art, and when he happened to examine the mastodon skeletal finds, he perceived an opportunity.

Peale ran a commercial art museum out of his home, exhibiting his portraits of revolutionary luminaries. The public's enthusiasm for American heroes never faded, yet to keep paying visitors coming, he needed something fresh and unprecedented—and still American. That something was nature. Personally captivated by the nonhuman world, he shifted his museum's focus from works of art to, as he and others put it, works of nature.*

When bones of what was still being called a mammoth turned up at another Hudson River valley farm in 1801, Peale did something extraordinary. He traveled to the farm, bought the bones for $200, and paid another $100 to excavate the property. To fund a crew and equipment, he borrowed $500 from the American Philosophical Society. Peale threw himself into the dig—figuratively, of course. Uncovering little at the New York farm, he moved to a new

---

* Source after source describes Peale's natural history museum, opened in 1784, as the first in America. Although Du Simitière's preceded Peale's by two years, Peale's museum, portrait art, and historical legacy would outshine Du Simitière's star.

site nearby where there were bones enough for him to become the first to recover a near-complete mammoth (which, remember, was really a mastodon).

Rarely one to overlook a picture opportunity, Peale immortalized the dig in what became one of his most famous paintings: *Exhuming the Mammoth*, as it was originally titled. In France, at the time Peale was committing his earthy excavation to stretched canvas, a young, brilliant, and influential naturalist named Georges Cuvier examined the big, fossilized mammals of Siberia and the Hudson and Ohio River valleys and determined that they were distinct from living elephants—and from each other. He gave the North American species its own name: "mastodon." Peale would eventually rename his painting *The Exhumation of the First American Mastodon*.[14]

By December, after his excavation, Peale and his eldest son, Raphaelle, had assembled a complete skeleton at his natural history museum. Visitors paid 25 cents for a ticket or a dollar for a yearlong pass. The ticket was itself a work of nature art. Printed on the front was a sampling of the museum's holdings: elk, bison, porcupine, turtle, paddlefish, crab, crustaceans, flying squirrel, tree squirrel, bats, ostrich, pheasant, woodpecker, toucan, horned owl, and bald eagle. Peale sold enough tickets to afford a comfortable life for his family, which included a succession of three wives and eighteen children (not all survived childhood). He named his children after distinguished artists and naturalists, and some followed their father in one or both of his professions.

Peale's youngest son, Titian Ramsay, a naturalist and painter, drew a bald eagle in flight that was adopted for a silver dollar that was minted for three years starting in 1836. Titian Peale's bird reappeared two decades later on the so-called Flying Eagle cent, a large pure-copper penny. Looking back at these coins, the art historian and curator Cornelius Vermeule identified Titian's flying eagle as the "first numismatic bird that could be said to derive from nature rather than from colonial carving or heraldry."[15]

Not entirely from *wild* nature, though. Residing today at the US

Mint in Philadelphia is Peter, a taxidermized bald eagle hanging by a wire in motionless flight inside a glass box.* According to the brass label next to Peter's box, he lived in the 1830s, foraging around the city by day and roosting at the mint by night, with full access to the facilities. One evening, a worker engaged a coin press without seeing Peter perched on top. The machine caught and broke one of Peter's wings. He never recovered from the injury. By then, as numismatic historians have written, Peter had served as Titian Ramsay Peale's model for the silver dollar.

Perhaps not the only model. Outside on display behind the Peale Museum was a menagerie of live animals: an elk, a lion, a tiger, and a baboon, as well as bears, snakes, hawks, monkeys, and birds of all sorts, among which was a bald eagle. Sometimes called family pets, each was fated to extermination and mounting. The bald eagle lived on a perch in a cage next to a second cage with a pair of chattering parrots, which sat on top of a bear house. Above the eagle cage a sign read, "Feed me daily for 100 years." The eagle was turned into a mounted display long before a century could pass.[16]

Twelve blocks from Independence Hall, at the Pennsylvania Academy of the Fine Arts, hangs *The Artist and His Museum*, a mammoth eight-foot-by-six-foot painting. The oil-on-canvas was another of Charles Willson Peale's promotional self-portraits, undertaken at the height of the museum's popularity. The portrait's subject, Peale himself, stands front and center of the painting, showman-like, holding up a red velvet curtain trimmed with a gold-tassel fringe. He is welcoming the viewer to enter his museum, where, behind the curtain, he has his sizable collection of mounted animals on rows of shelves. In the foreground next to Peale's right foot, a deboned feathered carcass of a turkey lies draped across a wooden box, beside which is the tool kit of the taxidermist. Directly above the turkey, on the uppermost shelf, is the bald eagle that did not live one hundred years. Interpreters of the painting have concluded that the raptor on its high

---

* Those working in the art of taxidermy loathe the expression "stuffed," so I shall not offend by using it.

perch represents the top tier of the animal kingdom. They have also said that its position above the turkey is a metaphor for its "triumph" over the gobbler, reflecting Peale's own ranking of the bald in American iconography.[17]

The naturalist community similarly held the museum at a level above others for "its scientific arrangement." In Peale's view, what made America America, what put it on par with European nations, was not its art but its biological and geographic attributes. Wisdom, he believed, came from knowing nature, which was the indelible pathway to truth. Because America was so richly endowed, its citizens had the opportunity to elevate civilized society to a level above others, like the bald eagle on the shelf above the turkey on the floor. The "birds and beasts," the museum's advertisements and tickets proclaimed, "will teach thee."[18]

This wasn't carnival hype. Suggesting the natural basis of American ideals, Peale wrote, "As this is an age of discovery, every experiment that brings to light the properties of any natural substance helps to expand the mind, and make man better, more virtuous, and liberal." For Peale, the nourishing struts of a healthy society—order, harmony, diversity, and freedom—were the same as those of nature. "In any one of the departments of animal Nature," he added, sounding like indigenous Americans of his time and ecologists of a later one, "the variety is so great of the wonderful provision [as] to sustain and keep in equal equilibrium each sustaining link to complete a whole system." Peale recognized what few who called themselves naturalists recognized: the interconnectedness—the web—of life.[19]

Not long after Peale erected the first mastodon skeleton, his second-oldest son, Rembrandt, gathered with twelve male companions to sip imported wine at a table beneath its expansive rib cage. To the side, near the great mammal's pelvis, a pianist played "Yankee Doodle." The men raised their glasses toward the belly of the beast while someone delivered a patriotic toast: "The American People: may they be as preeminent among the nations of the earth, as the canopy we sit beneath surpasses the fabric of the mouse!" Some have reduced the moment to a mere publicity stunt. Maybe it was, but the toast was

a true sentiment of the interconnected organic world delivering greatness to the American people.[20]

Little mouse and big pachyderm were, as it happens, part of
the same woven fabric that Charles Willson Peale perceived. In
"preeminence"—to borrow from the toast—the mastodon did not,
in reality, surpass the mouse. Each had its equally essential place in
the web of life, though the mouse had evolved earlier and it survived
long beyond the mastodon. In its position in the food chain, the little
rodent was a morsel with more than a morsel of importance to car-
nivores, bald eagles among them. Which is to say that the mastodon
at one time contributed to the feeding prospects of eagles. Big, tram-
pling herbivorous animals, when grazing, unwittingly flush insects and
rodents from hiding places in grasses and brakes, scaring up meals for
observant carnivorous birds to pounce upon.

⌒

ALONG WITH "YANKEE DOODLE," the pianist at the mastodon
soiree played "Jefferson's March," which had been composed for the
pomp and ceremony of the third US president's 1801 inauguration.
Jefferson and Peale were friends and letter-writing companions, and of
like minds on the subject of the mastodon and the universe. Whereas
many saw randomness and chaos in the nonhuman world, Jefferson
saw order and symmetry, as purposeful in form as an eagle's tail feathers fanning out in flight. He translated these natural aspects into the
architecture at Monticello and into the incisive words of the Declaration of Independence. Freedom, equality, and happiness, he believed,
were determined by incontrovertible laws that derived from nature
rather than a political authority—a philosophical understanding that
the colonies used against Great Britain.

Nature, wrote the political scientist Charles A. Miller, was an "engine
of attack on the authority of tradition." The Founders—who disavowed
women's rights, regarded Indians as uncivilized, and held Africans and
African Americans in bondage—failed to live up to their own philosophy, falsifying the freedoms embodied in their iconographic bird to the
benefit of white privilege. Still, they established a lasting ideological and

constitutional platform. Natural rights and laws persisted as a crucible for the nation, and over the course of its history, the socially and politically disfranchised would continually invoke Jefferson's words and principles to seek full access to the fruits of American society.[21]

Jefferson was more than a political philosopher, word master, inventor, and architect. He was a naturalist. Many of the admired thinkers of the day on both sides of the Atlantic could call themselves naturalists. Franklin, Adams, Madison, even Washington, were students of the physical world. Adams was proud to claim that in drafting the original constitution of Massachusetts, he was the "first" to recommend the promotion and study of "Natural History to his Countrymen."[22]

In the middle of the war against Britain, Jefferson wrote a book titled *Notes on the State of Virginia*, the only full-length book produced from his pen. Charles Thomson thought Jefferson should have called it "A Natural History of Virginia." Of its twenty-three chapters, Jefferson devoted the first eight to zoology, botany, and geography, and he did not limit his geographic range to Virginia. He put forth the continental environment as a national asset and source of energy and vigor. The wild and domesticated animals of his republic, he maintained, were taller, healthier, and more abundant than those in Europe, and there were places in the United States that were "worth a voyage across the Atlantic" to see "the most sublime of Nature's works."[23]

Jefferson's fascination with the subject dated to his boyhood among the peaks and valleys of the Blue Ridge foothills. He grew up long and lean and inquisitive in Shadwell, Virginia, on the Rivanna River, a branch of the James River, which in the nineteenth century would be straddled by a dam with the name Bald Eagle.

Jefferson admired clever mechanical devices and smarter aspects of the human-built environment. But no design anywhere was smarter than nature's. From atop his little mountain, Monticello, Jefferson wrote, "How sublime to look down into the workhouse of nature, to see her clouds, hail, snow, rain, thunder, all fabricated at our feet." He knew the natural histories of the plants and animals in the Blue Ridge foothills, often having learned from those who maintained intimate relationships with the land, his slaves. Birds were his favorite, the ser-

enades of mockingbirds a special pleasure. The outdoors was a con-
cert hall from dawn to dusk. Its music for this Enlightenment savant
was evidence of the exquisite rationality of nature. Jefferson wanted to
know how the nonhuman world worked and what physical laws guided
its enterprise. He wanted to know how these laws extended to human
behavior and civil society.[24]

Jefferson was a member of and tirelessly active in the American
Philosophical Society (APS). If Philadelphia was the American capital of intellectual ferment, the APS was its executive office, promoting
the exchange of "useful knowledge" in lectures it hosted and papers
it published from its proceedings. Jefferson sat on the organization's
Antiquarian Committee, sportingly known as the "bone committee,"
and served as president of the APS when he was also president of the
United States, as well as when the committee lent Peale $500 to dig up
bones in New York.[25]

One of the APS founders was John Bartram, a Philadelphia bota-
nist. When America was still British, European gardeners knew there
was something exceptional about its natural attributes. Bartram, who
lived and gardened beside the Schuylkill River, had much to do with
this perception. For several decades during the eighteenth century, he
sent millions of seeds and young green plants to Europe, inspiring a
horticultural craze. Bartram's vegetation advanced studies in botany
and gave impetus to Carolus Linnaeus, a Swedish naturalist, in developing a taxonomic system for naming organisms, which remains the
standard today. Gardeners in the British Isles and on the European
continent could not get enough of American plants. They were unique,
beautiful, robust. To have seeds from Bartram and to reproduce flora
native to America was to be a bona fide gardener. The ornamental
greenspace in England became America transplanted, such that by the
mid-eighteenth century, writes Andrea Wulf in her wonderful book
The Brother Gardeners, Bartram's plants were growing in "every landscape garden and shrubbery in England." Bartram and his good friend
Benjamin Franklin, another cofounder of the APS, foresaw their
homeland as a living laboratory where scientific achievements could
equal or surpass those in Europe.[26]

Bartram's son William brought that living laboratory to a broader demographic when he published a book recounting his one-man natural history expedition through the colonial South. Jefferson, Peale, and Bartram the younger were of the same generation, born within four years of one another and dead within the same number. A Quaker true to his faith, William Bartram lived simply and frugally in the stone house in which he had been raised, tending and expanding, in preferred solitude, the garden started by his father. "I am continually impelled by a restless spirit of curiosity," he wrote, "in pursuit of new productions of nature." When John began to retreat into retirement, his gentlemen-collector and -scientist clients took note of William's spirit, as well as his talent for drawing nature's productions. They turned to William to continue the flow of plants and seeds across the Atlantic.[27]

One of those gentlemen, John Fothergill, a wealthy English physician and botanical maven, agreed to finance William's trip through the southern colonies. Bartram spent nearly four years combing specimens in the Carolinas, Georgia, East Florida, and West Florida (all British colonies at the time), and across the western boundary in present-day Alabama. Setting out from Charleston in May 1773, the thirty-five-year-old Bartram ultimately covered twenty-four hundred miles on foot and horseback and by boat, traversing every conceivable terrain and enduring every imaginable weather condition, including a hurricane that pursued him like an eagle an osprey. From a Euramerican perspective, the region was still a frontier, although people had occupied it for thousands of years. Outside the principal British American cities of Charleston, Savannah, Pensacola, and Saint Augustine, a scattering of whites lived in small settlements in the backcountry, trading off moments between peace and conflict with Indians. The Quaker Bartram traded only peace.

His was a long walk through a living museum of the most remarkable plants and animals, so many of which he learned about from Indians. The novelties and "beauties in the bounteous kingdom of Flora" stood apart from the environment that most Americans knew, which itself stood apart from Europe's—distinct from the distinct, in other words. Bartram walked from the upper temperate zone to the lower, encountering much that was perennially green—pines and palms and palmettos, and a vigorous, expansive cycad that Indians called *koonti*. Along the Carolina and Georgia coast and across Florida, he was humbled by live oaks, sprawling-limb hardwood trees not yet assigned a botanical name. Rooted in acidic loam soil and watered by fifty-plus inches of rain a year, live oaks grew to eighteen feet in diameter. Stout was a regional affliction: Bald cypresses, swampland trees, shot up as straight as a ship's mast, some fifty feet to a feathery-green overstory. Their fluted trunks splayed outward as much as twelve feet. Grapevines could fatten to a foot thick.[28]

Bartram cataloged 358 species of flora and fauna, giving Linnaeus more to add to his system of taxonomy. Nearly half were unknown to Western science. Of birds, he ticked off 200 species. Those he rendered in ink and pencil are as he saw them engaged in the wild: hunting, feeding, calling, and, in the case of the elegant sandhill crane, strutting.

He came upon bald eagles, although he knew them quite well from the Schuylkill River easing past his garden in Philadelphia. The Creek invited him to witness an Eagle Dance, and he noted matter-of-factly the white tail feathers used in costumes and as ceremonial wands. He was innately drawn to Native peoples' correspondence with the indigenous land, and he waxed lyrical about most wildlife, but not the spiritual bird of the Creek. Sounding like a sardonic Franklin, Bartram called the bird with its "ferocious" brow and "fiery" eyes an "execrable tyrant: he supports his assumed dignity and grandeur by rapine and violence, extorting unreasonable tribute and subsidy from all the feathered nations." Ospreys contribute "liberally to [its] support" and vultures hold themselves "in restraint" when carrion is to be had but not shared by a bald eagle.[29]

Although Bartram was judgmental from the point of view of his culture, what he was observing was species coexisting in pyramidal relationships within their indigenous settings. He acknowledged such coexistence in other observations. Bears and turkeys, he wrote, "are made fat and delicious, from their feeding on the sweet acorns of the Live Oak." Dead trees were exhibit halls of living spaces. They attracted insects that attracted woodpeckers, which hammered out cavities to make homes, which, when abandoned, were bequeathed to bluebirds, owls, and squirrels. On or near the ground, insects fed frogs that fed snakes that fed hawks and owls. Or plankton fed little fish that fed bigger fish that fed bald eagles, sometimes courtesy of ospreys.[30]

All this activity and these exchanges happened within habitat, what today we often call an "ecosystem." An ideal habitat for bald eagles was longleaf and loblolly pine woodland. They ranged tall and forthright from southern British colony to southern British colony in unbroken expanses that awed Bartram. Rivers, lakes, wetlands, and estuaries patched and coursed the pine woodlands, where balds made themselves plentiful in the winter. In the sixty- and seventy-foot-tall pines, they found highly favorable nesting platforms. The wide spacing of the trees within the longleaf environment opened their view to small movements on the ground and gave space for six- to seven-foot wings that would cast down on prey. The eagle's nest at the crown of those trees was, literally and figuratively, the top of the woodland pyramid.

The specimens Bartram saw in the Southeast attested to the bald's habitat range, yet they were, on average, smaller than the balds familiar to him in Pennsylvania. Regional size and weight differences are common among the species along a north–south axis in North America, and differences generally diminish toward the middle latitude of the continent. If Bartram had hopped a whaling ship, rounded South America's Cape Horn, and sailed up to Alaska, balds larger than Pennsylvania's and significantly larger than Florida's would have greeted him. And he would have seen balds in the winter—hundreds, in areas where the fishing was still good. Stouter eagles can cope with temperatures several degrees below freezing as long as they can still eat. To explain the regional size differences, scientists often call up Bergmann's rule. Carl Bergmann was a nineteenth-century German biologist who argued that birds and mammals in colder climates grow larger and heavier to conserve body heat.

Gene pools can contribute to size variations too. Bartram was not likely to have seen a Pennsylvania-born eagle that had taken up with a Florida eagle. When eagles reach breeding age, they almost always return to their natal region to find a mate, and there they produce young. In the next century, ornithologists would start referring to two subspecies, the southern and the northern bald eagles—yet they are *Haliaeetus leucocephalus* all.

Bartram's eagle observations come from his book, compiled from field notes, drawings, and memories. No other of its type from the era exceeds this work's scientific contributions, and few its lyrical gifts. Delayed in its completion by the Revolution and a plodding pen, the five-hundred-plus-page book appeared in 1791 under a leggy fifty-word title. To publish a book in those days, the author or an agent often sold advanced subscriptions to prospective readers. President Washington, Vice President Adams, and Secretary of State Jefferson all subscribed. Bartram was admired, respected, consulted, modeled, even plagiarized, but for some inexplicable reason, sales among his countrymen and -women fell short of a thousand copies. Critics believed that his descriptions of the wild South were merely the flowery likes of his wild imagination. A reviewer in the widely circulated *Columbian Magazine* accused Bartram of "rhapsodical effusions," calling his "style so incorrect and disgustingly pompous." That comment must have sent Bartram out into his garden for quiet distraction.[31]

The *Columbian*'s was not a universal assessment. The English poet Samuel Taylor Coleridge, who knew a thing or two about style, embraced Bartram's *Travels* as a "work of high merit [in] every way." Coleridge and William Wordsworth paid Bartram the ultimate form of flattery by borrowing imagery from *Travels* for poems, most notably Coleridge's "Kubla Kahn" and Wordsworth's "Ruth." In the book's first chapter, Bartram put forth an evocative image: the "amplitude and magnificence of" many scenes along his route, he said, presented "to the imagination, an idea of the first appearance of the earth to man at creation." Upon reading this suggestion of Eden, the face of the *Columbian* reviewer likely twisted up and darkened.[32]

Those who could appreciate the analogy were readers abroad, where many were growing weary of the expanding swaths of denuded forests, the shrinking open spaces, and the filth, noise, and congestion of industrializing cities. To Europeans, America had always represented a new start, a rebirth. Whether or not European readers had ever been across the Atlantic, Bartram's effusiveness enabled them to remember or dream, to see, smell, taste, and hear America. *Travels* was a hit in Europe. It enjoyed nine printings, in London, Berlin, Paris, Dublin, Vienna, and the Netherlands.

⌒⌒

BY THE TIME BARTRAM published *Travels*, the Floridas had returned to the possession of the Spanish in compliance with the 1783 Treaty of Paris. The American gaze shifted momentarily away from the territory he celebrated and sharply westward toward the Appalachian range, out to the Mississippi River, and beyond into unmapped reaches (a shift in outlook that likely contributed to the poor US sales of his book). Jefferson's physical and philosophical view from Monticello was clearest in that direction. The year after finalizing the 1803 Louisiana Purchase, which doubled the size of the United States, Jefferson sent Lewis and Clark off with the Corps of Discovery. James Clark was thirty-three and Lewis's former military commander. Meriwether Lewis was twenty-nine and Jefferson's secretary, with valuable experience in field botany. Both had the requisite discipline and were ruggedly built to more than six feet tall.

The expedition's main objective was to find an uninterrupted water route across the Northwest to the Pacific Ocean. Seeking evidence of America's natural value was of no small importance, though. In his detailed instructions Jefferson highlighted the following: "Worthy of notice will be the soil & face of the country, its growth & vegetable productions; especially those not of the U.S., the animals of the country generally, & especially those not known in the U.S. The remains & accounts of any which may be deemed rare or extinct." Re the latter, he was of course thinking of the mastodon.[33]

The yield from the two-year expedition added up to some 80 plant and 122 animal species new to the taxonomy of Western science, although quite well known to indigenous peoples of the region. Among the "new" birds were Clark's nutcracker and Lewis's woodpecker, the first of which is black-and-gray handsome, and the second black-gray-greenish-reddish-and-pinkish vibrant. One of the expedition members shot a black, broad-winged soaring bird with a bare head. Clark identified it as a vulture. It was a California condor, another first.

Mistaken identity, even by specialists, wasn't uncommon when it came to birds. The expedition saw both bald and golden eagles. Goldens weren't foreign to easterners but they were more common in the West and not as familiar as the founding bird. The explorers called goldens "gray eagles." They also mistook full-size juveniles of either the bald or golden for adults of a third species. "The colors are black and white, and beautifully variegated," Lewis and Clark's published report reads. They addressed this bird as the "calumet eagle," since its feathers hung from the stems of ceremonial pipes used by Native groups that the Americans met. They additionally recognized the versatility of the feathers: Indians "attach them to their own hair, and the manes and tails of their favourite horses, by way of ornament. They also decorate their war caps or bonnets with these feathers." The expedition met Natives who captured eagles to "raise for the sake of feathers" and others who killed them for the same. So "highly is this plumage prized by the Mandans, Minnetarees, and Ricaras that the tail-feathers of two of these eagles will be purchased by the exchange of a good horse or gun, and such accouterments."[34]

Lewis and Clark delivered back east crates of natural samplings, with live animals as a bonus. Jefferson kept a few items for Monticello. Most of the botanical specimens went to the American Philosophical Society. Jefferson sent a live magpie and prairie dog, along with numerous animal skins and skeletons and Indian articles, to Charles Willson Peale. The magpie and prairie dog ended up in mounted poses.

No sought-after northwest passage was discovered, no hoped-for mastodon. In no way, however, was the expedition disappointing. The natural wealth of the West added to the biological and geographic treasures of the East. The West had breathtaking beauty, vastness, and multitudes, all properly delineated as extraordinary.

Many agreed: the monumental grandeur of Jefferson's annexation would give rise to a monumental American culture.

The consensus was shared by no small number in Europe, where despite a sometimes snobbish attitude toward the people of the United States, the general fervor for wild America remained strong. The diplomat and Romanticist writer François-René de Chateaubriand said, "[In] vain does the imagination try to roam at large amidst our cultivated plains" in France, while in America the "soul delights to bury and lose itself amidst boundless forests—loves to wander, by the light of the stars, on the borders of immense lakes, to hover on the roaring gulph of terrific cataracts, to fall with the mighty mass of waters, to mix and confound itself, as it were, with the wild sublimities of Nature."[35]

In a two-volume work on North America, François-Jean de Beauvoir, Marquis de Chastellux, caught himself writing so excessively about the "epochs," "powers," and "phenomenon" of "sublime" nature as an "indelible character" that he apologized to his reader. "I pass from one object to another, and forget myself as I write. . . . I must now quit the friend of nature, but not nature herself, who expects me in all her splendor."[36]

Alexis de Tocqueville, best known for the perceptive observations of national politics and character he made in *Democracy in America*, the first volume of which appeared in 1835, recognized that "Nature offers the solitudes of the New World to Europe." The "curse of the American wilderness," he declared, were its mosquitoes, "generally bigger" than European mosquitoes, yet the curse did not dissuade him from extolling America's endemic virtues. Starting with the Mississippi Valley, which stood out as the "most magnificent dwelling-place prepared by God for man's abode," Tocqueville extended his commentary to the "whole continent." It "seemed prepared to be the abode of a great nation yet unborn."[37]

Tocqueville, Chastellux, and Chateaubriand had each traveled the near and far corners of the United States between the 1790s and 1830s. Legions of others, both American and European, engaged in personal and official endeavors, trekking south and west about the broad land.

They invariably passed by or through places with "eagle" in their name: Eagle Pass, Eagle Point, Eagle Bluff, Eagle Bay, Eagle Harbor, Eagle River, Eagle Rock, Eagle Mountain, Eagle Station, or Eagle's Nest.

Zebulon Pike, famed for Pikes Peak, a precipice he never actually summitted, came upon several such places during two exploratory expeditions he led through the southern parts of the Louisiana territory. He also met Indians named Big Eagle, Black Eagle, and War Eagle. His objective was to access the commercial viability of the West, seeking in the indigenous landscape "great advantage to the United States." Yet in his expedition notes, he used the word "sublime" eight times, and on many occasions he was given to elegiac descriptions. The scenery from the scalloped hills above New Mexico's San Luis Valley inspired the following: "In short, this view combined the sublime and the beautiful. The great and lofty mountains, covered with eternal snows, seemed to surround the luxuriant vale, crowned with perennial flowers, like a terrestrial paradise shut out from the view of man."[38]

During an expedition in the 1830s to Oregon, crossing over mountains, passing through gorges, fording rivers, sinking into frostbiting snow, Nathaniel J. Wyeth of Boston abandoned words such as "paradise" and "sublime." The raw land to him seemed too often "clothed in gloom." His party lost at least seventeen members to drowning, disease, and Indian skirmishes. Still, there were occasional entertaining moments, as when a wily "bald headed eagle" flew off with a goose that had been shot for dinner. There was also the time when some in the party momentarily confused two balds perched in a tree across a river for Indian spies—a mistake that, once realized, aroused needed salutary laughter.[39]

It was the rare journal, diary, or letter composed in the backcountry that failed to see resplendence in the homeland and, in that, a pervasive glory in country and future. "The sublime became America's language of national identity," writes Wulf, "with artists scrambling up mountains to capture the spectacular sights and poets celebrating landscape." (In *Travels*, Bartram used the expression nineteen times, once as "grand sublime.")[40]

Washington Irving was yet another of the day who set out to behold the splendid country and tell the tale. Except he wasn't *just* another. He was Washington Irving, the first writer who consistently set fiction in the American landscape, not least of which are his enduring *Rip Van Winkle* and *The Legend of Sleepy Hollow*. He was a lodestar for Hawthorne, Longfellow, Melville, and Poe, widely respected on both sides of the Atlantic for his literary gifts; and, incidentally, he coined the phrase "the almighty dollar" and the nickname "Gotham" for New York City. Irving, being Irving, captured as well as anyone the deeper cultural meaning of America's extraordinary natural inheritance. When middle-aged, and a suitable physical replication of a pudgy avuncular character that might appear in one of his stories, he wrote, "We send our [male] youth abroad to grow luxurious and effeminate in Europe; it appears to me that a previous tour on the prairies would be more likely to produce that manliness, simplicity, and self-dependence most in unison with our political institutions."[41]

That nationalistic counsel comes from a memoir Irving wrote after his own tour of Europe. He was reflecting on a month he spent in the saddle out in the vast Oklahoma territory, where he had encountered bison, elk, deer, polecats, prairie dogs, wolves, wild horses, wild turkeys, cranes, owls, and crows. He took in a "vast and beautiful," "boundless and fertile" land, and he saw the future of America. On an elevated perch amid all this was the "monarch of the air," the bald eagle, "sovereign of these regions."[42]

⁓

THE HUDSON RIVER DAY Line named one of its steamers after Irving, and Gotham followed suit with a public school. Both did the same for Irving's contemporary DeWitt Clinton, who couldn't stand Irving as a writer, gentleman, or politician. Clinton served as governor of New York and subsequently one of its US senators, and he wore the bald eagle ribbon of the Society of the Cincinnati. He also once dueled with an outspoken supporter of Aaron Burr, a fellow donner of the Society of the Cincinnati eagle, and another whom Clinton loathed. One thing Irving and Clinton could agree on was that in all of America, "Nature had conducted her operations on a magnificent scale."

These are Clinton's words, spoken at the New York Academy of the Fine Arts (later the American Academy of the Fine Arts).[43]

The art historian James A. Craig says that Clinton's speech was "calling for nothing less than the codification of a national identity." What had encouraged his statement was the emergence of the first all-American art form: landscape painting. It had begun taking shape at the front of the nineteenth century with the Schuylkill River School, predecessor to the more famous Hudson River School.[44]

Most of the Schuylkill artists were from Philadelphia. Peale overlapped them, and some of his works incorporate elements of the genre. *The Exhumation of the First American Mastodon* captures more than the disinterring of fossilized bones from a marl pit. In the upper left corner of the painting, a drift of clouds blushes pinkish orange against a blue firmament; in the upper right, lightning strikes out of a dark overhead and reaches to the horizon. It's an exclamation point, a Romanticist statement of America's divine affirmation.

Landscape with Curving River, by Thomas Doughty, whose paintings would hang in the Metropolitan Museum and National Gallery, is one of the better-known Schuylkill works, completed around 1823. Three small female figures stand at its center in playful poses, dwarfed by hillside thickets and trees. They look down on a silent river, which draws out to the canvas's nether reach in correspondence with floating shoals of unthreatening clouds. The aesthetic force in the Schuylkill works is largeness of land and sky and solace of rivers, evoking a reverence for rather than a foreboding of nature. When they appear in Schuylkill paintings, humans are small in number and size, subtle immersive objects in blue-green vastness.

The fountainhead of the Schuylkill River School was the Pennsylvania Academy of the Fine Arts, the nation's first art school. In an 1810 speech on the fifth anniversary of its opening, Joseph Hopkinson, son of Francis Hopkinson, one of the would-be designers of the national seal, slipped in a comparison that occupied the minds of many: "Do not our vast rivers, vast beyond the conception of the European, rolling over immeasurable space, with hills and mountains . . . and luxuriant meadows through which they force their way, afford the most sublime and beautiful objects for the pencil of the Landscape?"[45]

The Schuylkill River School answered that question with a resounding yes. The school was not so much a movement about style as an ideological vision of a nation. Unspoiled North America was a new Garden of Eden, an Eden to nurture a people and their flowering republic. It opened the way for the new Hudson River School, relocating this vision, and the locus of creativity, to New York and beyond.

The artist commonly held up as the Hudson exemplar is Thomas Cole, who studied at the Pennsylvania Academy of the Fine Arts. Born in England, Cole grew up in the Ohio River valley—mastodon country. As he matured as an artist, he eventually settled on a farm in the Hudson River valley, that other mastodon country. An old storehouse built of barn-board siding on the farm served as his studio, and it was there that he completed the paintings that came to define the Hudson River School—paintings of the Catskill range and the White Mountains. Cole toured Europe on a couple of occasions and returned ever more convinced of America's exceptionalism in nature. In an 1836 article that he wrote for the debut of the *American Monthly Magazine*, he made individualized comparisons between the mountains, forests, rivers, lakes, waterfalls, wildness, and sky of home and abroad, and home came out on top. America had features "unknown to Europe," he said. They were "unrivalled," "unsurpassed," "everlasting," "boundless," "glorious," "most impressive," "more enchanting," "sublime."[46]

As in Schuylkill works, eagles typically eluded Hudson River canvases. One exception is a thirty-six-by-fifty-inch painting by Robert Havell Jr., the principal engraver for John James Audubon's *Birds of America*. He not only represented the Hudson River form; he lived on that very river, working in a cupola studio with unobstructed views of the riparian environment. Near the end of his career, Havell completed a work titled *Death of a Warrior: White Headed Eagle*. At its center is a mature bald eagle fighting off two canvasback ducks that are protecting their scurrying brood from the invading raptor. The eagle is on the losing end of the struggle. One wing is folded in, the other raised as if broken. The left eye seeps blood, and more blood reddens the underside plumage. The scene is set on the bank of a large river that flows to its own temperament between still banks, apart from the animated violence at the forefront, calmly toward a sylvan gorge and an umber-colored rocky mountain. A heavy darkness swabs the sky, and a precisely vertical sheet of rain falls to the painted horizon on the right. Toward the upper left, a small parting in the clouds allows brightness to shine through.

There is no certain way to interpret what this painting is trying to convey. On the back of the work, someone wrote in pencil, "Death of a Tyrant." This apparent alternate title was likely a more accurate reflection of Havell's perception of the demeanor of bald eagles, which would have been consistent with prevailing attitudes in the nineteenth century, including the attitudes of ornithologists (more on this in chapter 3).

Similarly unclear is why the ubiquitous bald eagle never established a formidable presence in early American landscape art. Creative minds may have thought that the founding bird was better suited for patriotic and decorative arts, where its presence had been established since George Washington's buttons. As the eighteenth century yielded to the nineteenth, the bald eagle, observed the art critic Philip Isaacson, was putting in an appearance on badges, bottles, buttons, butter molds, vases, flags, stoves, fabrics—in flight, at rest, with the US shield, without it, grasping olive branch and arrows, sometimes thunderbolts—on "virtually every type of tangible object made in this country."[47]

What additionally might account for the bald's rarity was the increasing scarcity of wildlife in the East—the fallout of commercial and recreational hunting and habitat devastation. Fewer and fewer of America's bird were around to be seen in the sky and over rivers as the Hudson River School matured (a subject of chapter 4).

It seems also that the masters of the two schools were not terribly interested in animals, period. If the artist focused on, say, a bird, the resulting work might appear ornithological, and as a consequence the eternal sweep of physical America would devolve into backdrop.

LANDSCAPE ARTISTS OF THE early and mid-nineteenth century were not painting backdrops. They were painting natural America's unblemished pageantry. Many US citizens and visitors wanted to immerse themselves in the sublime offerings of the country. Starting before the Civil War, they began taking nature excursions to exotic Florida (which became a US territory in 1821) and the unbounded West, responding to a second generation of Hudson River Schoolers, who helped bring wonders of the spreading-out nation to the people back east.

The earliest and perhaps most important among these artists was George Catlin, who in the decades before the Civil War followed Washington Irving's prescription for an exceptional experience on the prairie. Clean-shaven, with angular features and a fondness for wearing native buckskin shirts when on the Great Plains, Catlin was a friend of Charles Willson Peale and fascinated by artifacts from the Lewis-and-Clark expedition displayed at the Peale Museum. In 1830, Catlin joined a diplomatic mission that William Clark led up the Mississippi River into Indian country. He was treated to the sight of eagle feathers more than he might have imagined. Tribal members "denominated" eagle feathers as "first-rate plumes," he wrote. These adornments radiated from headdresses, at dances, during medicine ceremonies, from the hilts of lances, at the ends of pipes, amid the manes of horses, and among "objects of superstitious regard, and [were] made with great labor and much ingenuity." Probably no other artist has painted as many "brows plumed with the quills" of eagles.[48]

Native life enthralled Catlin; he gave more brush time to composing portraits than to painting landscapes. A substantive purpose grew out of his infatuation. Witnessing the unfolding of continental expansion, he wanted the government to create a "magnificent park," a "*nation's Park*, containing [native] man and beast, in all the . . . freshness of their nature's beauty!" It would be, he said with repeated exclamatory enthusiasm, a "beautiful and thrilling specimen for America to preserve and hold up to the view of her refined citizens and the world, in future ages!"[49]

The first national park was decades away. In the meantime, art would have to preserve for future ages. That's what a new fad in scenery albums did. Nothing else so overtly connected national consciousness with nature. Scenery albums brought the organic wholeness of the continent to people who could not afford to travel and inspired those who could to take in America's natural heritage. The earliest album was put out by a Philadelphia printer in 1820. The album's illustrator was a former bird scarer (essentially a living scarecrow) named Joshua Shaw, an English-born Philadelphian painter of the Schuylkill River School. In the book's preface he wrote, "In no quarter of the globe are the majesty and loveliness of nature more strikingly conspicuous than in America." Later in the century, in *Picturesque America*, the writer and poet William Cullen Bryant asserted in the preface, "We have some of the wildest and most beautiful scenery in the world."[50]

That was in 1872—after the US assembled all the territory that would constitute the lower forty-eight states, after steel rails and steam locomotives connected the Atlantic Ocean to the Pacific Ocean, after Americans boated down the Colorado River through the Grand Canyon, summited peaks in the Rocky Mountains and the Sierra Nevada, and set eyes on the blue geysers at Yellowstone, the bluer water at Crater Lake, the white snowcaps in the Grand Tetons, the gray rock face in Yosemite Valley, and the redwoods in northern California. That was the year of the founding of the first national park, Yellowstone.

As the nation swelled into lands new to Americans, their feelings for the "sublime" swelled. This favored adjective of so many early writers and naturalists emerged from a feeling or response deep within, not from intellectual discernment but from pride. For Bryant, the eagle's proper place amid the "most beautiful scenery" of his "beloved land" was that which the monarch of the wild sky already claimed everywhere Americans encountered it. In his poem "The Skies" he wrote,

> Far, far below thee, tall old trees
> Arise, and piles built up of old,
> And hills, whose ancient summits freeze,
> In the fierce light and cold,
> The eagle soars his utmost height.[51]

~~

BEFORE JEFFERSON'S DEATH ON the nation's fiftieth birthday (coinciding with Adams's), the polymath Founding Father started acquiring American landscape prints and originals. He owned two works depicting Niagara Falls, and others of the Blue Ridge Mountains and Natural Bridge. If he had happened to see it, Jefferson likely would have approved of Joshua Shaw's scenery album for how it portrayed his nation's identity nursed in the land, mountains, rivers, lakes, gorges, valleys, and sky beyond Jefferson's little mountain.

Monticello's entrance hall was a type of scenery album in full dimension. Mindful of every detail in the house that he never stopped designing, Jefferson filled the seven-hundred-square-foot hall, rising to an eighteen-foot ceiling, with illuminating morning light coming through two tall, arched windows. A prelude to the thirty-five-room house, the entrance hall was a pleasant anteroom with a polished parquet floor where visitors waited to be received by the host. Yet as much as guests occupied the room, it occupied them with its own logic and narrative separate from the house. One caller said it was "rather a museum." Another said, "[The] eye is struck and gratified . . . with objects of science and taste" on full display.[52]

There were paintings and sculpted stone busts of famous men, including a Sack Nation chief; engravings of the Declaration of Independence, and John Trumbull's depiction of its signing with Charles Thomson and the delegates; and maps of Europe, Asia, South America, and the United States—in the last instance, the one Jefferson sketched on to plan the Lewis-and-Clark expedition, apparently using an eagle-quill pen.

Those sunny 12,600 cubic feet were not only a museum; they embodied Jefferson's ample intellect and passions and, as much as anything, his vision of America's unique qualities. Visitors reported seeing on the wall to the right the mounted heads of a ram and a bison, and antlers of a deer, an elk, and a moose, American all. Spread across tables was an "array of the fossil productions of our country, mineral and animal." The most prized were the tusk and jawbone, complete with five-pound teeth—of a mastodon recovered from the Ohio River valley.[53]

There was a bald eagle in the hall too, yet less obvious than the other artifacts. To see it, visitors had to look up to a plaster relief molded into the ceiling. That plaster eagle is there today, restored and painted beige. Eighteen gold stars float around it, representing the number of states that existed at the time Jefferson had the relief installed, sometime between 1812 and 1816. It is decorative art, unlike everything else in this museum-like space, yet it is strategically positioned. The bald eagle is in flight on open wings, peering down over all in the room—king bird lording over the animals in the collection of Americana, just as it lorded over those of the broader land.

THREE

✦ ——— *Twice-Baked Turkey* ——— ✦

THE NAME "AUDUBON" IS A PROFOUNDLY FAMILIAR ONE, best known for the organization using it, the National Audubon Society, and the person from whom it derives, John James Audubon. Turn to the average man or woman on the street for the name of an American ornithologist, and you'll no doubt get Audubon—or maybe Peterson or Sibley, if your survey responder has one of their avian field guides at home. Names you are unlikely to hear are those of Audubon's ornithological predecessors and contemporaries. He had many: Mark Catesby; Georges-Louis Leclerc, Comte de Buffon; Marie Jules César Savigny; John Abbot; George Ord; and Alexander Wilson.

Some have called Wilson, who lived from 1766 to 1813, the father of American ornithology. He is the subject of four biographies, which isn't that impressive for someone ordained with such a distinguished title. And he'll likely never escape Audubon's shadow. A case in point: the first word in promotional copy for the latest Wilson biography, published in the bicentennial of his death, is "Audubon."[1]

Biographical interest in few American scientists—Einstein, Edison, and Franklin—exceeds that of Audubon. He has been immortalized in at least thirteen biographies (not to mention the scores of juvenile books dedicated to him). Three excellent works have been published since 2004, and one—William Souder's *Under a Wild Sky: John James Audubon and the Making of* The Birds of America—was a finalist for a Pulitzer Prize.[2] Towns, schools, parks, sanctuaries, conservation areas, wildlife refuges, neighborhoods, golf courses, recreation centers, streets, and quite a few bridges consecrate him. Twice, his image has graced US postage stamps. He has been honored with three birds bearing his name: Audubon's oriole, Audubon's shearwater, and the yellow-rumped warbler. (The scientific name for the last is *Setophaga coronate auduboni*.)

Is this idolatry excessive and blithe? Compared with Einstein, who has inspired more than a thousand biographies, no, but then probably so. Some observers say that ornithologists regard Audubon as a great artist, and that artists regard him as a great ornithologist. Each camp's willingness to claim him for the other is a telling equivocation. Audubon's chief contribution to science was in identifying and cataloguing birds and discussing a bit of their behavior (he spilled as much ink, if not more, talking about shooting them—to wit, spilling their blood). Of the species he included in *The Birds of America*, the big portfolio of a book that is the collection of Audubon's hand-colored etchings, 94.7 percent were already familiar to Western science. He added twenty-five species to its inventory of birds; Wilson, whose published ornithology also includes illustrations, added thirty-nine. So, did Audubon truly advance ornithology to a new level? Let's just say there is little purchase in being a temple destroyer.

What is unequivocal is that Audubon had a huge ego and a colorful life, and he compiled a sizable body of writing: letters, field journals, a memoir, and an ornithological text. Historians have been drawn to him like Audubon's hunting dogs to a venison thighbone. From his archived life, they have extracted a picture of an emergent endemic ornithology and an evocation of a particular early to mid-nineteenth-century white American male character.

He also left his impression of bald eagles. The praise that one might expect to come from his pen was very much measured. Not because this French-Caribbean immigrant who had been taught English by provincial Quakers lacked the necessary vocabulary. Not because he felt the restraint of scientific objectivity. Just the opposite. The mores of Audubon's civilized world weighted his analysis and dragged him down to a low opinion of bald eagles that he was unabashed, even militant, about sharing. *Haliaeetus leucocephalus* found no friend in *Homo sapiens auduboni*.

In the 1831 ornithological volume that accompanied *The Birds of America*, five words into the first sentence of his entry on the bald eagle, Audubon dropped in "noble," already then a cliché. The paragraph that this first sentence opens is a tribute to "our national standard" on which the "figure" of the eagle is emblazoned, honoring a "great people living in a state of freedom," the very life state of the bird itself.[3]

In the next paragraph, Audubon deployed additional gleaming superlatives: "daring," "great strength," "cool courage." Then, qualifying this garland of encomia, he wrote that if the white-headed eagle were to exhibit a "generous disposition toward others," it "might be looked up to as a model of nobility." Wait. So, the bald eagle isn't noble after all? After putting it up on a perch, was Audubon taking aim and knocking it off? Indeed, before he left the paragraph he called America's bird "ferocious" and "overbearing." It had, he said, a "tyrannical temper."

Audubon then used the ornithological essay as a chopping block on which to behead the avian monarch. In the next two paragraphs he offered anecdotes of bald eagles hunting their prey. They kill to live, in other words—as do most meat eaters in nature. But Audubon accused balds of having a "cruel spirit," of being the "dreaded enemy of the feathered race." He projected his own feelings and affixed the raptors with anthropomorphic traits. All scientists at the time anthropomorphized animals; everybody did. They judged other species as they would members of their church, though animals belonged to a different congregation that had thrived for millions of years by its own tenets. Nobody truly knows whether birds, à la Audubon, condemn bald eagles as cruel and tyrannical.

Audubon furthermore accused the "tyrannical" raptors of a "great degree of cowardice." When a human approached with a gun, he said in giving an example, the bird had a tendency to fly off "in zig-zag lines, to some distance, uttering a hissing noise, not at all like their usual disagreeable imitation of a laugh," which he described as that "of a maniac." Was he saying that an otherwise brave bird would sit still and take a shot of lead like a real man? It seems that the laugh was not that of a coward or maniac but of an adept, smart, zigzagging bird getting the last laugh on the outsmarted ornithologist with a gun. Two paragraphs after Audubon turned eagles into laughing birds, he revealed the reason for his animosity: they ill-humoredly stalked and robbed the innocent osprey, stealing the "hard-earned fruits of its labours."

In the essay's conclusion, America's titular birdman summed up his feelings with a final resolution. Somewhat apologetically, he wrote, "Suffer me, kind reader, to say how much I grieve that [the bald eagle] should have been selected as the Emblem of my Country." He went on to quote the "great Franklin on this subject," saying his opinion "perfectly coincides with my own." He followed with an excerpt from Dr. F's letter to his daughter in the form of his famous screed against the lousy, thieving, lazy, dishonest, immoral, and craven feathered national representative.[4]

Audubon was writing in the 1830s, more than two decades after Franklin's grandson published the letter. The ornithologist, famous on both sides of the Atlantic by this point, was canonizing Franklin's opinion of America's chosen bird of officialdom. Perhaps Audubon allowed himself to be unduly influenced by the views of the Founding Father and eminent scientist at the expense of independent ornithological assessment. Even to the extent that Audubon's favorite bird was the turkey.

The Birds of America, the remarkable volume of artwork that emanated from Audubon's tireless labor, opens—arguably defiantly—with a big, beautiful, color plate of the wild turkey. For no other bird did Audubon likely spend more time painting, and with a flourish of metallic pigment. A visitor to his studio commented that his devotion to capturing his dead model's image to his satisfaction continued "till it rotted and stunk—I hated to lose so much good eating." Audubon bestowed a generous 112 inches of text to the turkey's ornithological essay. To the bald eagle, he allotted only 48 inches, and the bald's image appeared thirty plates after the turkey's (more on this in the pages ahead).[5]

To a man disdainful of a raptor that fled for its life rather than submitting to its death, the turkey could do no wrong in his eyes, ears, and taste buds. He dressed it with a thick gravy of praise for its upstanding behavior, sounds, and delicious flavor. And his adulation did not stop with his senses.

Before Audubon assumed the endeavors of an ornithologist, his family kept a pet turkey, which he had taken from the wild when still a nestling. As an adult, it wandered the local village during the day and roosted on the roof of the Audubon home at night. Audubon's wife, Lucy, gently tied a red string around its neck to distinguish the bird from the feral quarry pursued by quick-to-shoot hunters, including her husband. The pet was likely the model for the gold-and-carnelian fob seal attached to John James's watch chain, which bore the image of a wild turkey. He had a calling card to match. The motto on both read, "AMERICA MY COUNTRY."[6]

Though something of an inconvenient fact, it seems that America's most exalted ornithologist hated America's most exalted bird, the bald eagle. But maybe not as much as he thought. As if he needed it to happen, as a salve for his unhappiness with the founding bird, Audubon discovered a whole new species of eagle. He assigned it a scientific name, *Falco washingtonii*, and found as much satisfaction in it as he did the turkey.

As it turns out, Audubon wasn't seeing what he thought he was seeing. The eagle fooled him, as birds so often fooled naturalists in the early days of ornithology, when men and women of science were distinguishing, sorting, relating, and cataloguing birds, when every identification and assumption and conclusion was provisional. When eagles and birds, to paraphrase Charles Willson Peale, had much to teach thee.

◆━━━━━━━━━━━━━━━━━━━━━━━━━━━━━━━◆

THE BALD EAGLE
IN EARLY SCIENCE,
OR ORNITHOLOGISTS
WITH GUNS

BIRD-INFATUATED NATURALISTS WHO IMBIBED THE MYS-
teries of the universe learned that bald eagles shared a lot about
themselves and withheld just as much. It was common knowledge that
songbirds, with few exceptions, flew south for the winter and north
for the summer, but bald eagles befuddled naturalists. Some of these
fishing raptors willfully stayed put year-round. Some left soon after
the end of nesting season in spring and summer, and appeared to go
off alone or with one or two others rather than with a gathering flock.
Before bird-banding became common in the twentieth century, the
peripatetic bald eagle did not divulge where it went, how long it might
sojourn at any given place, or how many miles—ten or a thousand—it
traveled to get there.

For a long time, scientists debated whether bald eagles even
migrated. The distances they travel pale in comparison with song- and
shorebirds, driven by tight breeding schedules, that cross oceans and
complete transhemispheric journeys. And there is more randomness
than uniformity in the balds' seasonal peregrinations. With few exceptions, they follow no standard flight maps. One will go this way, and
another that. As a rule, although not a strict one, northern bald eagles
fly south after nesting season—mid- to late summer—yet rarely on
transcontinental flights. When southern balds depart the nest—late
spring to early summer—they generally fly north and west, and many

clock hundreds, even a thousand or more, miles. Balds from southern Mississippi, for example, will wing to southern Canada. Sometimes they roost in areas where their departed northern counterparts had been nesting, and will hunt the same grounds and fish the same waters. Some relocate no farther than the state next door. Like other raptors, they migrate during the day—unlike most birds, which travel in the cool and safety of night, when predators are fewer.

Bald eagles choreograph their seasonal movements not to changing temperatures and daylight hours but to the availability of food. Whereas other birds will flee cold places, bald eagles will often stay put as long as they can feed themselves. Snow can cover the ground and if victuals are to be had, eagles will have them. If it hasn't frozen access, ice doesn't discourage them. Pouncing and pecking, they will break through thin layers on a lake or river to catch a live one, having perfected the technique of ice fishing long before humans walked out onto frozen sheets with a staff, ax, or auger. It was not unusual for early New York City residents to see convocations of eagles riding ice floes down the Hudson and East Rivers, pulling fare from the water around them. Some of these birds likely would have been nesters arriving from Canada to spend the deeper days of winter in the area—this was a just-down-the-river *south* to them. Eventually, renewed feeding opportunities and the nesting impulse would recall them to the thawing upper northland.

There were observable clues for early naturalists, yet still many questions. When breeding season circled around again—starting roughly in late November in the continental South, and January to early March in the North—mature eagles returned to domestic routines at existing nests, which can survive decades for reuse. Early naturalists watching for the eagles' return could not be sure whether they were seeing the same birds that had occupied the nest the year before or birds new to the area and nest. Males are roughly twenty percent smaller than females; that much naturalists knew. But it was hardly easier to tell the difference between those of the same sex than between ten-dollar golden eagles coined at the same mint, stamped by the same press.

What eagles made obvious is how busy they were during nesting season. Couples spent the first weeks constructing a new nest or refurbishing an existing one. Domestic sites were all about location, and the best locations were near water with a surplus of fish. By the time eagle hatchlings appeared in respective areas, fish were on spawning runs: shad up the Delaware and the Saint Johns Rivers, salmon the Snake, chinook the Columbia, bass the Elk, crappie the Green, herring the Rappahannock, and sturgeon up and eels down the Hudson; yellow perch spawning in Lake Michigan, alewives the Chesapeake Bay, mullet the Gulf of Mexico, and whiting the waters of Maine; rainbow trout running the Eagle River in Colorado, bluegill the Eagle River in Wisconsin, and salmon the Eagle River in Alaska; steelhead breeding in Michigan's Eagle Lake, bluegill Minnesota's Eagle Lake, and bass Texas's Eagle Lake. Nesting coinciding with spawning was an unassuming accord negotiated between predator and prey. Fish that were inevitably swallowed by larger fish evolved with the survival tactic to reproduce in generous excess, feeding others in the ecosystem—bird, bear, badger, raccoon—sustaining the inimitable circle of life.

A naturalist standing on the ground could easily see the seat of the eagles' continuing existence. Their nests are big and conspicuous, several feet high and several across. They suggest both a lack of concern for camouflage and shrewdness in construction by their top-predator occupants. Eagles integrate their nests into the uppermost candelabra crook of the tallest trees—sturdy ones that rise like watchtowers over hunting territory. From below, the naturalist with an understanding eye could admire a couple at work, flying out to dead trees and attacking small branches with the weight of their bodies to break off building sticks. After carrying each stick back one at a time in their beak or talons as they would a fish or rodent, they would fit each one in place, ultimately raising a structure that was as stout as an old warship. The couple lived in a veritable construction zone. The out-and-back ferrying of building materials continued after the female laid eggs, when the observer below would see the white head of a sitting bird above the thousands of assembled sticks.

What was not initially obvious is that place asserted a deep claim on the eagles' attachment. Partnered couples came back to the same nest or nesting area nearly every year. More than likely, they had not been together during the four to five months outside breeding season, each having risen skyward and gone separate ways in late spring to late summer. Unless tragedy befell one of them, the same two would meet up at their remembered nest year after year. Swans, cranes, and geese mate for life, putting bald eagles in good company. They maintain a fidelity to both spouse and home, which, given their life span of twenty to thirty or more years in the wild, is a significant commitment. As far as science knows, one partner does not cheat on the other. Lying behind their enduring attachments could be the expeditiousness of not having to build anew or not having to clumsily court a different mate each season. Or it could be love. No one can say definitively.

What scientists have learned is that the neurotransmitters in many wildlife species are wired the same as our own and excite emotions we know well. Love is one of them; so, too, jealousy, frustration, and grief. For a long time, the experts maintained that just as animals did not exhibit intelligence, they did not experience emotions—two falsehoods that kept animals at a psychic distance from humans, who frequently harmed them. Instinct—not thought, not intelligence, not calculation—was supposed to be the all-powerful driver of animal behavior. Not until well into the twentieth century did the scientific community let go of these canards, even though there had long been evidence to dispute them.

John James Audubon gathered such evidence in late fall 1820 while traveling down the Mississippi River on a flatboat to New Orleans. Rivers are among the most biologically diverse ecosystems in the world, and just downstream of Audubon and his party were the coastal marshes of Louisiana, fifty-five hundred square miles of the richest estuarine environment in the country. Above and beside the wild smorgasbord of flowing waters and wetlands were kites, cranes, vultures, egrets, herons, geese, and pelicans engaged in perennial quests for sustenance. On December 7, with nesting season underway, one of Audubon's companions took aim and winged a bald eagle. It dropped into the water. They brought the bird aboard alive. Audubon later wrote, the "Noble Fellow Looked at his Enemies with a Contemptable Eye."

The men tethered one of its legs to a push pole, only to have it jump over the side and start swimming, using its wings "with great Effect." If they had not retrieved it, according to Audubon, it would have made shore some two hundred yards distant, while "dragging a Pole Weighing at Least 15lbs." The bird wanted to survive, and its lifetime partner watching from above wished the same for it. The "femelle," Audubon wrote, "hovered over us and shrieked for some time, exhibiting the *true sorrow* of the *Constant Mate*."[7]

One of Audubon's patrons described the birdman's time in Louisiana as a "Tour through the Extensive forests of Western America for a Scientific Purpose." Audubon was constantly collecting specimens to complete what would become his life's work, *The Birds of America*. Published in sections between 1827 and 1838, the book contains 435 hand-colored prints and represents, many say, the most important achievement in early American ornithology. It is also nothing short of an extraordinary artistic endeavor. At twenty-first-century auction houses, original copies have drawn bids exceeding $8 million, and in 2010, one copy commanded the second-highest-ever sum for a book. In its earliest form, *The Birds of America* was a work of science intended for more than scientists. It gave Americans a close-up picture of what their country was made of—one that highlighted the avian diversity and affluence of nature. If the mountains, valleys, canyons, prairies, deserts, and unfolding scenery in every direction made America special, its birdlife did as well. Outside their colonial possessions, England and France could not claim the same species range or the sheer number of birds. Scientists would later discover that America had four major migration flyways. Billions of birds traveled them twice a year. So large were some flocks that they were essentially cloud banks on wings.[8]

These wondrous feather eclipses diminished over time, however. For birds, the expansion of American ornithology and the American population brought about the opposite. Americans had hit the mother lode with nature, landed upon a windfall. Abundance was intoxicating—and deceptive too. It was the rare mind, scientific or otherwise, that could consider anything other than eternal bounties, that could envision depletion, destruction, and disappearance. There was much that science did not know and much that it dismissed. Coming from a deep time, among the oldest animals on Earth, birds had fully evolved in ways that naturalists could not fathom, with senses superior to their own, communicating in ways that naturalists could not hear or decipher, perceiving realities and truths they could not conjure. Feathered species had established symbiotic associations with other life that had persisted for longer than anyone could imagine, and birds thrived in numbers that could only be imagined. *Look at how they fill the sky and blot out the sun* was an utterance repeated across the land.

But science was not ready for ideas such as ecological communities and equilibrium. It was interested in what was out there. Bounties, plentitudes, myriads, slews, superabundance were out there. Species were coming afresh into scientific view—more and more of them. Recording and cataloguing exceptional natural wealth, science went along with the acquisitive standards of Western culture, thus not emulating the lifeways of birds or of the Native cultures that lived in a stable relationship with those birds. In a land of plenty, conservation for the most part seemed unnecessary.

In the middle of his sentence describing the tethered bald eagle desperately swimming for its life, Audubon the artist inserted a clause that reveals his priority in the moment: "I am glad to find that its Eyes were Coresponding with My Drawing." Two paragraphs later, with the flatboat carried farther downriver, he wrote, "Went to an Eagle's Nest; busily Employed Building. Shot at the femelle . . . the Male was also asitting—Killd."[9]

ONE OF THE OLDEST types of animals on Earth, birds have always fascinated humans. Dating to the age of rock and cave paintings, they have stirred in the featherless, wingless *Homo sapiens* a longing to fly, to experience the transcendent freedom and spirit of the air, and to see the world with a bird's-eye view. Birds are masterpieces of nature. The fluid beauty in their colors and physical form is living art. Their every subtle and conspicuous movement—the undulating traverse of the wren, the high step of the heron, the dance of the crane, and the contemplative blink of the owl—is poetry. Wheeling, pitching, pivoting, swooping, and swerving are an aesthetic. "The sailing eagle many a circle sweeps" is how the ornithologist and poet Alexander Wilson once composed the bald in verse. There has probably never been a time when birds did not capture a place in the literature of Western culture. They establish mood and setting, stand as metaphor and simile, signal suspense and direction, connect readers to characters, and appear as characters. Homer's Trojans in the *Iliad* advance on the Pygmy tribes with a "din" recalling the "clamorous cries" of cranes rising above a "measureless rainfall." In Sappho's "Ode to Aphrodite," the goddess of love and beauty arrives in a chariot drawn by "lovely sparrows." Seventy-one species, from cormorants to nightingales, show up in Shakespeare's works.[10]

In and out of literature, no other undomesticated animal makes its presence so conspicuous. From the thirty-thousand-year-old carbon-dated long-eared owl engraved in the Chauvet Cave in France to the Paleo-era eagle of Seminole Canyon in Texas to the present day, birds have captivated the curious mind. Observing their habits is one of the oldest of human preoccupations. Homer, Herodotus, Aristotle, and Pliny the Elder kept notes on them and were mindful of their seasonal flights. The books of Job and Jeremiah speak to migration. In ancient Egypt, birds were symbolic of the human soul. Even when religion discouraged, inquisitive spirits pondered similarities between birds and humans. In sixteenth-century France, the naturalist Pierre Belon included an illustration in his *Book of Birds* that compares their skeletal structures. If Pelon intended to court astonishment, he must have succeeded. The likeness between the human and avian forms evokes some distant nonbiblical ancestry, surely eerily enough to elevate the hackles of devout society.[11]

If you were faithful to the teachings of the Bible, you excluded from your diet certain feathered beings: pelicans, ravens, hawks, owls, falcons, vultures, gulls, cormorants, and eagles. Even storks and ospreys. These were not to be eaten, but not because they were sacred as in indigenous cultures. They were, as confirmed by Leviticus 11 citing God counseling Moses and Aaron, "detestable." Some of the offending birds eat other repudiated creatures, such as mice, lizards, and eels.

Since you are what you eat, you—bird or human—can be found sinful by consumption. Many of the blacklisted are incorrigible scavengers. Scavenging was a detestable habit in itself—one that could, and usually did, lead to eating detestable things. The bald eagle was therefore in double jeopardy, since it was predisposed to both sinking its beak into the flank of a rotting corpse and wrapping its talons around a fresh eel or field mouse.[12]

The first known notes jotted down by a European about North American birds date to Christopher Columbus and his 1492 reach into the so-called New World. Seafarers had always heeded the presence of birds. Song and coastal species coming into sight of an ocean-bound ship indicated land nearby, and certain behavior gave signs of fair or foul weather. A mess of kestrels, gulls, puffins, or frigate birds fishing offshore forecast good weather. When any of these flew inland, an approaching storm had likely persuaded their retreat. Mariners believed that birds had a sixth sense about weather, and they were more or less right. In 2013, researchers learned that white-throated sparrows have the equivalent of internal barometers that enable them to adjust nesting and migration behavior according to weather expectations. As the writer and naturalist Scott Weidensaul points out in an insightful book with an elegant title, *A World on the Wing*, birds possess an extraordinary array of physiological navigational tools related to memory, sight, smell, "quantum mechanics," beaks, and even a nerve that might perform as an internal magnetic compass.[13]

No species is beyond fallibility, of course. Birds can fly astray, and they can get thrashed around in storms just the same as luckless mariners. Still, any sea captain worth his salt knew that the sight of shorebirds was good for the morale of a crew long at sea, particularly on a voyage like Columbus's, sailing toward an uncertain destination.[14]

Columbus's New World destination turned out to be home to an almost unfathomable number of new bird species, new to the newcomers. Among them was a bird that would one day achieve fame as the symbol of a nation yet also draw the scorn of one of that nation's most esteemed ornithologists. Before that nation and ornithologist, a succession of observers described that bird from afar and on its home turf.

⌣

TWO THOUSAND OR MORE years before Columbus, the sight of an alert raptor pursuing live quarry from the wing prompted some-one—early Assyrians, historians believe—to come up with the sport of falconry. The popularity of hunting with trained, quasi-domesticated birds of prey quickly spread across western Asia and into Europe. The favored birds have historically been falcons and hawks, followed by eagles and owls.

The earliest book on falconry written in the Western world, *The Art of Falconry*, was composed in Latin in 1241 by Frederick II of Hohenstaufen. Frederick was the Holy Roman emperor who some-times dressed as an Arab; it was likely his fashion modelers who taught him falconry. His book is fairly expansive, dipping into what today we would call ornithology. By the seventeenth century, books devoted solely to ornithology had been published in Germany, England, and France.

When Europeans began settling the Eastern Seaboard of North America, they had known the golden eagle and the sea eagle from their side of the Atlantic. But they had little familiarity with the bald and for some considerable time did not know that only the bald and golden lived in North America. Colonial endeavors brought a growing variety of birds to European attention. Naturalists felt a greater imperative in constructing a taxonomy, a standardized system for identifying species and making nature more knowable. The task was daunting, and grew even more so as previously unknown birds came to light.

The first to attempt to place the bald eagle in scientific nomencla-ture was most likely made by Francis Willughby. A persevering young seventeenth-century British naturalist, he spearheaded the earliest ambitious effort to inventory birds worldwide in a book of ornithology. Willughby was well-to-do and well educated. Precocious in youth and boyish-faced in adulthood, he wore his blond hair below the shoul-ders, with bangs over the forehead. He never matured out of his boyish looks, though he grew a mustache, a rather feeble one, according to drawings. Either a lung infection or pneumonia cut his life short, yet he made enough of that life to leave a legacy. While a student at Trinity College, Cambridge, he latched onto the scientific revolution sweeping Europe in the 1600s, when natural science moved to the center of intellectual inquiry. In animal studies, the insurgency meant basing classification on observed anatomical features rather than on accepted characterizations handed down from Aristotle and the Bible.

The naturalists of this new era started with the understanding that no bird is solitary; each is joined to the vast avian community, yet each kind is distinct. In compiling his catalogue, Willughby tried to grasp that community by identifying an individual bird as a member of a species and assigning that species to a group of related species. A group was called a "genus," and each genus would belong to a named family of many "genera." So, it went (and still does) individual to species to genus to family.

Just as Willughby was completing his life's work fitting birds into categories, illness took him, at age thirty-six. His friend, housemate, and former tutor, John Ray, put the final touches on Willughby's composition and in 1676 published *Ornithologiae libri tres*. The book is far from a complete inventory of the world's birds, but at four hundred pages, it exceeded anything like it at the time.[15]

Open the stiff, yellow pages of *Ornithologiae* and the first birds you encounter are eagles. Willughby began with *Aquila*, the golden, and then several lines down referred to "*Haliaeetus Aldrov*, The Bald Buzzard; by some called the Sea-Eagle." *Aldrov* is the species name and was apparently chosen to honor the man many considered the father of natural history, Ulisse Aldrovandi. Willughby's bald buzzard that "some called the Sea-Eagle" may be the first mention in an ornithological text of a bald eagle.[16]

In seventeenth-century Europe, hawks were often called buzzards, and "hawk" was used as a catchall term for falcons, kites, ospreys, and eagles, meaning that "buzzard" could refer to an eagle. *Haliaeetus* was the genus Willughby ascribed to the European sea eagle, the genus to which the bald eagle now belongs. According to some sources, the genus name *Haliaeetus* wasn't introduced until 1809—by a French naturalist with his own swanky name, Marie Jules César Savigny. But there on page 17 of Willughby's *Ornithologiae*, published 133 years earlier, is *Haliaeetus*—"The Bald Buzzard; by some called the Sea-Eagle." The Eurasian sea eagle has a dark head, so the *bald* buzzard in Willughby's text is presumably the white-headed eagle of America.

Despite Willughby's precedent, the name *Haliaeetus* didn't fly with other ornithologists. On their home turf, bald eagles would witness humans fighting in four major international wars* and one civil war before science settled on its name for them. One naturalist would try something, and another would come forth with something different. The number of names, noted the twentieth-century writer Polly Redford, became "as richly varied as Americans themselves—a bouillabaisse of Latin, French, and Greek mixed with various birds of prey."[17]

Before ornithologists could reach a consensus, they had to agree on what was out there in the wild and what they were looking at. Bird identification was not, and is not, always easy. With balds, male and female differences weren't the challenge as with many other species. Juveniles were the issue (aren't they always?). They reach the size of adults within a few months, yet for a number of years the colors in their feathers, feet, legs, beaks, and eyes are different from those of their elders. All the while, juvenile balds resemble an adult of potentially another eagle species. Those similarities threw off a long succession of ornithologists. A case in point is John Brickell, who published his *Natural History of North-Carolina* in 1743, the year Franklin, Bartram, and others founded the American Philosophical Society. A physician and resident of the colony of North Carolina, Brickell was known to allow racoons he studied to sip rum. A witty Redford speculated, "Some of his bird lore suggests he may have had a drop or two himself."[18] Brickell wrote,

> The Eagles being accounted the King of the Birds, I shall
> therefore begin with them. Of these are three Sorts, *viz.*
> the *Bald*, the *Black*, and the *Gray Eagle*. The *Bald Eagle* is

---

* The Seven Years' War, the Revolutionary War, the War of 1812, and the Mexican-American War.

the largest, and is so called, because his Head to the middle
of the Neck is covered with a white sort of downy feathers,
whereby it looks very bald.[19]

In identifying a black eagle and a gray eagle, Brickell was not mistak-
enly looking at goldens; his gaze remained well within nearly exclusive
bald eagle territory. Full-size young bald eagles had fooled him.

They fooled William Bartram too. In *Travels* he wrote, "I shall name
the eagle, of which there are three." His included the gray, the bald,
and the "fishing hawk," which he assigned the scientific name *Falco
piscatorius*. Here, he was obviously referring to the osprey, the bird that
"contributes liberally to the support of the bald eagle."[20]

Bartram was writing at one end of the century and Brickell at the
other, and neither was a bird specialist, so they might be forgiven
for their mistakes. Joining them, but adding color—literally—to the
enterprise of classifying, was Mark Catesby, an Englishman who produced a work that overlapped chronologically with Brickell and geographically with both Brickell and Bartram. Catesby was a naturalist
too (meaning a generalist) rather than an ornithologist (meaning a specialist). A resident of England, he visited the colonies twice, spending
approximately eleven years total compiling information for a natural
history of Carolina, Florida, and the Bahamas.

No one before had attempted to classify North American flora and
fauna on such a scale. Printed in 1731 and 1743, and supported by the
Royal Society and 166 presold subscriptions, Catesby's leather-bound
two-volume book sold for 22 guineas, a considerable sum. Yet so eager
were Europeans for more information on American flora and fauna that
his subscribers ranged from royalty to the landed gentry (who owned
parcels in North America) to the humble gardener. The expense was
in the 220 hand-colored images, printed from engraved copper plates,
that infuse the book. Alongside them, Catesby wrote page-length
summaries of the habits and physical characteristics of his subjects,
which were fish, reptiles, mammals, and birds. As to be expected of
a text of the type and time, Catesby's scientific theories and empirical
observations were not without flaws. Still, with its color images the

book was essential enough to the study of homegrown nature for Jefferson to eagerly pay 10 guineas for a used copy in 1783.

Dedicated solely to birds, and following the precedent of Willughby and Brickell, volume 1 of Catesby's natural history opened with eagles. He made no mention of gray, black, or brown eagles and began with the eagle best known to people on the Eastern Seaboard, the bald eagle. He named the bald eagle *Aquila capite albo*, a scientific designation doomed to failure once ornithologists agreed that bald eagles were not a part of the *Aquila* genus, to which the golden now belongs. The eagle in Catesby's illustration is diving for a fish dropped from the talon clench of a fish hawk. The osprey is pictured in the background in its unhappy state. The bald is reaching to grab the fish, a quite large one, in midair with its beak. In real life, the eagle would use its talons. The image is nevertheless dramatic, and for those who had never seen the eagle's tactical ploy, it told a story.[21]

When Americans won their independence and looked to nature as a source of national pride, birdlife still had endless stories to teach naturalists. There persisted the general belief, for example, that swallows hibernated beneath pond ice or self-interred in marshes. As late as 1809, the American Philosophical Society published a letter that gave an eyewitness account of the swallow's alleged winter habit. As for bald eagles, the birds themselves had set naturalists straight on at least two attributes: their dining preferences ran to fish, and with a feathered rather than unfeathered scalp, they were more suited to being called "white-headed eagles." Among all the species names that the lordly raptors tried on for naturalists over the years were several references to their alabaster crown: *blanc*, *albicilla*, *capite albo*, *pygargus*, *leucogaster*, and *leucocephalus*, the one that ultimately fit.[22]

To pull the best fit out of the bunch, birdmen needed answers to some basic questions. For one, was the bald related to *Aquila*, the golden? Or was it, with its white tail feathers and fishy appetite, related to the sea eagle of the same attributes? There was no agreement. And who and what were these other eagles, the gray and the black, that observers recorded? With the world and nature growing conceptually larger and more complex, some naturalists were leaning toward specializing in a concentrated area of study, such as astronomy, entomology, or ornithology. A generalist, the influential eighteenth-century French naturalist Georges-Louis Leclerc, Comte de Buffon, whom Jefferson admired, had recorded roughly twelve hundred birds, but it seemed that every other day scientists were finding errors in Buffon's work. Furthermore, the method of classification he had devised, which rejected the binominal system, was something of a disaster. The avian population was wanting for a more complete identification and standardized classification—which Alexander Wilson bemoaned as a "source of great perplexity." The system needed to reflect at least an elementary understanding of the behavior of birds and relationships with their surroundings, including other birds.[23]

One trait of bald eagles that stuck out in scientific minds was their peculiar rapport with the osprey. Catesby highlighted it in his short description of the bald and then again in the description of the osprey. He also briefly noted the less familiar domestic habits of balds. "They always make their Nests near the sea, or great rivers," he wrote, "and usually on old, dead Pine or Cypress-trees, continuing to build annually on the same tree, till it fails."[24]

Visual images were as important as written ones in learning about birds. Naturalists did not always draw their subjects within their environment—that old dead pine or cypress. Birds were often suspended against the white (actually yellow) of the page or placed upon a generic unidentifiable perch, made more object than subject, presumably to emphasize identifying colors and features. Catesby departed from this tradition. He drew most of his birds with a specific tree, shrub, or flower that he associated with their species—a groundbreaking practice for which he is often not credited. Eight of his subjects are plucking a seed, fruit, or bug; the red-crested woodpecker (now called "pileated") is digging out bugs with a stick in its beak; the kingfisher is swallowing a fish; and the fish hawk is standing on one.

To picture a bird interacting with another species, as Catesby did with the bald eagle partaking in plunderous activity against the osprey, was a bold departure from convention. Some conventionalists, preferring the blank background, could not abide it. One blunt critic maintained that Catesby's "sole objective was to make show figures of the productions of Nature, rather than to give correct and accurate representations." The critic could not have been further from the truth. Catesby's very objective was representation.[25]

EMPTY BACKGROUNDS REMAINED THE standard when Alexander Wilson made himself into a dedicated ornithologist, and he followed it. Still, a poet at heart, his nature was to be creative and to allow his subjects to speak to him. Wilson was born in 1766, the son of a poor working family in Paisley, Scotland. After growing into a young man with a long and narrow face, sloping nose, and geometrical jaw, he tried his hand as a cowherd, a weaver, an itinerant peddler, and a published poet. Sympathetic to the plight of weavers exploited by factory owners, he decided to become their voice of protest, using satirical verse. His creations got him arrested, and authorities once forced him, as in an *auto-da-fé*, to burn some of his censuring poetry on the steps of the Paisley courthouse. Persistently a footfall away from the jailer's grip and a pauper's life, Wilson emigrated to America in 1794, the year Charles Willson Peale announced he was quitting portrait art for natural history.

In America, Wilson supported himself in assorted vocations, none of which satisfied his creative spirit. What stimulated him was the new country and its natural history, and the inventive and intellectual genius that surrounded him. He had the good fortune to land at the southern outskirts of Philadelphia, America's Edinburgh—two cities that were centers of the Enlightenment. America introduced him to freedoms and opportunities he had not experienced before. In the nation's promising future, he saw his own, and in its living, star-studded political leadership he recognized political sensibilities he had long embraced. When Jefferson ran for president, Wilson composed turgid verse celebrating him as the "enlightened philosopher—the distinguished naturalist—the *first statesman on earth*—the friend, the ornament of science."[26]

Those words appear in a letter to William Bartram. By 1804, Wilson had taken up residence at Gray's Ferry, a short walk—on the Floating

Bridge over the eagle-fishing Schuylkill River—away from Philadelphia. The Bartram farm was just up the way over a hill where, Wilson once wrote in a poem, "happy millions" raised "voluntary songs" to "Nature's universal God." He became a regular visitor of "Bartram's hospitable dome" and "fairy landscapes," and before long he was calling Bartram "my venerable friend"; seven times he called him that in his eventual book. Wilson also befriended Peale and willingly surrendered to the allure of Peale's museum. The two veteran naturalists encouraged Wilson's budding interest in natural history, which flourished beyond what they probably anticipated. In a letter to another Philadelphia acquaintance, engraver and fellow Scotsman Alexander Lawson, an irrepressible Wilson announced, "I am most earnestly bent on pursuing my plan of making a Collection of all the Birds of this part of N. America." Apparently feeling a little embarrassed for his boldness, he added, "Now I dont want you to throw cold water as Shakespeare says on this notion."[27]

Dousing Wilson's ambition to compile the most complete ornithology of America would have been a justifiable action. His knowledge of birds was not very extensive. He was hardly more qualified as an artist, skills of which were essential for any successful naturalist. But those who assumed roles as Wilson's mentors did not go Shakespearean on him. When he was three years into his effort and frustrated with streaking in his watercolor paint, Wilson wrote to Bartram, "You know the whole affair ten times better than I can pretend to." Beyond artistic advice, Bartram shared his knowledge of southern species, and Peale made his inventory of nearly 760 mounted birds available. The two directed Wilson to other private and public collections, and to published ornithologies, library references, and the roads to follow on research trips.[28]

Wilson eventually traveled twelve thousand miles, crossing through and over the Appalachian range and staying primarily east of the Mississippi River, moving steadily in all seasons, using whatever transportation was immediately accessible: horse, stage, boat, his well-shod feet. The venture came with risks. In the water at Cape May, New Jersey, he was nearly swept out to sea when he tried to retrieve a downed American pied oystercatcher (pied with a white belly against a black-gray body). Following a long and trepidatious stretch in the southern back-woods in 1810, he wrote his brother, "I have slept for several weeks in the wilderness alone . . . with my guns and my pistols in my bosom." In Georgia, where he often thought of Bartram's own excursions through the "sultry clime," he met John Abbot, an English-born artist-naturalist who had lived in what was then the southernmost state since 1775.[29]

Abbot was a refreshingly pleasant entomologist with sheeplike eyes and a sensitive brow—quite the opposite of the bald eagle. He was best known for his insect sketches, some three thousand of which he would complete during his life of eighty-nine years—forty-two more than Wilson lived. He also amassed some thirteen hundred bird images. "Abbot appears to have been an ornithologist," said one of that scientific species at the shank of the nineteenth century, "but without the name." Abbot rendered several balds in watercolor drawings. Unlike European naturalists who waited for skins, taxidermized samples, and descriptive letters to arrive from commissioned collectors in the New World, Abbot spent endless hours studying his subjects in their natural setting. He and his visitor from the North undoubtedly found common ground with this approach. "I have been on several excursions with him," Wilson wrote to Bartram. "He is a very good observer." Wilson relied on empiricism in its purest form—or "personal intimacy," as he described it—to lead him to "truly ascertain the character of" each among the "feathered race, noting their particular haunts, modes of constructing their nests, manner of flight, seasons of migration, favorite food, and numberless other minutiae."[30]

Dogged curiosity and a shining, wide-eyed sense of wonder that never turned off were essential to a naturalist seeking to prosper in the profession. The willingness to question entrenched gospel could elevate one above others. These very qualities enabled Wilson to debunk the myth of the ice-hibernating swallow. They also lent credibility to a novice fledging into a professional. When starting out on his quest at age thirty-eight, Wilson hardly knew a swallow from a martin or, as Shakespeare's Hamlet said, a hawk from a heron, and of none did he know the proper nomenclature. Wilson compared the passion he afforded his ambition to the excitement of an eight-year-old boy he met on his travels. In woods near his home, the boy had recently discovered the "beautiful simplicity of nature." As soon as he came out of them, he couldn't wait to go back in.[31]

⌒

IN 1808, FOUR YEARS after entering the thickets of avian studies, Wilson published the first of nine volumes of *American Ornithology, or the Natural History of the Birds of the United States*. The Boston *Columbian Centinel* wrote, "Every gentleman of easy fortune, who has any taste for Natural History, and every Public Library throughout the U. States, ought to be possessed of a copy of this splendid national work."[32]

Wilson's colored illustrations are exceptional in quality and detail (Alexander Lawson was the principal engraver for *American Ornithology*)—and quite remarkable for someone who began as a beginner. They are at the same time austere. Unlike Catesby, who had the birds in his paintings engage with their botanical habitat, Wilson's subjects mostly sit on nondescript unleafed twigs and branches unattached to trees, or on bare ground without background. He made up for the emptiness by devoting an average of four pages of written text for each bird. (Catesby, by comparison, stuck to a single column on a single page; a second column duplicated the first in Latin.) *Blackwood's Edinburgh Magazine* wrote of Wilson in 1831 that "he is the best painter in words of birds that the world has yet seen"—a compliment that presumably pleased the poet. Wilson ultimately wrote on 262 species, 39 of them new to Western science. This vast landscape required nine volumes, the last of which was published six years after the first.[33]

The bald eagle makes its appearance in volume 4, which came out in 1811. Wilson did not form a Franklinesque opinion of the bird. He embraced it as worthy of its status as the proud symbol of the nation. The "adopted *emblem* of our country," he wrote, "is entitled to particular notice. . . . Among all our Birds, none has so noble an exterior as the Bald Eagle."[34] Wilson agreed with Peale and Jefferson that a nation—its character and quality of ideas—corresponded with its natural environment, and he believed the bald eagle exemplified this correspondence. Wilson gave the American raptor eleven pages of descriptive narrative, with its own color plate and poem. Both painting and poem are set at Niagara Falls.

> High o'er the watery uproar, silent seen,
> Sailing sedate in majesty serene,
> Now midst the pillar's spray sublimely lost,
> And now, emerging, down the rapids tost
> Glides the Bald Eagle, gazing, calm and slow[35]

Wilson was less poetic than strategic with the caption he chose for the section: "White-Headed, or Bald Eagle." Note the conjunction "or." He cared little for the name "bald eagle." "Bald," said the tousled-haired ornithologist, was an "epithet," an "improper and absurd" description. The "most beautiful of his tribe in this part of the world" deserved better.

Wilson found satisfaction in using the classification system of Carolus Linnaeus. The species name that the revered taxonomist assigned to the bald eagle was *leucocephalus*, meaning "white head." Linnaeus listed the bald eagle for the first time in a 1766 revision of volume 10 in his *Systema naturae*. But he opted not to use the golden eagle genus, *Aquila*. He did not think the bald, which he had never seen in the wild, belonged with the sea eagle, *Haliaeetus*, either. He assigned it to the falcon genus. So, for a time the bald eagle, according to supreme ornithological authority, was *Falco leucocephalus*, the name Wilson used.[36]

No other eagles appear in *American Ornithology*—no black or gray or brown ones, only the white-headed. Someone had given Wilson a young eagle, he explained in the text, and his captive treasure aged into its mature feathers, into a bald eagle, revealing to him a common mistake made by fellow birdmen. The "gradual metamorphosis," he wrote, "will account for the circumstance, so frequently observed, of the Grey and White-headed Eagle being seen together, both being in fact the same species, in different stages of color, according to their difference of age." Charles Willson Peale, having witnessed his family's pet bald eagle grow into an adult, had previously come to that conclusion, which he shared in a 1792 meeting of the American Philosophical Society. To bolster his authority with firsthand empirical evidence, Wilson may have, in fact, claimed an observation and bird that belonged to Peale from years earlier, before Wilson disembarked in America. One way or the other, Wilson underscored the gray eagle's false identity.[37]

Wilson also underscored nesting behavior, of which uncaged eagles taught him much. "On some noted tree," he observed, "the Bald Eagle builds, year after year, for a long series of years." The enduring fealty to nest, he noted, matched that to life partner. Wilson then offered an observation that had apparently emanated from unseemly scientific methodology (unseemly for an observation of the twenty-first century, but not one from the nineteenth century). "When both male and female have been shot from the nest," he wrote without further comment, "another pair has soon after taken possession." Eagles were not only loyal mates but also caring parents. "No bird provides more abundantly for its young," said Wilson. "Fish are daily carried thither in numbers." Those numbers were so great, he claimed, that the "putrid smell of the nest may be distinguished at the distance of several hundred yards." Contributing to the pungency was a mishmash of moldering construction materials: "large sticks, sods, earthy rubbish, hay, moss, &c." Writing some years later, a naturalist and historian of North Carolina, John Lawson, picked up the same sniff on the right wind, reporting that a nest was "commonly so full of nasty bones and carcasses that it stinks most offensively."[38]

Wilson added to scientific understanding, yet sometimes allowed his imagination to take over. The ornithologist-cum-poet all but referred to the bald as a "spirit bird." It exhibited "powers of flight capable of outstripping even the tempests themselves," ascending "till it gradually disappears in the distant blue ether . . . along the face of the heavens." Wilson also got some things wrong. Balds, which in the wild can live well into their thirties, reach a "great age," Wilson wrote, as much as "sixty, eighty, and as some assert, one hundred years."[39]

Wilson's *Falco leucocephalus* is one of the few birds he painted not against an empty background but actively engaged with the wild. For a setting, he chose the Niagara River and its precipitous falls, a "noted place of resort for the bald eagle." The top predator stands on the riverbank, with folded wings, sweeping tail feathers, and thrusting neck—a continuous axis parallel to the ground that gives it the appearance of forward motion, though it is committed to staying put. Death was common at the falls—death that fed the living—and Wilson committed himself to portraying it. Staring outward into its surroundings, his eagle has its right foot planted on a lifeless fish. Blood reddens the raptor's beak.[40]

Sometimes, in the white-headed raptor's interactions with ornithologists, the blood was its own, which came by way of the sharpshooter eye of the ornithologist, epitomized by no one more so than John James Audubon.

⁀

IN OCTOBER 2018, ONE of Audubon's prized possessions, smelling of black powder and oil, sold at auction for $192,000. The auction house called it a "Long Tom," the common name for a particular type of English percussion fowler, a gun. "Fowler" is for fowl, and "fowl" is for birds—waterfowl and wildfowl—all of which is to say that the gun was made specifically for shooting birds. This long shooter measured nearly five feet from brass-plated butt to brass bead, a length that increases firing distance and enhances accuracy. The barrel's smooth bore and gauge size had been designed for birdshot, a compaction of several score to several hundred lead pellets. The number and size of pellets used depend on the size of the quarry. Upon firing, the pellets scatter to improve the chance of hitting a moving target. Ideally, the target is scored with only a single or a few pellets, to avoid an explosion of feathers and body parts, to bring the bird down whole. As guns go, Audubon's Long Tom was a fine-looking one and in fair shape, for a piece that spent most of its working days in all corners of the American backcountry, from Florida to Labrador, from Louisiana to the Great Lakes, enduring harsh environments and heavy use.

If you've seen a portrait of Audubon, you've probably seen the Long Tom. He customarily cradled the instrument in his arms when he sat or stood for artists. His son John Woodhouse Audubon painted at least two portraits of his father with said gun. In one, the subject holds his shooter as if preparing to level its barrel on something with feathers. At the time Audubon arranged to publish the first installment of *The Birds of America*, the Scottish portrait artist John Syme composed what has become the most familiar image of him, which hangs in the White House. Audubon is forty-one with an aquiline nose and dark hair waving down to his white shirt collar. His aspect is serene, his eyes set in a gaze to his left—eyes that, despite their deep-brown color, Audubon thought resembled "more those of an enraged eagle than mine." He is wearing a wolfskin jacket and heavily buckled leather ammunition straps crossing over his shoulders. Clasped between his hands, pulled against his chest like a favorite book, is the Long Tom.[41]

As he posed for others, so also did Audubon depict himself, yet more daringly and dramatically. In the color plate of the golden eagle in *The Birds of America*, he slipped in a tiny detail of himself straddling a downed tree while crossing over a perilously deep ravine. Strapped to the audacious adventurer's back is a gun and a dead bird.

At age sixty-five, some moments before his death in late January 1851 at home in New York City, when he was uncommunicative, immobile, and lost in the recesses of his mind and flashes of old memories, Audubon did not utter but declaimed his last words. They came out when his brother-in-law, William Bakewell, walked into his room: "Yes, yes, Billy! You go down that side of Long Pond, and I'll go this side and we'll get some ducks."[42]

Observing, admiring, painting, and shooting birds filled the arc of Audubon's life. With his work, tireless self-promotion, and longer life, Audubon pushed Wilson, whose own life and efforts were cut short by dysentery, irreverently into a deep chasm of the dead and forgotten. Audubon's images opened up America of the feral past to latter-day viewers and became synonymous with the study and articulation of avian life. Those heavily thumbed modern-day field guides to birds lying at the ready on side tables next to windows all across America—David Sibley's and Roger Torey Peterson's—are descended from Audubon. They are based on the theory that drawings and paintings like his—but also Wilson's and Catesby's—can teach us more about birds than photographs can.

Birds began to shape Audubon's life, and his consciousness, early on. So, too, did revolution, which framed the circumstances of his vulnerable youth. His birth in 1785 in Saint-Domingue, soon to become Haiti, was something of an insurgency. It resulted as a presumably unintended consequence of a liaison between a chambermaid and a sugarcane planter/French naval officer who had fought with the Americans during the Revolution. But the boy was not an unwanted child. His name at birth was Jean Rabin Audubon, which married the names of his unmarried parents: father Jean Audubon and mother Jeanne Rabin. His mother died when he was a few months old, and not long after, the revolution soon to erupt from legitimate discontent among the enslaved, some of whom were owned by the senior Audubon, prompted the exodus of the family—father, son, and a half sister from another liaison. They hastened to France, where Jean Rabin was raised by his father's French wife. Perhaps in consideration for her, they changed his name to Jean-Jacques.

After the French Revolution, with the country spiraling toward the Napoleonic wars, the senior Audubon spirited his son out of France to avoid military conscription. The younger Audubon crossed the Atlantic to his father's plantation north of Philadelphia on Perkiomen Creek, near its unassuming confluence with the Schuylkill River. In America, the eighteen-year-old Audubon began operating by his third and final name, John James.

It wasn't long before he met and fell for the daughter of a neighboring planter. Lucy Green Bakewell was well-read, musically talented, delightfully attractive, and blissfully tempered, which he was not. Audubon had ambition, masculine good looks, and a French accent sprinkled with quaint idioms, such as "thee" and "thou" (remember, Quakers taught him English). He also possessed a powerful sense of self, to say nothing of vanity and self-absorption. John and Lucy married in 1808, the year the first volume of Wilson's *American Ornithology* appeared—a book that Audubon would not see for a couple more years (1808 was also the year Peale completed his painting *Exhuming the Mammoth*).

By then Audubon had forsaken Pennsylvania, traversed the Appala-

chian range, and settled in Louisville, Kentucky, where he and a partner opened a dry-goods store. The country was still young—as, too,
its population. More than half of those the national census counted,
excluding Indians and the enslaved, had yet to reach the age of twenty.
Youth was ever on the move westward. Audubon wanted to meet way-
farers as they passed in that direction, and unload from them some of
their saved-up coins.

Two years into Audubon's business, a customer, trim and cultured
yet a little worse for wear, walked into the store and up to the dry-goods
counter. He spoke with a Scottish burr, which didn't make him stand
out. In the backcountry, almost everybody seemed to be from somewhere else—including John James and his business partner from
France and Lucy from England. The customer from Scotland wasn't
a customer. He was Alexander Wilson, who, for the moment, was an
itinerant peddler with one product to sell: subscriptions to *American
Ornithology*. He had samples of the volume, which he gladly shared
with the storekeeper. Taken by them, Audubon very nearly made a
purchase, but either he decided he could not afford the $120 subscrip-
tion, or he thought his own paintings were as good, if not better.

Audubon had been drawing birdlife at least since his years in France.
Everywhere he traveled, he made notes and sketches. His interest in the
avian scene had heightened in America, where the "feathered beings"
existed in sight and song in conspicuous numbers. Certain species
could turn water, land, and sky into a crowded natural canvas—egrets,
herons, and ibises retiring to roosts at sunset in the Everglades; white
pelicans lake- and bayou-hopping around the Gulf coast; and passenger pigeons, often called wild pigeons, fording migration routes across
the North. Nothing is and was more renowned than the spellbind-
ing flyovers of passenger pigeons. They brought people out of houses,
workplaces, and taverns with a sense of awe. Many ran back inside to
retrieve guns—too many, ultimately. For "every old woman in the village," as a character explains in James Fenimore Cooper's 1823 novel
The Pioneers, the sky proffered a pigeon potpie.[43]

By the time Audubon moved to Louisville, he had assembled a
sizable portfolio, part of which he brought out to show Wilson, who

later confided to his diary that the work was "very good." Audubon related the moment by maintaining that Wilson grew nervous upon seeing them, fearing he was looking at the endeavors of a competitor. Each had discovered a kindred spirit, but they became rivals, and they remained so even after Wilson's premature death three years later, fourteen before the first installment of Audubon's *Birds of America*. Wilson left Louisville without having sold a single subscription, embittered that "science or literature has not one friend in this place."[44]

Soon after, Audubon left town too. He moved with his family to western Kentucky to open another store. Prosperity bloomed and then withered. An ill-fated business investment and a national economic crisis, flaring after the War of 1812, sent him to jail for unpaid debts. Sitting behind bars with time to think, Audubon conceived of an opportunity. Ever since meeting Wilson, he had craved to do his own book of America's birds. As it took shape in his mind, his ornithology had to be bigger in physical size—so, too, in its inventory of bird species—than Wilson's. And the art would be better. Declaring bankruptcy, Audubon secured his release and settled his affairs the best he could. Leaving Lucy and their two boys in Kentucky for the time being, he went to New Orleans to paint portraits for hire and birds for his reconceptualized future.

Like Wilson, Abbot, and Catesby, Audubon studied birds in their environment and acquired most of those he would use as models, completing acquisitions primarily with the Long Tom. To ornithologists, guns and powder were as brushes and paint. Shooting skills were as important as a sharp eye for detail and steady hand with the horsehair. You didn't go out into the field without a collecting gun. Rendering one image required several, a dozen, or a score of dead specimens. Birds that escaped your sight you sought from someone else, who would shoot, skin (leaving the feathers), and deliver them to you (not the best alternative for trying to paint a bird's living likeness). Being a serious ornithologist epitomized loving to death.

Let's be fair, though, and consider the times, the environment, and the wildlife numbers. Death was a daily occurrence not only in nature but also in human society. People were accustomed to witnessing it and instigating it. They commonly killed to eat, whether slaughtering a pig or wringing a chicken's neck on the farm, shooting dinner in the woods, or catching it in the river. And there is that word "abundance"; habitat remained largely unmolested and handsomely supplied with wildlife. There were *birds birds* everywhere, singing, screeching, chirping, hooting, flying, feeding, hunting, nesting, and fouling the roof, stoop, and fence rail. Taking a dozen or so robins for a drawing would hardly, seemingly, put a chink in their tribe. Besides, what choice did one have other than to use a gun? Birds don't sit still like Washington did for Peale. A camera suitable for the field was still two decades away from development, and binoculars even further in the future. In the meantime, science had its demands, and the alternatives to loading and firing a shot of pellets weren't any better: clubbing a bird to death or lassoing and drowning one.

These mitigating circumstances lose something of their thrust, however, when viewed beside Native cultures that regarded the eagle as a spiritual being and whose people witnessed as much death in their daily lives as anyone. Indigenous traditions, for example, would not have tolerated certain persecutions at Casco Bay, Maine, in the 1660s. The historic feeding fortunes of wintering bald eagles on the rocky shore took a fatal turn when white settlers began shooting them to feed their hogs. That ominous turn in fortune portended the future that bald eagles would endure on a continent culturally and ecologically transformed by Euramerican occupation. Part of the transformation included superseding indigenous spiritual rituals with alien savage deeds. Of the latter, shooting was probably the least violent of violent takings of balds. Those alternatives—clubbing, lassoing, drowning—happened, and characteristically without a grain of compunction.

Things weren't any more pleasant on Louisiana's Lake Pontchartrain in the winter of 1821. Audubon was the instigator. Lake Pontchartrain is an ocean-like 630 square miles of estuarine brackish water with sedge marshes, cypress swamps, bottomland hardwoods, and blackwater lagoons. From an aerial perspective, the lake suggests the giant head of an octopus on the squiggly body of New Orleans. It's the most conspicuous geographic feature in the area—more conspicuous than the big river that corkscrews through the usually uncorked city.

The wet environment attracts restless populations of residential and migratory birds, and they duly attracted Audubon. He wanted both adult and young models. Audubon could not use the Long Tom with a pair of eaglets he happened upon at Lake Pontchartrain, fettered to their nest high in a tree. Neither he nor a couple of helpers cared to climb the tree either, risking confrontation with protective parents armed with saber talons (although such attacks are rare). So, he devised another strategy: he had the helpers cut down the tree. Audubon admitted that it wasn't his most propitious tactic. It "caused great trouble to secure" the unfledged. Once on the ground, with nest shattered, the chicks scrambled about until finally they "became fatigued, and at length were so exhausted as to offer no resistance, when we were securing them with cords."[45]

Audubon never explained what scientific use these hapless eaglets served him. No visual or ornithological account shows up in his books, yet he continued to take other nestlings from the wild. A decade after the chopped-down home on Lake Pontchartrain, he was on the Saint Johns River in Florida hunting eaglets, after shooting five adults in one day. The deaths of two of the young were drawn out in a long torture too excruciating to repeat here. In this instance, Audubon shared what use their demise served. The "young birds were skinned, cooked, and eaten by those who had been 'in at the death.'" He assessed them as "good eating, the flesh resembling veal in taste and tenderness."[46]

Nesting tree and eagles: sentenced to death by the ornithology of the day. If you could not be an executioner, you could not succeed. If the ornithologists with guns harbored ambiguous personal feelings about admiring the domestic routines of birds and then obliterating them, they did not openly share those feelings. Emotionally and philosophically, at least as manifested, collecting animals was no different from collecting plants. You spoke as matter-of-factly about one as about the other. Note Wilson's reference, offered earlier, to shooting an eagle couple from their nest. He wrote about that incident as if a scientist taking their lives were as natural as lightning striking them, although if he had witnessed the latter, he might have denounced cruel nature. Audubon shared the tree-chopping vignette in his *Ornithological Biography*, a companion volume to *The Birds of America*. The author obviously had no worries that disclosing the sanguinary realities of his practice would trigger a disgusted reader into flinging *The Birds of America*, a twenty-year endeavor with 435 color plates, against the wall (never mind that it was too big for that).

Four hundred and ninety-seven species make appearances in that large book. If Audubon needed an average of five birds to model for each, a veritable battalion of nearly twenty-five hundred birds would have to have given up their lives for his science and art. Why so many? No matter the level of marksmanship, not every shot brought down a bird undamaged. If the drop was bad, you shot another and hoped for better results. The living color of a dead bird can fade ever so quickly too, so you kept a supply of fresh ones coming. But is twenty-five hundred really so many? It depends on how you look at things. For argument's sake, let's double the ante, to five thousand. Spread across two decades of shooting, Audubon's kill rate per year would then average 250. In a reedy blind in the Louisiana coastal marshes, a lazy duck hunter with a good aim and using a breech-loading rifle, which came along in the 1830s, could bag that many birds in a single day. In that context then, the ornithologist's death toll seems relatively insignificant, except that Audubon was part of that larger context. And so, too, was Wilson.

Both men fancied hunting for sport. When Wilson was visiting Louisville, the two went out together to shoot birds for amusement. Given the alpha-dogging that went on around their art back at the store, one cannot help imagining that it followed them out into the field, where birds paid the bloody price.

THREE YEARS AFTER STEPPING off the flatboat in New Orleans, having undertaken his share of commissioned portraits and collected and painted a sizable number of birds, Audubon set out for Philadelphia to find a sponsor to publish his work on America's avian life.

The country would soon turn fifty. Franklin, Bartram, and Benjamin Smith Barton were dead. Peale was not three years from his last breath. Wilson was eleven years in the ground at Philadelphia's Gloria Dei Church cemetery. But Philadelphia was still the epicenter for natural history and the logical place to seek a financial backer—except it wasn't, not for Audubon. He made his bid to the board members of the Academy of Natural Sciences but was stalled before hardly getting started. A reputation for being something of a blustering, backwoods boor and pseudonaturalist didn't help, yet character wasn't his worst enemy. George Ord was. Ord was the chairman of the proceeding and throughout was eager to quash Audubon's overtures. In the midst of overseeing the posthumous completion of Wilson's *American Ornithology*, Ord stood unstintingly as the keeper of Wilson's legacy—a duty he performed to the point of crowning the dead poet the father of American ornithology. Wilson was long in his grave, but the city remained his paladin.

Rebuffed, Audubon ended up following Bartram's publishing route for *Travels* and crossed the Atlantic to England. His engravings were also done abroad, some of them in Wilson's homeland. Audubon wanted his feathered subjects portrayed in life size. To fit a nearly five-foot-tall bird like the pink flamingo and great blue heron on a page, he had to show it feeding with head and neck turned down or low and twisted around. The book was necessarily large, a size that printers called a "double-elephant folio." The original *Birds* measures twenty-six inches wide and thirty-nine tall—larger than an average tombstone and, at sixty-plus pounds, about as heavy. (If you had servants, and presumably you did if you could afford a copy, you summoned one or two whenever you wanted the book moved.) New plates were published and delivered to subscribers every few months between 1827 and 1838, unbound and without text. The original is less a book than a portfolio, with each image appearing on one side of a page. Stunning in mass and artistry, *Birds* was fit for a king, and it cost a king's ransom. One subscription lightened your liquid assets by $1,000 (equal to $28,000 in 2020 dollars).

Along with the size of each painting, Audubon's use of pastel, graphite, and ink enabled him to bring out intricate details his predecessors could not—every barbule in a feather, serration in a beak, pigmented layer in an iris. People immediately made comparisons between the works of Wilson and Audubon. Wilson's birds were proportionally incorrect, many said, and rigid in form. Detractors also argued that Wilson's drawings lacked the botanical backdrops common to Audubon, implying, while ignoring Catesby, that backdrops originated with Audubon.

Not surprisingly, passionate charges of artistic theft rolled out of the Wilson and Audubon camps. When volume 6 of Wilson's *American Ornithology* appeared with the small-headed flycatcher on a color plate between pages 52 and 53, Audubon was sure Wilson had copied one of the drawings he had seen in Louisville. Ironically, such a bird probably didn't exist, proving either Audubon right or both men wrong. One or both of the rivals had likely misidentified an immature warbler of some type, awaiting its mature colors.[47]

Audubon was standing on boggy ground when he made allegations about being the victim of an art thief. A number of his own birds seem lifted from Wilson's brush—most especially the Mississippi kite, red-winged blackbird, and bald eagle. There was no logical reason for him to have copied Wilson's bald; Audubon had seen and shot his share of them. Yet their balds are virtual twins in their features and streamlined poses. Audubon painted the tail feathers up, Wilson down, but both accentuated the wing feathers at the shoulders, called "coverts," which in nature resemble perfectly jeweled pieces of an elegant tapestry. Their birds also identically stand with right foot on a dead fish and look outward with a proprietary glare and beak open. Only Wilson's is bloody.*

Narratives of Audubon's subjects came out in a separate bound volume of standard size, *Ornithological Biography*, first printed in Edinburgh in 1831. Audubon's narrative of the bald eagle ran nine pages in volume 1, three short of Wilson's narrative. Then in volume 2, he added five pages. Was he competing with Wilson? The entry in the

---

* To this viewer's eye, Audubon's bald looks more cartoonish—a criticism that has become fashionable.

second volume amounts to little more than anecdotes about hunting down—shooting, roping, clubbing, even poisoning—young eagles. What stands forth in the entry is that eagles had a strong will to survive, that they knew, or were learning, to be wary around humans. In volume 1, Audubon's perception of the eagle's characteristic cruelty and cowardice dominate the pages, but those pages also offer a glimpse into what people were beginning to learn about the representative of their nation.

Audubon honored Linnaeus's classification, *Falco leucocephalus*, for the bald eagle. Audubon also rejected the name "bald," as had his dead rival—more so apparently than Wilson, though, since Audubon refused to use it altogether. Audubon was no more able than Wilson to get a handle on the longevity of his subject, saying he had heard of one that lived to be one hundred years old, though he doubted his source. Also like Wilson, Audubon attested to the eagle's devotion to its mate and nest. He as well acknowledged the good care that parents gave to feeding and protecting their young. (He blithely related shooting a female "as she sat on her eggs." What happened to those eggs he did not say.) When the environment provided, the young did not want for food delivered by their parents. Scraps littered about and below a nest were a feast in themselves for other foraging animals.[48]

The scavenging ways of balds may have disappointed Audubon's own professed standards, but eagles impressed him with their shrewdness when hunting "honestly." "The bird now and then procures fish himself," he wrote, sometimes by herding a school into water shallow enough for the eagle to wade in and, much like a heron, strike "at them with his bill." He witnessed couples that hunted as a team. The exhibition of two pursuing a duck amounted to well-nigh genius for the nineteenth-century observer: One eagle swooped in and frightened the prey into a deep dive beneath the water's surface. When the duck reappeared elsewhere for air, the other eagle forced it down again. Back and forth this tandem continued until the duck tired and ceded its life.[49]

Audubon also related the takedown of a swan. It all happened in flight. The male chased a swan while the female stood on call to assist. Audubon said the eagle's greatest weapon was its "awful scream"; it was as deathly terrifying to the swan as the "report of the large duck-gun." The eagle then revealed its talents as a flyer by approaching the swan from below, rolling sideways, and striking underneath the swan's wing with its talons, forcing it to the ground—flying talents that many other observers have witnessed since. On the ground, the eagle drove "his sharp claws deeper than ever into the heart of the dying Swan." Feeling its prey's "last convulsions," wrote Audubon, the eagle "shrieks with delight." The victor then shared the spoils "eagerly" with its mate.[50]

As much as Audubon disliked bald eagles, he discovered a new species that seemed to have redeemed the *Falco* genus. He noted that he first spotted this unknown bird at night on the upper Mississippi River near the Great Lakes in the winter of 1814, and again on the Green River in western Kentucky. He managed to acquire only one specimen, knocking it off its unlucky perch with a sitting-duck shot from the Long Tom. "Had the finest salmon ever pleased him as he did me?" Audubon asked of his trophy and himself. "Never."[51]

The trophy was an eagle like no other seen—much larger than the bald, he said—with a ten-foot-plus wingspan to the continental bald's seven. He painted it big, filling in the entire double-elephant sheet for plate number XI in *Birds*. As the discoverer, he had the privilege of naming it. He called it the "Bird of Washington" and assigned it *Falco washingtonii*. A brown bird—dark brown, blackish brown, chestnut brown, yellow brown, grayish brown—it had unbooted dirty-yellow feet and toes, and a bluish-black beak. It had no pure white. The tail and head feathers were dark. It was a noble bird, the same as the upstanding George Washington. The Founder, war hero, president "was brave," wrote Audubon, "so is the eagle. . . and his fame extending from pole to pole, resembles the majestic soarings of the mightiest of the feathered tribe."[52]

Like the small-headed flycatcher, the Bird of Washington was another species of improbability, with an ironic twist. Even though Wilson had sorted out the identity confusion around young balds, scientists continued to claim the existence of brown and gray eagles.

The eagle that Audubon took to be a noble bird and worthy of Washington's name was probably a newly fledged bald eagle. His *Falco washingtonii* was a fish eater, but it was also significantly larger than the bald—by Audubon's measurements, that is. How, then, could the two be the same? Did Audubon measure incorrectly? Fed so well by the parents, the brown-feathered young balds tend to be somewhat larger than the adults when they leave the nest, weighing a pound or so more and wearing juvenile wing feathers that are slightly longer. They are not as big as the purported *Falco washingtonii*, but Audubon may have been so keen to discover a new species and pay a patriotic salute to a man he revered that he lost his objectivity.

Audubon loved his adopted country, and his allegiance to men who personified that country, such as Washington and Franklin, was greater than his allegiance to a symbol—a flawed symbol, in his opinion. One could argue that Audubon was blinded by his patriotism.

Of all the birds he wrote about, Audubon reserved his harshest words for the one chosen to stand for his adopted country, which is to say that his assessment, his adjudicating, of wildlife came from a human-centered place. He went out on the river, into the woods, down to the shore, but he never truly left behind his own world when he contemplated birds and nature. He failed to acknowledge or humble himself to the idea that there was a life bigger than his own and beyond his ability to fully comprehend.

Wilson, by contrast, approached nature with deference. Perhaps that quality came from the poet within, or made of him the poet. Yes, he, too, anthropomorphized, pinning human standards on birds like so many merit badges or demerit marks. The bald was "notorious" for its "injustice and rapacity," he wrote. Yet taking from the osprey was only one of "its whole organization" of habits it had "adapted" for survival in a wild world that necessitated periods of fasting and feasting and, of course, killing. Wilson regarded nature on its own terms, discovering the existence of an intelligence, an aged wisdom, that was larger than his own consciousness and his capacity for understanding. Here is an excerpt from his *American Ornithology*:

[There are no] animals so minute, or obscure, that are not invested with certain powers and peculiarities, both of outward conformation and internal faculties, exactly suited to their pursuits, [possessing a] character solely and exclusively their own. This is particularly so among the feathered race. If there be any case where these characteristic features are not evident, it is owing to our want of observation [and to] a morose unfeeling and unreflecting mind.[53]

Indeed, the feathered race—most especially the ones within it that had the white-feathered head and tail, might have important truths about earthly life to pass on to humans, whom he called "the lords of the creation." In the closing sentences to his long narrative on the "distinguished bird," Wilson broke the boundary between animals and humans by questioning the hierarchy of nature that members of his species had constructed, putting themselves above all else.

[The bald eagle's] food is simple, it indulges freely, uses great exercise, breathes the purest of air, is healthy, vigorous and long lived. The lords of the creation themselves might derive some useful hints from these facts, were they not already, in general, too wise, or too proud, to learn from their inferiors, the fowls of the air and beasts of the field. Those fowls and beasts are not in their ways inferior, only so in the human mind.[54]

Humans had made the world confusing for the bald eagle. They had saddled the top predator with an undeserved reputation for being a tyrant and shameless coward, and ornithologists had affirmed that reputation. Yet, paradoxically, Americans had also put the bald eagle up on a symbolic perch, where it asserted the ennobling virtues of a great nation.

Nineteenth-century taxidermists with immature bald eagle (on top of bureau, right side). (Preston Cook Collection)

Part Two

PREDATOR, SYMBOL, AND DIVINE MESSENGER

FOUR

— Perches —

WHATEVER UNSAVORY PROCLIVITIES LORE AND ORNITHOL-
ogy presumed of *Haliaeetus leucocephalus*, the bald eagle continued to
persevere as a national symbol, possessing a charisma that was ideal for
the imagery of an emerging nation.

With George Washington having rebuffed his presence on official
iconography, Lady Liberty and the American eagle made an effective
alternative pairing. Edward Savage validated the potential and potency
of their partnership with a 1796 stipple engraving titled *Liberty in the
Form of the Goddess of Youth*. Young Liberty is an image of angelic poise,
with sheer dress, flower garland, spilling hair curls, and an offering cup
held high. A bald eagle on the wing descends in front of a thunderstorm amid shafts of sunlight to sip from her cup as she treads on the
implements of monarchy: a scepter, a key, and chains.

Instantly a fitting pair, they came together again when a new Capitol was being built after the War of 1812, during which the British burned most of the public buildings in DC. Lady Liberty was a
descendant of the goddess Libertas of the late Roman Republic, the
much-vaunted model for America's own. Positioning this nonnative
figure alongside the native bird attested to the reputation of both and
enhanced the authority of each.

Their first pairing occurred with a fourteen-foot statue positioned
in a niche high on the south wall overlooking the Speaker's desk in
the House of Representatives (which is now the National Statuary
Hall). There, a classically draped Liberty held a scroll of the Constitution in her right hand, with the eagle standing at her feet to that side,

its wings open and head turned toward something in the distance. Before the two were lifted into place, the Capitol's superintendent of construction, Benjamin Latrobe, experienced a bit of a scare. He sent Charles Willson Peale an urgent request for a drawing of the indigenous species. The observant superintendent worried that the Italian sculptor modeling the statue was creating an "Italian, or a Roman or a Greek Eagle," an *Aquila*. Latrobe told Peale he wanted an "American bald eagle" and asked him to send a drawing of the proper bird. Peale obliged, sending not only a drawing but also a taxidermized "head & neck of the American White head Eagle," and in the end the sculptor portrayed the native species.[1]

In a subsequent pairing three decades later, during Capitol renovations, the eagle went head-on-head with Liberty—an abutment that calls for elaboration, as does the excitement that erupted during the planning of a new cast-iron dome for the Capitol.

Placing Liberty atop the dome was a given. She was to be called the "Statue of Freedom," yet the meaning of "freedom" bore the weight of the times. It was the 1850s and the eve of the Civil War. The commissioned sculptor, Thomas Crawford, an Irish immigrant who had married into a family of Yankee abolitionists, opposed slavery. The man he answered to in Washington was the secretary of war, Jefferson Davis, who supported slavery. Indeed, the Mississippian would in a few years become president of the Confederate States of America. Davis in the meantime supervised the Capitol renovations, and he did so with an unwavering vision of the architectural and artistic features that would define the building and the national sensibilities it stood for.

Responding to one of Crawford's early drawings of the proposed sculpture of Liberty, Davis said, "[She] impresses me most favorably." *Most*, but not altogether. She wore a "freedom cap," symbol of the emancipated Roman slave. Historians have said the sight of the cap had Davis spitting with anger. Whether or not that was true, he was tempered and generally thoughtful in correspondence to his chief engineer, writing, "Without pressing the objection, it seems to me [the cap's] history renders it inappropriate to the people who were *born* free and who would not be enslaved." Davis suggested an alternative: "Why should not armed Liberty wear a helmet?" Crawford went with that, and either he or Davis decided the helmet should be crested with the head of an American eagle, along with an arrangement of its body feathers.[2]

The plaster model with proper adornment was shipped from Crawford's studio in Rome, Italy, in 1857. Delayed in transit by a leaky cargo ship requiring unscheduled ports of call, the model statue arrived in Washington two years later, in five puzzle pieces. To assemble it, the Capitol crew commissioned a professional sculptor. After completing the initial job, the sculptor insisted on overtime pay before breaking down the model for transport to a nearby foundry for casting. A slave named Philip Reid, who worked for the foundry, stepped in and devised a tackle-and-pulley system to separate the parts.

Cast in bronze and platinum, the twenty-foot, fifteen-thousand-pound statue was ready for installation in the fall of 1863. Piece by piece, the crew hoisted the figure to the top of the dome. The head with eagle headdress went on at precisely noon on December 2, followed by an earth-shattering thirty-five-gun salute.

Reid was by then a beneficiary of the recently passed Compensated Emancipation Act, which had abolished slavery in Washington, DC. A free man, he could appreciate the fuller meaning of the Statue of Freedom, a meaning that Davis, who by then was sitting in Richmond, Virginia, in his own capitol, had tried to abridge. Seeing the irony, a *New York Tribune* correspondent reported that the face of the statue was "turned rebukingly toward Virginia."[3]

The joke was actually on the news correspondent. The statue faced east toward Maryland, a Union and slave state. And there is more. What Davis had liked when he was US secretary of war, the current architect of the Capitol construction, Thomas Walter, hated: Liberty's headdress with a head on it. Walter, a designer of churches and prisons, didn't understand the point of it. The disembodied eagle's head and splay of feathers struck him as odd and odd-looking. In a letter to a colleague, Walter said of the statue, "Mr. Crawford has made a success of it except so far as it relates to the buzzard on [Liberty's] head." Walter disliked the impertinent ornament enough to support a resolution in the House of Representatives to prune Liberty's helmet clean. Having succumbed to cancer six years earlier, Crawford could not defend his design. The resolution, nevertheless, failed on a technicality.[4]

Walter's irritation with the statue wasn't entirely unjustified. When looking 288 feet up to Crawford's original on the Capitol dome, the observer is challenged to understand what Liberty has on her head. From one angle, you could imagine an Indian headdress (except only Native men and the imaginary Indian princess wore feathered headdresses). From a different angle, Liberty looks to be wearing a mohawk, which is problematic in the same gender way as a headdress. From yet another angle, she might have a whole bird up there. Then maybe it's a palm tree: the burst of wing and tail feathers float outward like fronds, and the eagle head beneath hangs like a coconut.

At the Capitol Visitor Center, guests can stand close to Crawford's plaster model. The eagle feathers and head are apparent, but not their reason for being. Even if one were to find some conceptual meaning in the eagle's decapitated state, the search for meaning in the attachment of body feathers to a disembodied eagle head leaves one baffled. Walter was less baffled than aggravated: "It is an eagle so disposed as to constitute a cap, without being a cap," he grumbled half a decade after the statue's installment, "a helmet without being a helmet."[5]

Walter's complaint was never with the eagle itself but with clipping the wings of a national symbol. Yet in another way, the disembodied figure spoke to a deeper truth. At the time that the eagle's star was rising as a popular emblem of greatness, the real bird in its real world was getting more than its wings clipped. There were animals—beasts of the field, many of them identified as undesirable predators—that Americans prescribed for eradication while marching westward across the grand canvas that defined their country. As it happened, bald eagles got caught in the scatter shot of eradication and frequently ended up condemned to the fatal bull's-eye. As with Audubon dispassionately cutting down the forsaken tree of eagle nestlings in Louisiana, the story isn't a pretty one.

BIRD OF PARADOX

WINGS HAVE ALWAYS FASCINATED HUMANS ENVIOUS OF bird flight. They are a truly extraordinary feature of the extraordinary avian anatomy. They are exquisite in function and movement, in the way that birds use them for lift and drag, and how birds can twist and turn, lower and raise, fold and unfold them, origami-like—although living wings are the original and everything else replication. They are beauty perfected in their motion and mien. Dark gray or gray brown, the bald's wings may not be as colorful as those of a Lady Gouldian finch or lilac-breasted roller, but they are exceptional in their articulation of strength. Feathers are densest on the forebone of the eagle's wing, making the front thicker to enable lift. Airplane wings are modeled after this organic design. It is said that, pound for pound, the eagle's hollow-boned, feather-clad wing, which weighs less than two pounds, is structurally stronger than an airplane's. Large and powerful, eagle wings move a lot of air, propelling the flyer on a horizontal plane at speeds of up to thirty-five miles an hour—more if the bird has a tailwind. If you are ever close to an eagle taking off or flying low and flapping its wings, the force of the air it moves can feel like an explosion. Exaggerating slightly but making a point, the naturalist Neltje Blanchan noted that as eagles "swoop earthward, the tops of the trees over which they pass sway in the current of air they create."[6]

Although people have long been infatuated with the art of flying, they have not always been so enamored of the bearer of the wings. This was true for many birds in America: hawks, crows, owls, bobolinks, starlings, sparrows, vultures, and eagles. While the country as a whole celebrated its allegorical founding bird, untold numbers of Americans condemned its living representative—a fatal bias that was legitimized by the opinion of John James Audubon and other ornithologists of the nineteenth century. Americans forced the bald eagle to exist in virtual parallel universes. In one, it was a heart-beating species: chick, juvenile, parent—a dynamic of nature. In the other, it was symbol, metaphor, icon, avatar—a manufactured article, such as a stamped button, bronze casting, or painted image. In one universe people hunted it down; in the other, the Americanization of popular culture raised it up. The bald eagle was object and ornament, alternately perceived as cowardly and courageous, cruel and caring, villainous and virtuous. The intertwining universes gave birth to the "bird of paradox." On rare occasion, a living, feather-and-blood bird thrived in that raised-up symbol universe. That's what happened with a young bald eagle that went to war and lived the paradox.

~

In a sepia-toned *carte de visite* made for the 1876 Centennial Exposition in Philadelphia, Sergeant John F. Hill is a US military veteran of twelve years. *Cartes de visite* were photographs mounted on card stock cut to wallet and business-card sizes. They were all the rage at the time of the centennial. People exchanged them as gifts and used them as calling cards. Shops sold *cartes de visite* of celebrities, and enthusiasts collected them like baseball cards (some of which used the same technology and had come into vogue).

For his *carte de visite*, Hill put on his old Union frock coat with brass buttons, gold epaulets, and three wide chevron stripes on the upper sleeve. Turned with middle point down, the chevrons were sometimes called "wings," and complementing them is Hill's felt cavalry hat with a spread-eagle infantry insignia on the front. The coat fits him well. The veteran is still trim, and sporting a gunslinger mustache and what today we would call a "soul patch" over a pointed chin. He was making a celebrity *carte de visite*, but he wasn't the celebrity. A smart-looking bald eagle sitting on a fabricated wood-and-metal perch next to Hill held that status. The eagle was fifteen years old and named Old Abe.

He and Hill had served together in the recent war, the Civil War, and seen a lot of action—nearly two score battles and numerous small skirmishes.

Old Abe is without a doubt the most beloved bald eagle in American history. At least eight biographies, the first published in the year of the centennial, tell the remarkable story of his experiences. His fabled journey began early in life near Lake Superior after Chief Sky, an Ojibwa of the Flambeau band, removed him from the wild.

Old Abe was born in Wisconsin in spring 1861, around the time recently seceded South Carolinians were shelling Fort Sumter. Chief Sky* had been out hunting when a large nest high in a tree attracted his attention. Above it, two adult eagles "circled and swooped about." He knew from their behavior that eaglets were in the nest, and he "wanted"—as he told an interviewer for the *Indian Leader* decades later, when he was peeking into his eighties—the "little ones." In Ojibwa culture, eagles were spirit birds, and their feathers were used in ceremonies and medicine rites. But Chief Sky's desire for the eaglets doesn't appear to have been related to custom and tradition. He took them by violence.[7]

Unable to climb up to the nest, he chopped the tree down with a steel ax, fighting off the parents as they tried to protect their young. When tree and nest crashed to the ground, he discovered two eaglets. One had died on impact. The other he took to his camp, where women and children fed it "meats and scraps from the camp kettle." A few weeks later, Chief Sky joined others who paddled eighty miles down the Flambeau River to trade with white settlers. The bird went with him. At the town of Eagle Point, Wisconsin, he made a deal with a fiddle-playing tavern keeper to exchange his living bundle of feathers for a bushel of corn. "Why should I not?" the interview quotes Chief Sky. "The eagle was no larger than a chicken, and a bushel of corn was a bushel of corn."[8]

The tavern keeper was forty-year-old Daniel McCann. He had a toss of dark hair, along with a rugged face and heavy-shouldered build

* Some sources refer to him incorrectly as "Chief Big Sky."

suited to the backcountry life he shared with his wife and seven children. The eagle was ostensibly a pet for them. But the pet grew to full
size and into its sharp talons and biting beak. Likely at the protest of
the children, McCann took the bird to Eau Claire to sell or trade. He
met soldiers there who were forming a company that would join the
Eighth Wisconsin Voluntary Regiment and partake in the great sec-
tional conflict. McCann tried to convince them of the need of a mascot, a war eagle.

The war eagle is one of those long-ago concepts that took root separately on opposite sides of the globe. The ancient Greeks and Romans
used it as a mark of distinction, and so, too, did indigenous North
American cultures. "War eagle" could refer to one who was courageous
or bellicose or both. The Wisconsin volunteers were skeptical about the
need of a war eagle, and then, as the story goes, McCann pulled out
his fiddle. He played a lively rendition of "Napoleon's Retreat," prompt-
ing the eagle into its own lively rendition of a dance. The men loved it
and passed around a hat to come up with $2.50, and a local threw in a
Quarter Eagle, a gold piece equivalent to what the men had raised. The
company captain suggested naming their war eagle Old Abe, an idea
that was more appropriate than he might have thought. "Bald eagle" was
a tag given to smart, stalwart politicians. When Abraham Lincoln first
entered public service, supporters dubbed him a "young eagle." By the
time Company C mustered with the Eighth Regiment at Madison, the
members of the former were calling themselves the "Eagle Company."

Old Abe the bird was inducted into service along with the men—an
event highlighted with red, white, and blue ribbons draped around his
neck. One of the men built a perch mounted to a replica of the shield of
the United States, and fastened it to a staff. In an 1863 photograph of
Olde Abe on his perch on its staff, surrounded by the color guard of the
Eighth, he sits higher than the heads of the soldiers and does not yet
wear mature feathers. He is probably two years old or less at the time.
In marching columns, he rode next to the national colors at the head,
held aloft by one eagle bearer in a succession of many. During battle,
the bird and his bearer moved to the center of the regimental assembly,
where Old Abe sat in the line of fire.

Indeed Old Abe often found himself in the thick of things. At the Battle of Corinth in Mississippi, wrote David McLain, Confederates aimed "to take the Eagle dead or alive." They targeted Old Abe as they would a commanding officer on the battlefield, to strike at the heart of the enemy. The Confederates launched three charges, and each time Old Abe rode his staff-borne perch beside the flag and into a valance of gray uniforms and gun smoke. A Minié ball cut Old Abe's tether, and he made a break for it. After he had flown some fifty feet down the line, McLain caught him. When the fighting ended, McLain and Old Abe had both been winged, literally—McLain clean through his blouse at the shoulder, and Old Abe through three wing feathers that were lost. Neither suffered a flesh wound. The Eagle Company, tragically, fared less well; it lost half its men.[9]

The next major theater for Old Abe was at Vicksburg, situated atop the tall eastern bank of the Mississippi River. Again, the Confederates were out for the blood of the mascot. A shot grazed his breast but did little more than ruffle his feathers. Then a tree behind which he and his new bearer took cover exploded from enemy cannon fire. Both survived, but others with them did not. Later in a ravine, where Union troops were temporarily pinned down, Old Abe drew the soldiers' attention momentarily away from the surrounding carnage by devouring a rabbit a soldier had caught. It was a prelude to Union victory.

When the troops marched triumphantly into Vicksburg, their commander, General John A. Logan, was saddled at the head of the procession. He was known as the "Black Eagle." To his left was Old Abe, the more popular eagle.[10]

In the occupied towns through which the Wisconsin troops traveled, Old Abe had the power to both thrill and torment white southerners. He had a set of instinctual and miming gestures that entertained locals, children especially. He shook the hand of his bearer with his talon, fiercely stretched out his wings and opened his mouth with a chitter, raptorially stalked and killed barnyard chickens when prompted, and drank water poured into his mouth from a canteen.

Like the sight of blue uniforms, though, his presence could unwittingly divide occupiers from occupied. Many of the latter would have preferred to see his body laid out stiff and cold amid flowers as a moral victory in defeat, confirming the persistence of their cause. Instead, deflated white southerners settled for slinging epithets, such as "dirty crow" and "Yankee buzzard."

For Black southerners, the sight of Old Abe must have affirmed the arrival of coveted freedom, and he was always a good one for lifting the spirits of federal troops who had lost their comrades. "I have seen Generals Grant, Sherman, McPherson, Rosecrans, Blair, Logan and others," McLain wrote of inspiriting moments, "raise their hats as they passed Old Abe, which always brought a cheer from the regiment, and then the Eagle would spread his wings as he always did when the regiment cheered, and he did look magnificent at such times."[11]

In September 1864, after witnessing years of violence on a level he would not have in the wild, Old Abe was honorably discharged from service along with seventy of his regimental comrades. He was not yet four years old, although fully dressed in white head and white tail feathers. Had the stress of war matured him early? A commanding officer maintained that Old Abe "enjoyed the excitement" of battle and "would scream with wild enthusiasm." A writer with *Harper's Weekly* said he had been "in his element."[12]

The truth is that Old Abe had been *out of* his element, the natural environment, and would remain so for the rest of his life. He would not become a benefactor of the freedom he symbolized and risked his life for. Among the possibilities of what to do with him upon his honorable discharge, liberation does not appear to have been on the table. As Daniel McCann's children learned, certain wild animals, no matter how cute, adorable, or majestic—pandas, tigers, black bears, opossum, prairie dog, raptors—don't domesticate well. You can train raptors to hunt from a glove, but they evolved in nature to live in nature, to keep their distance from humans. Unlike a dog or cat, they don't seek to interact with people. Interaction is largely forced on them when they are in captivity, and captivity is a form of forced interaction. Wildlife trainers don't regard bald eagles as highly trainable (goldens, slightly more so). If Old Abe could truly perform the tricks his biographers and handlers claimed, he was an exception among his species, probably because he had lived with only humans since being kidnapped as a nestling.

Remaining in captivity was the only future the men of the Eagle Company could imagine for Old Abe, and they unanimously voted to donate him to the state government in Madison. He had seventeen more years in front of him and would live them as an adored yet tethered celebrity who charmed, enchanted, humored, and inspired.

When Sergeant Hill delivered him to Madison, Old Abe was formally received by the governor and state quartermaster. His new home was two rooms in the basement of the capitol building where the ceiling had been painted blue with white stars, and the bars of his apartment-size cage in alternating red and blue. He depended on a succession of attendants, eight of whom were veterans, to feed and exercise him. The press said he was fussed over and living in luxury. A writer for *Audubon Magazine* claimed that Old Abe was "in no danger of trying his wings for a wide excursion; he had adopted the folds of the flag of stars." It's true that Old Abe wasn't flying anywhere, even behind the red and blue bars. Ever since his untethered maiden flight at Corinth, he had been subjected to having his wing and tail feathers routinely clipped. The shingling robbed him somewhat of that flawless, fluid appearance of a free bird, making him look thinner than natural and perhaps a little underfed, because sometimes he was—because sometimes his handlers were neglectful. The clipped wings also suggested that he was to be more symbol than living species.[13]

Old Abe's fame by war's end, one of his biographers said, was "in everybody's mouth." When he was seen in real life, people realized the bird of liberty was not just an illusion; it was a legitimate battlefield veteran, and it reminded them of the cause for which their country went to war. The state made use of Old Abe's popularity by sending him out to participate in events around the country to raise money to support Civil War orphans and widows, and medical care for veterans.[14]

He and Sargent Hill attended the Centennial Exposition, where Wisconsin sponsored the Old Abe Museum of Ornithology. The "museum" was essentially a place to meet this most famous of birds, known nearly as well among European visitors as among Americans.

You could also buy, for 50 cents, the first bird biography, titled *The Soldier Bird*. Its 126 pages were written by Joseph O. Barrett, who, in addition to being a prolific author, was a medium and spiritualist. According to one source, thousands of copies were sold, netting $16,000 for a veterans' fund. If you didn't have the 4 bits for the book or wanted something a little extra, you could buy a photograph of Old Abe, who would sign it on request by piercing a hole in the picture with his yellow beak. P. T. Barnum wanted not a photograph but the bird itself—so great was the public adulation that Old Abe excited. Through a third party, Barnum made a $20,000 offer to the state. He was turned down.[15]

Old Abe almost didn't make it to the centennial. The year before, his caretakers had gravely shirked their duties, leaving him too long without expired mouse or raw fish to eat. He fell critically ill before a veterinarian restored his health. He was looking "well and hearty," said a Chicago newspaper, by the time he appeared at the 1875 convention of the Grand Army of the Republic (GAR).[16]

The GAR was a fraternal organization of Union military veterans similar to but not quite as elitist as the Society of the Cincinnati. Members were given a chest ribbon that had a five-point star dangling from a silk flag in miniature, above which was an eagle with wings open and standing on crossed cannons. The GAR eagle was similar to the silver eagle insignia, adopted in 1832, that was worn by army colonels, who became known as "full-bird colonels," except the colonel bird is fixed on a bundle of arrows rather than cannons. A natural before the camera lens, Old Abe never posed with arrows but did on cannons—that year, in fact—for a landscape photographer who had fought at Vicksburg with him.

During the GAR convention, Old Abe was a celebrated guest at the Grand Pacific Hotel and the "object of much flattering attention from the citizens of Chicago," noted the Chippewa Falls *Weekly Herald*, which was more or less Old Abe's hometown newspaper. His last public appearance was a reappearance at the GAR convention five years later, this one in Dayton in 1880. In another year he would be dead. As yellow fever, cholera, and tuberculosis were to defense-

less society, fire was to buildings and whole cities. All were suscepti-
ble. The story that traveled down the years is that Old Abe sounded
the alarm when a fire broke out in the capitol. He and the building
were saved, but not before he inhaled too much smoke. For weeks he
remained listless, and then he died in the arms of his last caretaker,
nineteen years and ten months after he had hatched in a nest in a tree
that no longer existed.[17]

The state naturally could not let him go completely, so he became a
mounted bird placed in a glass box displayed upstairs in the restored
capitol building. He continued to generate the interest of newspapers,
magazines, poets, and, of course, biographers, many of whom wrote for
young adults. Inevitably, wild stories were told of the exploits of this
allegedly tame bird. He was often the hero of the battlefield, alerting
the troops of enemy movements, leading the charge, and inspiring tactical maneuvers (a 2010 children's book by Patrick Young is titled *Old
Abe, Eagle Hero*). The contributor to *Audubon Magazine* who said Old
Abe would never fly away wrote, "I have seen him once in a procession
at Chicago, perched high over a wagon . . . looking in wise, solemn
attitude as if he would rebuke the whole great land of freedom that ever
it had warred against itself."[18]

Then, in 1904, the capitol burned again, like an aboveground purga-
tory, in a fire lasting eighteen hours and reducing everything to ashes,
including Old Abe's mounted remains. It's as if fate had decided he was
not supposed to have ever existed as a living symbol, even with clipped
wings. Nor was he supposed to be preserved as a mounted memorial,
where he dwelled on the borderline between the symbol and species
universes. It is true that after the incineration of his taxidermized likeness, the memory of the real bird persisted. That memory, however, is
colored by the myth of a glorious bird dedicated to a noble cause and
the unity of the nation. The memory suggests that the reverence for
Old Abe was accorded to living bald eagles generally. It was not. In
Old Abe's day, people did not leave them undisturbed, and their life
span was significantly shorter than the war mascot's.

Old Abe may have dodged Minié balls on the feverish battlefield,
but outside of captivity he would not likely have escaped the shot of

someone like Arty Kerker, who went to a city park one day to shoot a bird. There were a lot of someones like Arty Kerker.

WHAT KERKER PERPETRATED IN Highbridge Park was not uncommon, untoward, or unlawful—not in his day. From the time it opened in Upper Manhattan in the late 1860s, the park was a leafy and granite-rock haven—all but a wilderness enclave in an expanding metropolis. The park flanked the east side of the Harlem River and got its name from the High Bridge nearby. Spanning 1,450 feet across the river on sixteen stonework arches, the High Bridge resembled a Roman aqueduct and, in fact, portaged clean freshwater for the city from the Croton River in Westchester County.

Families and friends picnicked in that stretch of green park along the Harlem. Lovers held hands, children played, and babies were pushed in strollers. But the park was not an entirely innocent place. Twenty years into its serene existence, Kerker visited Highbridge on a hard winter day in 1888, when Old Abe was seven years as a mounted form, for a specific reason: to kill a bald eagle.

According to the *New York Times*, which portrayed the moment as thrill-of-the-chase excitement, Arty Kerker and his "blonde whiskers" were "well known on the road" and river. He was thirty-one, the son of German-immigrant parents living in Harlem. In right-slanted longhand, the census taker at the time put down his occupation as "boatman." He worked for his father, a saloon keeper, who, as a side business, rented out watercraft at the foot of East 131st Street at Harlem River Park. It must have been from there that Kerker set off with three others for Highbridge Park on an icy January day on the steam launch *Van Cott*.[19]

Two balds had been spotted in the area a few days earlier, and the city was abuzz over their appearance. Bald eagles had once been, but no longer were, regular winter denizens of the woodlands and city parks, where they picnicked on prey caught on the grounds and from the surrounding rivers. They had been commonly seen mustered aboard ice floes, captains and crews contemplating their seasonal maneuvers among the larger avian fleet. Other times, and at night, they had roosted in leafless trees in wooded areas.

A group of eagles—as in a gaggle of geese, murder of crows, and charm of goldfinches—is called a "convocation." In the wintertime, arriving at a particular destination from different parts, some far away and some not so, bald eagles convene in roosts numbering in the tens, scores, even hundreds. Typically located in woodlands and near water, roost trees, despite their deciduous undress, help provide protection against seasonal winds. As important as anything, roosts are close to food sources. Much as they seek out spawning runs, eagles often trail migrating fowl—movable feasts, one might say—to their winter habitat. Finding fish and fowl together in the same place is ideal, and if one bald has figured out where the twain meet, others surely will too. Their winter communal inclinations are a departure from their solitary seasonal travels and nesting places. All ages come together. In any given roost, there are mature white heads alongside immature dark heads, yellow eyes alongside brown eyes, yellow beaks alongside gray. In 2018, the Center for Conservation Biology's eagle-roost registry compiled reports on sightings of 1,538 communal endeavors in the US and Canada, with the highest number, 622, in Maryland.[20]

Why are these otherwise semi-loners forming a winter convocation? It turns out that when eagles come together, they make a lot of chattering noise, and what erupts in high decibels seems like some grand repartee. Are they swapping big-fish lies, the braggart making a claim only to meet the loud retort of others? Scientists have never asserted as much. Instead, they think that the younger, unattached balds in the roosts prioritize communal gatherings to find a mate. Some of the racket then may be about who is choosing whom and who will ultimately go with whom. Scientists also believe that roosting eagles use their numbers for safety, sounding warnings of an approaching predator. That, of course, would be humans. Adult eagles tend to be more cautious than young ones, so a roost might be a learning place too. If you ever see a tree-bound convocation, the eagles perched on branches look like upward-growing fruit—or, if you're armed with a ten-gauge, double-barreled shotgun, they can look like a bunch of sitting ducks.[21]

That's the gun Arty Kerker took to the park, and before it was too late, his unsuspecting target might have been saved by the warning signals of a communal roost. But none was about. Without explaining why, the *Times* said eagles weren't showing up in Manhattan in the numbers of earlier years. Timidity probably had something to do with their absence. Arty Kerker wasn't the first person to shoot a bald eagle around New York, or the last. Balds are smart survivors—to wit, the stealing and scavenging—and disinclined to visit a place where they're greeted with bird shot.

Another reason to avoid New York was the declining availability of food sources. New Yorkers—every single literal body, two million of them—had fouled local rivers with their feculent waste. On top of that, the city was powered by horses, nearly three million clomping through the streets, and each dropping fifteen to thirty pounds of manure every day. What the shrinking number of farmers in the area did not broadcast on fields, draymen dumped in rivers along with street rubbish, shrouding the city in a noxious air day and night. Adding to the riverine blight, industries released untold tonnages of effluent. All these extras killed off aquatic grasses that were habitat and feeding grounds, as well as oyster beds that had been unparalleled in vastness, sending away any fish and fowl that avoided dying in the foul. And more death: bad water took some five thousand human lives in the form of cholera outbreaks in 1832 and 1849. That's why the city built the aqueduct that rode the High Bridge and delivered water from the unspoiled Croton River ten miles to the north.

So, the two lone eagles weren't in New York for the dining opportunities. The newspaper said a blizzard had blown them east from the Rocky Mountains. How the paper knew that it didn't make clear. Maybe weather pushed them to New York, but likely from up north rather than out west. Whatever the case, their presence stirred excitement, and most likely Kerker wasn't alone in thinking of taking a gun to the park. He just happened to be the first to get a clear shot at an unlucky, wayward eagle.

The High Bridge is nine miles upriver from where Kerker set out on the *Van Cott*. As he and his mates were approaching the park, they saw one of the eagles circling above the two-hundred-foot-tall, rock-faced granite water tower on the Manhattan side of the bridge. By then, the *Van Cott*'s propeller had broken, and the four were walking along the bank, shooting anonymous "birds as they went." Then, a few hundred yards south of the water tower, they came upon a dead tree, and Frank Holland, the "colored pugilist," as the *Times* called him, whispered to Kerker to lie low; "there goes a rabbit." Kerker, with florid mustache "frozen fast to his whisker," couldn't open his mouth to speak, but he could cock both barrels of his gun. As he scanned the grounds for the rabbit, his eyes met an eagle in the dead tree—head down, wings partially open, ready to dive for the prey. Kerker drew a bead and pulled one of the triggers of the double barrel. The bird's head flinched up before its entire body went down.

The report of the gun attracted a park policeman, who ran in its direction, although not to arrest Kerker. He gave advice on how to free the eagle from a crotch of the tree where it had dropped, still alive.

The wounded bird was full of fight as the men struggled to take it down. They knocked it unconscious and took it to a nearby hotel. While waiting for it to die, they warmed themselves. A few young women gathered around and looked at the eagle "with sympathy," but the hotel proprietor showed none of his own. The *Times* reported that he'd been losing two to three chickens a week since the eagles had come to town. He must have thought he was watching a thief die. When it did, someone weighed it—fourteen pounds, a female's weight—and opened its wings to seven feet.[22]

Kerker took the corpse to a taxidermist, who discovered that the bird's stomach was "entirely empty" and that "it must have been very hungry." So much for being a chicken thief. Kerker then granted the dead eagle a second life mounted on a globe of polished blackwood, captured in its last dignified act of survival—head down, wings unfolding, eyes in a death stare, preparing to pounce on a rabbit, a rabbit saved by Arty Kerker's bird shot.

Kerker's exploits, and the collaborating attitudes of the time (the cop helping him to remove the flailing bird from the tree), were instruments in the making of the bird of paradox. Like Old Abe, the Highbridge eagle crossed universes, from living object in Kerker's gunsight to a mounted corpse on an ornamental globe, where Americans most welcomed it.

Two months after the shooting, an editor at the *Times* apparently thought that Arty Kerker's quest for a trophy made an interesting enough story to give it 1,075 words in a column that drops three-quarters of the way down the page. The two-line deck beneath the article head refers to Kerker as a "sportsman," and the article's anonymous writer portrays him as neither honorable nor dishonorable. The tone of the piece would have been no different if Kerker's prize had been a goose or turkey, except to say that someone was looking forward to a feast. There was no offense or acknowledged irony in Kerker putting a bullet into America's vaunted bird, no reference to the bald eagle as a national symbol, idolized on hard coin and paper money, in patriotic art, and, however obliquely, on the US Capitol dome. Now it was on Kerker's polished blackwood globe.[23]

FAST-FORWARD FOR A MOMENT to a twitchy silent film starring a very un-Abe-like eagle that helps explain the prevailing disdain for the living bird. The 1908 production came out of the Edison Studios—Edison, as in Thomas. It's an eight-minute single-reeler titled *Rescued from an Eagle's Nest*, and a poor representation of the art form. *The Moving Picture World*, a magazine, called *Rescued* a "feeble attempt to secure a trick film of a fine subject." "Feeble attempt" refers to poorly executed special effects in the noticeable wires that flap the wings of a prop eagle. The "fine subject" is a father's courageous rescue of his baby. The father is played by D. W. Griffith, who years ahead would become infamous himself as the white-supremacist director of *The Birth of a Nation*, a phenomenally successful film that pioneered camera technique while also glorifying the Ku Klux Klan of the post–Civil War years. The film also stimulated the Klan's second coming in 1915. *Rescued* was Griffith's debut playing the lead in a film and, according to critics, not a terribly triumphant debut.[24]

Griffith's character is a woodsman who lives with his wife and child in a cabin crouched at the edge of a pleasant valley. In the opening scene, the father steps out into a bright morning, followed by his wife holding their infant girl. The morning light and yawning valley create a happy, picturesque mood. Kissing his family goodbye, the father bounces off toward the woods with a lunch pail in his hand and an ax on his shoulder. The mother returns to the cabin, leaving the child outside on the swept-dirt ground to play with a pot and wooden spoon.

Before long, a large eagle makes a pass over the valley and then disappears off-screen. The mother is still inside and her baby at blissful play. The audience knows what's coming. There is no music backdrop in the preserved version of the film, but one can imagine the day's theater organ in the orchestra pit escalating to a suspenseful crescendo. The worrying bird reappears, swerving clumsily on its prop wires from around and behind the cabin. The eagle is nearly as big as the child, which it quickly latches onto, and the two hurl away. The wires are obvious here, and the film editing a little jumpy. Still, it's an uncomfortable moment when the wild animal takes off with its human prey, which is not a prop.[25]

The camera moves in for a close-up as eagle and baby cross the valley, deeper and darker than before, no longer so pleasant. The predator bird is scary ugly, more like a flying ogre than a feathered being. The prop guys have outdone themselves with this one. The innocent child is hanging from the bird's talons (though really her own set of wires), kicking her feet and contorted in either pain or fear, or both. Her struggling creates a good effect, if not a good sign of the infant actor's comfort. The scene then cuts to the cabin. The mother rushes outside, grabs a shotgun that happens to be leaning against an exterior wall, takes aim, and thinks better than to shoot an eagle ferrying her child over an abyss. Throwing down the gun, she is frantic, but not so frantic that she forgets to retrieve her hat from inside before running for her husband.

He and three others have just felled a tree when she delivers the horrific news. The men point to the feathered kidnapper flying overhead, toss aside their tools, and race through the woods to the valley and the edge of a cliff. Using a rope that materializes out of nowhere, they lower the father over the side to a rock spur jutting out below. The baby sits quietly in the eagle's nest, content, as if a director's assistant had given her a treat. When the father reaches her, the eagle swoops in and knocks him over. They wrestle, and the father takes the bird in a choke hold. He raises a club, which, like the rope, magically appears, and bludgeons his attacker. It collapses.

With a flourish of celebratory arm gestures, the father kicks the dead predator over the ledge. He takes up the child in one arm, kissing her, and with his free hand grips the rope. He gives a signal, and the two are pulled smoothly and effortlessly up along the escarpment face, his one hand still holding the weight of the two. Wires for father now. When they reach the top, the men whip off their hats, wave them around in the air, and let out a silent whoop and cheer.

The director of this feeble trick of a film was James Searle Dawley. In another two years he would make the first Frankenstein movie (a hatchet face, dark feral eyes, and briskly combed hair gave Dawley something of the look of a mad scientist himself). With *Rescued*, he was dramatizing a familiar warning to mothers: never leave your infant unattended outdoors. People had been hearing stories of eagle kidnappings for generations; adults, since childhood. In the fifth edition of the popular primer *McGuffey's Reader*, published in 1844 and used in almost every grade school in the country, children encountered a story of an eagle bearing a girl back to its nest on a rock ledge, where it placed her next to an eaglet and amid the blood and bones of previous takings. People took the warnings seriously, but *McGuffey's* and the film were perpetuating a myth. Eagles don't prey on human babies, and if an eagle decided to, any child over preemie size would be too heavy for it to lift from the ground.

The myth was entrenched and powerful, deepening fears in a society in which one-third of all children died before the age of five (the majority from disease) and when wild animals were a threat people preferred to live without. Even those who presumably knew better bought into the myth. In *American Ornithology*, Alexander Wilson shared a story of a woman weeding in her yard with her child nearby when a "sudden and extraordinary rushing sound, and a scream from her child alarmed her, and staring up, she beheld the infant thrown down and dragged some few feet, and a large Bald Eagle bearing off a fragment of its frock." Thomas Nuttall, a naturalist who liked the bald eagle about as much as Audubon did, wrote in his 1832 *Manual of Ornithology*, "So indiscriminate indeed is the fierce appetite of this bold bird, that instances are credibly related of their carrying away infants."[26]

Nearly a century later, under the sway of the myth, the nature writer and historian of science Neltje Blanchan offered the same. "Scientists raise their eyebrows at tales of children being borne away by eagles," she wrote, "yet it would seem that some rare instances are well authenticated." In 1929, the *Auk*, the journal of the American Ornithologists' Union, noted that "even today the eagle which carries off babies has not been forgotten." In a congressional hearing the next year, a Georgia representative looked to Theodore S. Palmer, an ornithologist with the Biological Survey Unit of the US Geological Survey, for the truth about avian abductions. The congressman pointed to the "old picture books we used to study in school" showing "eagles carrying off children." He asked, "Is there any basis at all for that?" Palmer said he could "not recall any case at the present time," but that he "would not say there are none."[27]

What Palmer and the others were saying is that the living species was not the vaunted symbol bird; the former was a threat to society. Like George Washington in American culture, bald eagles were mythic figures. They were real, but people with straight faces ascribed false realities to them, casually parsing their attributes into objectionable and acceptable ones. Just as Americans slotted Native peoples into good- and bad-Indian categories, cultures have historically classified plants and animals similarly. Valued were those that provided food, companionship, pleasure, ornamentation, usable power: a dog lying at one's feet, a canary singing melodies in a cage, a Clydesdale drawing a beer wagon. Unacceptable were those that harmed the dominant species: starlings descending on an orchard, mice raiding a cupboard, a porcupine chewing out a barn—anything that challenged the well-being or self-determination of the human species.

Old Abe landed in the good category. That's because an Indian, a tavern keeper, a military regiment, a state, and a tether kept him from doing what wild eagles naturally do: hunt, scavenge, and take.

TAKE, AS IN THEFT. A greater offense than stealing a fish from an osprey was stealing livestock from people. Saying "bald eagle" paradoxically summoned the image of a free nation yet also that of an antagonist to American progress. A free nation in the Jeffersonian vision of the independent family farmer depended on reducing or eradicating purported nuisance animal populations. To protect livestock and commercial game, every territory and state instituted predator control programs, which often included the payment of bounties. The programs typically targeted wolves, coyotes, wolverines, and mountain lions, as well as smaller animals like jackrabbits and prairie dogs. In the nineteenth century, safeguarding so-called natural resources, whether grown on the farm or in the wild, was called "conservation," and early conservationists tended to support predator reduction. William T. Hornaday, director of the New York Zoological Park (now the Bronx Zoo) and a recognized conservationist, labeled the wolf a "black-hearted murderer." Theodore Roosevelt, the first conservation president, regarded it as a "beast of waste and desolation."[28]

Here's how the rationale went: The integrity of American freedom depended on expansion into new territory—the West and deepest South. Expansion depended on protecting the means of living, and thus the individual freedoms of the independent farmers moving into new territory, who were the industrious foot soldiers and validators of the American system's legitimacy. The farmers' liberty, in turn, depended on safeguarding the return on their labor, not unlike the necessity of the country defending its shores against hostile invaders. Safeguarding the farmers' interest depended on fending off crop-destroying and livestock-killing animals.

A lot of dependence was wrapped up in this formula for independence, and sometimes the nation, made possible by nature, could not abide nature, or certain aspects of it. This was especially true as the unfettered country expanded geographically and demographically.

In Old Abe's lifetime alone, the American population swelled from thirty million to fifty million—a nearly seventy percent increase. In the twenty years after, the population surged another fifty-five percent. In the twenty after that, ending in 1920, it grew by forty-one percent. By the time Old Abe was settled in his stars-and-stripes capitol apartment, the Ohio River valley and the Erie Canal's farther reach were no longer the American West. The new West ran all the way out to the Pacific coast. Chicago, on Lake Michigan, was the brawny shipping and processing hub between the old and new. Cattle, corn, and wheat were steadily filling in the Midwest and West Texas. The Transcontinental Railroad opened when Old Abe was but eight, and Indian country was joylessly shrinking. It was as if, after winning the war, Americans had excess energy to work off, and did so by multiplying, acquiring, building, and conquering.[29]

For some, when the national centennial came around, the experiment with constitutional government by the people was working. Which is not to say that it benefited everyone, or that the United States was not still discovering itself, still figuring itself out as a nation (that's in part what the war was about). Patronizing European powers were no longer trifling with it so much. The nation's preoccupation had shifted to domestic issues, mainly trying to mesh its founding values with its cultural diversity, to find some agreeable space in society for women, who were organizing for suffrage rights; black men, who were trying to exercise, too often in vain, new rights that an amended Constitution finally granted to them; and immigrants, who in profound ways were putting a new face on the social and cultural landscape. The country was also trying to balance rural places with ever-expanding urban industrial places. In expansion was relief, as in a high-pressure valve on a steam boiler. Rural America offered natural resources to bolster industrial-urban America, and in those resources was the promise, although often an empty one, of a greater sense of freedom for people who felt cheated out of opportunities or confined in the crowding cities.

While earthly forces were compelling the dispersal of the population, there was also an otherworldly calling to the West. Prophets, policymakers, and opinion makers especially subscribed to the divine purpose laid out before them. John L. O'Sullivan distinguished the purpose as such: it was white America's "manifest destiny to over-spread the continent allotted by Providence for the free development of our yearly multiplying millions." O'Sullivan introduced the term "manifest destiny" in an 1845 article in the *United States Magazine and Democratic Review*, a periodical he edited. He repeated the expression in the *New York Morning News*, adding that manifest destiny was a God-given "right" of the American people—white men with their women as helpmeets—"for the development of the great experiment of liberty and federated self-government entrusted us." That experiment translated into taming the land by deciding how it should be used and who and what should remain on it.[30]

It didn't matter that much of the area Americans were moving into was cold and semiarid (thirty percent of the contiguous US) or already occupied. They believed their chosen-people presence would bring rain, moderate weather, and conquest. This wasn't mere folklore in their minds. A century before, the esteemed French naturalist Georges-Louis Leclerc, Comte de Buffon, had maintained, erroneously, that the climate of the New World was unsuitable for civilized society, that it had stunted the development of indigenous cultures. Yet civilized people, he asserted, possessed the capacity in mind, body, and spirit to rectify it: "Some centuries hence, when the lands are cultivated, the forest cut down, the courses of the rivers properly directed, and the marshes drained, this same country will become the most fertile, the most wholesome, and the richest in the whole world, as it is already in all the parts which have experienced the industry and skill of man." Americans believed that the West would deliver an idyllic climate, by the grace of God or Mother Nature, not after centuries but immediately upon their subjugation of the lands and the waters in the very ways Buffon prescribed.[31]

The most famous artistic reference to manifest destiny is John Gast's 1872 oil-on-canvas titled *American Progress*. It depicts the full glory of the white people of a nation traversing the continent—a buoyant parade of them on foot, horseback, and iron rails; in covered wagons and stagecoaches; and full of anticipation. Their guiding spirit in a silken form of golden locks, ivory skin, and gossamer tunic is Lady Liberty. She is at the painting's center, celestially aloft in a golden sky. In one hand, she clutches a "School Book" to her bosom, and with the other, she is stringing telegraph wire across the land. Civilization is on its decisive march. Indians and wildlife are, as Gast painted them, in definitive retreat. If the perceived character traits of the bald eagle were supposed to mirror the nation, then those of aggression and predation would have to be included with strength and courage. Yet doing so would be unfair to the bird of prey, since its organic behavior was not destabilizing.

Less well known is an image on a map of the United States produced by Joseph H. Colten that predates Gast's painting by twenty-three years. A mapmaker of international stature, Colten used color shading to distinguish between the states and the territories. The US had nearly filled in the portion of the continent that would eventually add up to the lower forty-eight states, after the admission of Arizona in 1912. In a vivid illustrative inset in the upper right-hand corner is an eagle gripping a shield of the US lying at an angle—an illusion portending Old Abe and his shield a dozen years forward. (Was the maker of Old Abe's perch familiar with this map and illusion?) The illustration's setting is an unidentified East Coast seaport, and behind the shield is a plow and cornucopia of food crops. The artist extended the bald's wings fiercely behind it. Its neck is thrust hard forward, its mouth is open with tongue protruding aggressively, and its brow is stretched out in fierce exaggeration. It is the projection of an unwavering and impatient being, ready to soar and seize. The bald eagle is eyeing the continent—doing so for the people.

This is only a symbol, though, and a prophecy of the future of the United States. It does not speak for the species, which is all but doomed in this future. If it did, the living bird's look and posture would represent not aggressive conquest but repulsion for what the unfurling advance was doing to its species. While the United States grew more secure as a nation, the bird of the wild grew less secure. While the American population flourished, the bald eagle population withered.

It withered alongside many other nonhuman populations. Animals suffered for being beasts of prey or for the marketability of their meat, hides, or feathers. When the country was entering its second century of existence, a demand for bird feathers for fashionable women's hats, big and broad and showier than many of the species sacrificed for style and vanity, began consuming the globe. The profit yielded from hunting desirable plumage compared with that from digging for gold, and the pursuit of that profit was written in the plummeting numbers of certain bird species. In some areas of the United States, at the roosts and rookeries of egrets, herons, ibises, spoonbills, and flamingos, hunters conducted clean sweeps. The eagle was never as popular a target of the fashion industry as were wading birds with silky, vibrant feathers. The bald's brown-and-black plumage didn't add splash to desired fashion. Snowy-white feathers did, but the mere dozen from an eagle's tail were hardly worth the effort of chasing the raptor. According to one source, a bundle of one hundred eagle feathers netted a hunter a mere 9 cents. It was the guileless literal appetite of the bald that slipped it into deadly crosshairs.[32]

"The practice of killing predators on sight," the historian Thomas Dunlap writes, "had a long and respectable history" that dated to Old Europe and crossed the Atlantic to the New World. Nobody knows how many bald eagles were caught up in this respectable roundup. But the number taken in the contiguous territories and states from the year the Great Seal was adopted, 1782, to the end of the nineteenth century was surely staggering. (Using the date range 1850–1920, a search in Newspapers.com using the phrase "bald eagle shot" yielded 183,959 matches.) What's similarly extraordinary is that this massive slaughter transpired and that hardly anyone knows about it—at least, not as people know about the extermination of beavers and wolves in many parts of the US and the annihilation of bison, elk, and even birds for hats in some of those same parts and beyond.[33]

With ornithologists and popular culture portraying eagles as inveterate kidnappers, the myth became a green light for ranchers and farmers to shoot and poison bald eagles in the name of predator control and economic security. If a bald can steal away children, it can make off with livestock. That was the perception anyway. The indictment of the bald was close at hand on page 1 of Mark Catesby's *Natural History*, available for consultation since 1729: "Tho's it is an Eagle of a small size, yet has great strength and spirit, preying on Pigs, Lambs, and Fawns." Audubon completed a painting in 1828 of an eagle dropping from a grim, gray sky to take a lamb, pure white and innocent, into its talons. It was a long time before ornithological wisdom evolved beyond these fantasies.[34]

The observant Polly Redford noted in her keen little book *Raccoons and Eagles* that because the farmer has "been brought up on tales of legendary eagles he thinks the birds are much stronger and more dangerous than they really are." To the birds' misfortune, their native land had transformed into a place where someone always seemed to have a firing piece at the ready when a barnyard raider circled overhead. Flying, nesting, feeding, and migrating grew into a deadly game of chance for eagles. Opportunities for feathered couples to endure into old age waned. As early as 1888, the year Arty Kerker took his trophy from Highbridge Park, a biologist in Indiana made the stunning observation that "scarcely does an eagle come into our state now and get away alive, if he tarry more than a day or two."[35]

In the late nineteenth and early twentieth centuries, on just about any given day, some newspaper somewhere in the country ran a story on the slaying of a bald eagle. Not all of these stories were presented with the generous space given to the bald that got caught up in Arty Kerker's sport. More common were eight- to twelve-line briefs introduced by a lead-in on the top line. This in the Baltimore *Sun* in 1857 is one example: "An Eagle Killed by a Lady." The twelve lines that follow reveal that the lady was Mrs. Mary Taylor, who "came to the rescue" of geese belonging to her employer. When the eagle turned "for battle with her," she raised a club and "struck it a blow." Eleven lines in a Vermont paper are prefaced by "Eagle Shot." Another goose snatcher, this one fell to the gun of Mr. Philo Graves, who "had the luck . . . to shoot and capture alive a large American eagle . . . the king of birds." A first-person account from Ironton, Missouri, in 1899 read, "I, as quick as a flash, brought old Betsey to my face and pulled the trigger. I tumbled one of [the eagles] and the other left in a hurry." In Roxbury, Massachusetts, when a bald lighted on the steeple of a local house of God, the "sportsmen of that locality hurried for their guns." And this from the Boston Globe in 1892—words that collided the two universes of the bald eagle: "Bird of Freedom Winged."[36]

In any of the thousands of reports in the big, inky dailies and local sheets, one is hard-pressed to find an executioner chastised for slaying the bird of freedom. The killer might as well have been shooting bison out on the plains. Exceptions to this standard were rare, as when in 1858, a newspaper in Maryland castigated a farmworker who had killed a bald feasting on a dead sheep. The worker was a "negro," most likely a free man rather than a slave (newspapers as a practice identified the owners of slaves who made the news, and no owner is named in this article). The question of slavery's future was dividing the nation, and bloody conflict had already erupted between pro- and antislavery groups in the territory of Kansas. Whites in slave states like Maryland were tormented by the possibility of slave rebellions. To prevent the unthinkable, legislators imposed tighter restrictions on free Blacks to encourage them to leave for the North and take with them an image of Black freedom, as illusory as it was, that might invite the enslaved to revolt. Those who stayed behind were constantly reminded of their place in society.

The Star Democrat, reporting on the actions of the Maryland "negro," suggested that killing the bird of freedom was a privilege exclusive to whites. Instead of accepting the eagle's execution as prevention against livestock destruction, the newspaper asserted that "sheer revenge" had motivated the killer. "The scamp deserves to be skinned for it." Under the stimulus of race hatred, the Black life of the executioner mattered less than the avian life of a marauder, a life that generally mattered very little to begin with.[37]

In most instances, judgment, when rendered, was reserved for the avian kind. A 1905 article in the New York Sun highlighting the habits of Haliaeetus leucocephalus opened with some particularly harsh words: "The great American bald eagle is an illustration of the truth that fine feathers do not make fine birds. For, sad to relate, the original of our national emblem is a scavenger, a coward, and a thief." Mostly, the journalistic tendency was straightforward, favoring neither condemnation nor celebration, like in a hometown news item spotlighting a local angler who landed a rather nice bass down at the lake or river. A slain eagle was often described as a handsome, rare, fine, splendid, noble, beautiful, and perfect specimen. As the weight of the bass was certain to be mentioned, so, too, was the wingspan of the eagle. The victim of Mary Taylor's clubbing "measured six feet or a little more." A bald taken in Salem, Maryland, reached "five feet six inches from tip to tip of the wings." Another, in Trenton, New Jersey, "6½ feet from the tips of its wings." Near Holman, Indiana: "7 feet 4 inches from tip to tip."[38]

Sometimes conflicts between man and eagle happened in the presence of its companion, and newspapers, as did ornithologists, noted the companion's protective nature. A headline in the *Boston Globe* said it all: "Eagle's Fight for Its Mate." The event happened at the mouth of the Chicago River. When an eagle lighted on a ship coming into port sometime after sunset, "there was a rush for guns" among the crew. During the salvo that followed, the eagle came "tumbling down to the deck" of the ship. Another eagle, previously unseen, broke unannounced out of the dark and "attacked the marksmen." For two hours it circled around, according to the newspaper, and came in for repeated strikes. Each time it "braved the shotguns leveled at it," until finally it "flew lakeward and was lost in the night."[39]

Loyalty between mates made good copy. Yet absent was any acknowledgment by newspapers of what the bird couple experienced when gunmen ripped asunder their union for life. There was no comparing with humans the domestic habits and the strong desire of eagles to survive and rescue. The anthropomorphizing pretty much stopped at comparing the behavior of feeding birds with common thievery. The overriding factor was that eagles were a menace and controlling them a recognized necessity. The fact remained that, even if eagles weren't carrying away babies, they were lifting chickens out of barnyards and backyards.

No chicken wrangler wanted to sacrifice a single bird from the poultry stock. Chickens and pigeons were food on the table and money in the pocket. You could keep the weasels and foxes out of the yard by posting a varmint-chasing dog with the chickens, and you could frustrate the night-seeing owls by rustling the chickens in and out of a coop between sunset and sunrise. But these defenses weren't much in the way of obstacles between daytime raptors and grazing poultry. So, you took out a hawk or eagle with a gun whenever you had a chance, even if you weren't a chicken wrangler. You were doing a public service. Farmers weren't the only ones who raised poultry. Townsfolk commonly had a few laying hens and broilers pecking around their yards. That eagle in Roxbury, Massachusetts, was knocked off the church steeple because dutiful citizens perceived it as a robber on the road to Jericho—preparing to fall upon a man's chickens. And their perception was confirmed. The eagle dropped, wounded, from the steeple but managed to fly away. The next day someone found it lying dead on the top of a pigeon house.

BEING A FLAGRANT CHICKEN filcher didn't help to dispel the eagle's image as serial livestock thief. No less frequent than news stories of bald eagle shootings were stories of eagles snatching away lambs and pigs. A bald eagle in Wabash, Indiana, was accused of stealing some sixty pigs before it was captured and "ushered out of the pork business." Referring to the Profile House, a sublime wilderness resort looking out to Eagle Cliff in New Hampshire's White Mountains, William Cullen Bryant noted in his popular *Picturesque America* that there were "various traditions of children and lambs being snatched away and borne up to their lofty eyries" (nests). A bald eagle that apparently escaped from the Philadelphia Zoo in 1900 was shot "just as it was about to soar away with" the lamb of a local stockman. Another escapee, this one from the police zoo in Huntington, West Virginia, "carried off" a pig from a nearby farm. The stories of the zoo-breaking eagles were reprinted in newspapers across the continent—as were so many stories about raptor robberies and wingspans.[40]

As powerful and exquisitely designed as the eagle's wings are, they have limits to what they can lift. At best—*at best*—their cargo capacity is five pounds, and that capacity can be reached only if an eagle takes its prey in continuous flight with momentum behind it. Like children, lambs are, frankly, too heavy, and the scrawniest newborn beef calf weighs more than ten times what an eagle can go airborne with. A three-pound piglet falls within range, but a sow is likely to put her bulk between an eagle and her litter, and piglets pretty quickly grow beyond the alleged thief's lifting capacity.

Still, these realities dissuaded hardly anyone—even professionally trained ornithologists such as Albert K. Fisher. In an 1893 landmark book on birds of prey, *The Hawks and Owls of the United States in Their Relation to Agriculture*, he repeated the story of a bald having "carried off Bodily a large lamb and returned the following day after another." Even those who made the defense of wild birds their life's calling, such as T. Gilbert Pearson, the president of the National Association of Audubon Societies, talked of impossibly outsize payloads of mutton. In an educational leaflet that his organization distributed, Pearson claimed to have seen an eagle, a species he charged with having "no inconvenient scruples," ascend with a lamb after making a "downward plunge." Pearson also said that he knew of a "bird-student" who had witnessed an eagle absconding with a lamb that weighed more than the raptor, ultimately bearing it a distance of five miles.[41]

Being habitual scavengers made matters worse for balds. There were surely many times when an eagle with blood on its beak paid the consequences for the crime of another. Polly Redford wrote that false convictions often resulted when a stockman observed a "bald eagle tearing at the carcass of a lamb. He has caught the eagle in the act, and, as in the first chapter in a detective story, whoever is first seen bending over the corpse is immediately assumed to be the murderer." A four-legged predator likely would have been the culprit. After the killer got its fill, the scavenger bird would move in and take its turn—and the rap.[42]

Execution wasn't only by bullet, and it wasn't only carried out reactively. By the late nineteenth century, livestock farmers and the US Department of Agriculture were finding that the diffusion of poisons was an effective means of exterminating wolves, coyotes, and their marauding counterparts. From its beginning, the Department of

Agriculture was beholden to farming interests, and predator round-ups were an assumed part of its mission. Similarly, the US Fish and Wildlife Service, which was supposed to be "dedicated to the preservation of wild species," the Pulitzer Prize– and National Book Award–winning author Wallace Stegner once reflected, also caved to "pressure from stockmen." Fish and Wildlife promptly forged a history of waging "war on predators, especially coyotes, and the poison baits that it used to distribute destroyed not only coyotes but hawks, eagles, and other wildlife."[43]

Those poison baits were dead animals. To set a trap, the farmer or a worker used strychnine or another lethal substance to douse the body of a deer or elk shot specifically for bait, or livestock previously killed by a predator, and waited for an offending wolf, wildcat, or bear to feed on it. Cowboys, writes the historian Dunlap, "seemed to have regarded it almost as a social duty to 'lace' any carcass they encountered on the range." Like a steel trap, poison was not precision execution. To their peril, nonthreatening species partook in the deadly smorgasbord and got snared. Farmers might apologize for collateral casualties, explaining them as the cost of taking care of business, but they didn't apologize for the death of an eagle that fed on poisoned meat intended for some other predator.[44]

Then, along with the poison minefields, was Frank "Lefty" O'Doul, a baseball slugger who retired from the New York Giants in 1934 with a sterling .349 career batting average, 113 home runs, and 542 runs batted in. O'Doul went up in a plane with pilot Ben Torrey to shoot coyotes and eagles and attain more ribbies, of the wildlife sort. Torrey noted the occasion in the sports page of the *San Francisco Chronicle* in 1936. A burly, ex-military flight instructor from the First World War who had been shooting game from the air since 1929, Torrey took sportsmen like O'Doul up for a little sky fun in northern California, recommending that his clients use a shotgun with number 2 shot.[45]

Torrey, who once posed for a photojournalist between two spread eagles nailed Jesus-like to staked crosses, the butt of his shotgun against his hip, also hired himself out to poultrymen and sheep ranchers. The year before the *Chronicle* article, a group in Colorado paid the "birdman" $25 for every dead coyote and $10 for every eagle. Typically, he set out poisoned carcasses of cows and horses as bait and made the "rounds of his territory" to check the traps. And he got many of his eagles on the wing—not unlike the aces who shot down Germans during the war. "At times," he said, "I am able to fly within 50 feet of the bird by getting behind and slightly over it." Leaning out the door of the biplane, he'd take dead aim at the feathered being, and pull the trigger. Torrey claimed that he pursued only goldens and "not the bald variety," but the integrity of his claim is doubtful. A sheep farmer was quoted as saying, "We have lost lambs from our corral. We feel certain that they were taken by eagles. We do not know what eagles killed them, so we kill all the eagles we can." One winter and spring, Torrey's kill count, combined with that of another air hunter, peaked at more than two hundred eagles—of both varieties.[46]

Summer followed that record season, then another summer passed, and then a third. Then one day in 1938, a livestock farmer who was out in his field looked up at the high-pitched scream of an engine, and watched Torrey's plane take a nosedive into the ground, killing the forty-three-year-old and a teenage student pilot.[47]

WOULD ANYONE HAVE REGARDED Torrey's Icarus fall from the sky as poetic justice for eagles? Few, if any. There was no social stigma attached to killing a *predatory pest*. That's how newspapers, stockmen, agricultural officials, and, perhaps most frustrating of all, ornithologists and National Audubon egregiously categorized bald eagles. That's how they were seen by the guy with the gun ready to shoot them or the woman with the club ready to wallop them. The delineation *pest* was a far cry from *bird of freedom* or *emblem of America*. "Pest" justified expendability. The peculiar loss of balds in the wild, however, wasn't so grave to people, since the species was preserved in the form of a symbol, which, as the nation matured and gained respect internationally, became decidedly more popular.

As if in praise of that popularity, the Great Seal and its eagle got a facelift—a Tiffany's facelift. In 1883, the State Department asked the high-end jeweler of New York, Tiffany & Co., to correct flaws that were holdovers from previous modifications, most glaring of which was the cluster of only eight instead of thirteen arrows in the symbol bird's talon. The company's fifty-year-old muttonchopped chief designer restored the proper number of arrows and smartened up the eagle, shortening its neck so that it no longer appeared to be looking over heads in a crowd. He also cropped the hint of a crest that had persisted on the back of the neck, fattened up spindly legs, and opened the wings, giving them a harp-like profile, with the feathers drawn in rigid uniform rows that are anatomically inaccurate yet heraldically sound. Two years later, Tiffany's delivered the new die for a three-inch seal, the size used today for official documents.[48]

Kept close at hand in pockets and pocketbooks were similar yet slightly different, and somewhat shinier, renditions of the avatar bird on coins. Like those who conceived the statuary of the US Capitol, designers at the US Mint thought the bald eagle and Lady Liberty made a good pairing, and they debuted them together in 1792, the tenth anniversary of the Great Seal, on a half dime and a ten-dollar gold piece. The latter was known simply as "the eagle." Over the years, the mint struck quarter eagles ($2.50), half eagles ($5), and eventually, beginning in 1849, double eagles ($20), the diameter of each corresponding to its value. A final, and the most dazzling, incarnation of Liberty and the bald eagle appeared in 1907.

It evolved from a White House dinner a couple years earlier, when President Theodore Roosevelt shared his disappointment in the artistry of American coinage for falling short of the eminence of the nation. Seated with him was the acclaimed Beaux Arts sculptor Augustus Saint-Gaudens—bearded, bushy haired, and visiting from his green hilltop estate and studio in New Hampshire. In Roosevelt's estimation, "no greater artistic genius" existed, and he invited Saint-Gaudens to lend his talents to elevate the US Mint. Saint-Gaudens agreed with the president's assessment and immersed himself in the project, working closely off Titian Ramsay Peale's flying eagle on the 1836 silver dollar. When the Saint-Gaudens series debuted with the double eagle, with the bald on the reverse flying before lancing rays of the rising sun,

Roosevelt declared the coin "more beautiful than any since the days of the Greeks." The collectors' market today pretty much agrees. The limited-edition high-relief version is valued at over $100,000.[49]

Another showpiece of bald eagle adulation was the bronze bird that dangled from the red-white-and-blue ribbon of the Eagle Scout medal. When the Boy Scouts of America (BSA) was conceiving its highest rank of achievement, the group initially planned to designate the honor as a "Wolf Scout," based on the Silver Wolf award of the original Boy Scouts Association, founded in Great Britain a year earlier. But the leadership quickly decided that the BSA needed something more distinctively American and chose the bald eagle to replace the wolf. On Labor Day 1912, a neatly coiffed seventeen-year-old named Arthur R. Eldred became the first to wear the Eagle Scout medal, pinned above his left breast pocket and sash crowded with merit badges. His hometown newspaper, the *Brooklyn Daily Eagle*, reported that the "sturdy, well-built, keen-eyed, little fellow" remained "very modest" about his achievement. What neither Eldred nor the *Eagle* knew was that within a few years, nesting eagles would disappear from Long Island, as well as his hometown.[50]

In the column next to the Eldred story, the *Eagle* announced the start of the Junior Eagle Baseball League, which included fifty teams. Among them were the Eagle Midgets. Halfway across the country, Green Bay, Wisconsin, was organizing a semiprofessional baseball team called the Eagles. San Francisco already had a semipro ball club by that name. Pittsfield, Massachusetts, and Wichita, Kansas, had both baseball teams and daily newspapers named "Eagle." Butler, Pennsylvania, had a squad styled the same. So did York, Pennsylvania; Muscatine, Iowa; South Bend, Indiana; Butte, Montana; Santa Cruz, California; and Asbury, New Jersey. Along with its newspaper and baseball league, Brooklyn had the "Eagle" football league, as, too, did Pittsfield, Massachusetts, and Monongahela, Pennsylvania. Long Branch, New Jersey; Buffalo, New York; Akron, Ohio; Moline, Illinois; Minneapolis, Minnesota; Lafayette, Louisiana; and, across to the other side of America, Oakland, California, had "Eagle" football teams. In Richmond, Virginia, the "colored" football squad—The

Richmond Eagles—was looking for fresh competition in 1913 after five years straight without a loss. That despised predacious raptor inspired the most popular nickname among college and university sports teams, beating out bears, bulldogs, cougars, wildcats, and, handily, ducks and stormy petrels.[51]

Racine, Wisconsin, had an entire Junior Eagle football league that traced the lineage of its name to Old Abe through the J. I. Case Threshing Machine Company. Based in Racine, and one of the largest agricultural-machinery manufacturers, Case used Old Abe as the model for its company logo. In 1921, after the 101st Airborne Division of the US Army relocated its headquarters from Mississippi to Wisconsin, Old Abe's legacy inspired the image for the sleeve insignia of the infantry division, which nicknamed itself the Screaming Eagles. During the century and more that followed Old Abe's death, "eagle" landed as the name of or on logos for countless businesses and on the official seal of one federal agency after another, as well as eleven of the then sixteen offices of the US cabinet. It graced military uniforms, the Congressional Medal of Honor, and multicolored M&M's dispensed in bowls on Air Force One. It showed up on the emblems of the National Wildlife Federation and the National Rifle Association, and on soft-drink, beer, and oil cans. In 1931, eight eagle-head gargoyles in chrome-nickel steel began their exalted tenure staring out from the sixty-first floor of New York City's Chrysler Building, the tallest building in the world at the time.

As a symbol, the bald eagle conveyed a masculine quality that appealed mainly to the male ego. A women's organization displaying the bird in its logo or emblem was (and remains) an uncommon species. If historical documentation exists that explains a feminine aversion to the raptor, it has not passed beneath this writer's eyes. It seems not that women lacked an appreciation for the bald eagle as a representative of the country.* Apparently, the symbol embodied attributes that failed to conform to Victorian womanhood, which birthed a pro-

---

* The Girl Scouts of the USA started out in the 1910s with an eagle logo, which was replaced in the 1940s with a gold lily framed by a green shamrock.

liferation of all-female organizations. Society expected women to be songbirds, doves, and nesters, not raptors (and certainly not bald), and women generally agreed. In creating its emblem in 1889, for example, the Sons of the American Revolution positioned a bald atop a Maltese cross with a bust of Washington in the middle. The Sons' sibling, the Daughters of the American Revolution (established a year later) opted for its own logo, which displayed no animal or person but a golden wheel that signified, while upholding the separate roles of the sexes, a spinning wheel.

Gendered as it may have been, you couldn't beat the eagle as a symbol, but you could nullify its model. All this is to say that while Americans were taking it out of the wild, they were also putting it on uniforms, medals, currency, newspaper mastheads, and organizational insignia.

YOU MIGHT THINK THAT the living, flying bird of America would have experienced the freedom that it represented, the freedom that had belonged to it before Western cultures settled its domain. You might also think that the species that generated pride in country would generate trust in the people of that country. You might think that, instead of indifference toward slayings, newspapers would be filled with stories of compassion, as with one that appeared in several dailies in John James Audubon's day.

In the context of the deadly times, the exceptional article's two-word title, "The Eagle," feels like a statement. The story was shared by an unidentified correspondent on a "pedestrian tour" in the Missouri territory. He was admiring a bald eagle in a tree, looking into the "sun with an unblinking eye." When a turkey fluttered into the scene, the eagle attacked it, and then someone fired a gun and dropped the bird of prey. The correspondent wrote,

> I might too have killed the eagle, but admiration and awe
> prevented me. I felt he was the emblem and inspiration of my
> country; and for that moment would not for ten thousands

worlds like ours, have cut a feather from his wing. . . . It was a half hour before it died, and during that time my heart was filled with mingled emotions of regret and awe. I felt as though I was witnessing the last moments of some meritorious hero, who had fallen upon the hills of his fame.—This noble bird fixed his eyes upon me, and without a single blink, supported the pangs of death with all the grandeur of fortitude. I could not endure his aspect. I shrank into my own insignificance, and have ever since been sensible of my own inferiority.[52]

Writing in the 1830s, the correspondent merged the symbol and species into one, when the American people as a whole were doing the opposite. Confessing his "own insignificance" next to the bird echoes Alexander Wilson's recognition of this larger world that he believed humans could not comprehend in its entirety. It also suggests a third universe, in which the bald eagle dwelled—a spiritual universe, or world, known to Native cultures of North America. It was a world where the bird with the white head and tail was neither pest nor predator.

FIVE

◆————— *Feather Straight Up* —————◆

THE "INDIAN," WROTE LUTHER STANDING BEAR, "WORE the feathers of the eagle long before the white man came to this land. . . . There is a language of feathers for the Indian." Sharing that language in a 1931 book, Standing Bear reflected on how Lakota braves had traditionally affixed an eagle's tail feather to their headband. Erect and pointed to the left, a feather signified a courageous engagement with the enemy of the third level—the least life-threatening engagement. Pointed to the right, the feather signified bravery of the next level. Straight up, the highest level, meant the warrior had physically touched the enemy, "counting coup," as the act was known. A red stripe painted across the feather indicated that the warrior had sustained an injury.[1]

Luther Standing Bear's father, George Standing Bear, a member of the Brule Lakota of the Dakota Territory, wore an eagle feather straight up with a red stripe. He earned the feather as a young man while fighting territorial rivals. He had been with a small band that surprised a party of Pawnee skinning several bison killed on Lakota hunting ground, a green and amber prairie land west of the Missouri River where the herds ran in swelters of dust and grass. The Lakota pursued the Pawnee, and all escaped except one. The lone Indian had been stranded without his horse, and the Lakota quickly encircled him on theirs, sitting taller than he stood, although he was a big and tenacious man. Each time a Lakota's arrow pierced the Pawnee's body, he broke it off and threw it to the ground. George Standing Bear charged the man. An arrow from the captive's bow glanced off Standing Bear's shield and entered his left arm. Still, he managed to tap the Pawnee with his lance, purposely not killing him. Three other young Lakota followed, each similarly counting coup and taking an arrow.

The Lakota ultimately admired the Pawnee's courage and spared his life. Withdrawing, they turned for a long ride back to camp. The pain of George Standing Bear's injury felt crippling when the party stopped to rest the horses. He passed out and slipped into a dream that never fully explained itself. In it, an eagle appeared high in a seamless white sky. The great bird circled, and its shrilling cry rang down. It descended toward the dreamer, with wings outstretched and fixed on unseen currents. Standing Bear then awoke, opening his eyes to the medicine man performing a healing ceremony, circling him in a trot and whistling through the hollow bone of an eagle wing. The wounded young man carried on, and when the party re-joined its clan, the people celebrated with a victory dance and honored each of the newest warriors with an eagle feather.[2]

Some years later, Luther Standing Bear would himself wear feathers. He was born in December of the year of the Treaty of Fort Laramie, 1868, a month after the last Native signatory put his mark on the agreement between the US government and the Lakota, Dakota, and Arapaho of the northern plains. Out on the prairie there existed a body of time that seemed not to move; only the cycling seasons within it were in motion. Big land lay beneath big sky, with no end to either in any direction. In the season of his birth, "when the bark of the trees cracked" and colors had depleted to umber and gray, only the glistening of frost on cold mornings distinguished earth from heaven, with a wandering bird sometimes connecting the two.[3]

This was ancestral land of the Lakota. The prairie and unmoving time, like the rituals and stories of Luther Standing Bear's people, would always be bound to his identity. "We did not think of the great open plains, the beautiful rolling hills, and winding streams with tangled growths, as 'wild,'" he wrote. "Only to the white man was nature a 'wilderness' and only to him was the land 'infested' with 'wild' animals and 'savage' people."[4]

The undulant prairie was the landscape and capacious study hall of Luther Standing Bear's youth. He learned to ride a horse on it, to capture wild ones and race "sure-footed" ponies against other boys. He was taught how to stalk bison and elk. In the late fall, he hunted "fine to eat" migrating prairie chickens (pinnated grouse) that stopped to feed on the ripening buffalo berries and rosebuds. He kept the nicer wing feathers to fletch his arrows. Sounds of meadowlarks, crows, prairie chickens, and eagles were like the sounds of Sioux words. The "animals had a way of talking to one another just as we did." In "true brotherhood," the Lakota and "their feathered and furred friends . . . spoke a common tongue." As it was with the animal voices, so also was it with the wind, the seasons, and the night: each expressed the numinous energy mediating between all things. Human intelligence, Standing Bear knew, counted less than the enduring prairie.[5]

For Luther Standing Bear's elders and ancestors, and for their own elders and ancestors, the seasons of the plains melded into reiterating events. Nothing was forever on the prairie, yet everything was repeated, reawakened, regenerated, and renewed. That changed with Standing Bear's generation. He and others faced the onrush of Western society, when guns proclaimed conquest and silenced the thunder of bison runs; when the all-encircling land was no longer so open or governed solely by natural systems; when Christian missionaries, the US military, and government agents intervened with separate ideas of who the Indian should be and with common ideas of how the once timeless prairie should change. Felt by the prairie ecologically, change undermined historic Native relationships with animals and the land, as well as the organic rhythms that the Lakota people had always lived by.

The Treaty of Fort Laramie ushered in the era of reservation life and the bloody conflicts that arose around the government's first recognizing and then breaking up and appropriating tribal lands, as when it seized the Lakota Black Hills for gold miners. Indians were increasingly forced to abandon the prairie and pursue a double life—to be purveyors of opposing conventions, to split their personalities, to choose not one direction in a forked path but both, and to travel them simultaneously, similar to how bald eagles had been compelled into a double, parallel existence.

Cultural survival turned into an unending search for Indian identity, and ultimately Luther Standing Bear shrewdly crafted one that was as fluid as the seasons. His father had raised him to be one with the plains but then in 1879, sensing a strategy in the unavoidable, he put the boy, at age eleven, on a train that took him east to the US Indian Industrial School at Carlisle, Pennsylvania. The school was nothing less than a reeducation camp, where the students were required to cut their long hair short, dress in military-like uniforms, read the Bible, and speak only English. Luther Standing Bear quietly resisted yet believed that there was something inevitable and useful in learning the ways of the white man and how to read and write in his language, one different from the language of feathers.

Throughout the rest of his life, Luther Standing Bear retained a strong sense of tradition and history. He wrote books about his boyhood and his people's ways. He wanted others to know who they had been when they were as free as the birds that traveled the sky and the bison that once ran the prairie. After school, he eventually made a name for himself as a Buffalo Bill Wild Wester and a silent-movie actor who was Hollywood handsome. He and Jim Thorpe, Native athlete and actor, founded the Indian Actors Association in 1936. They pressed for dignified portrayals of tribal peoples on-screen and for Native actors to play the roles of Native characters, challenging Hollywood's established practice of casting whites as Indians (Burt Lancaster nevertheless played Thorpe in a 1951 biopic). When Luther Standing Bear performed in the Wild West shows and Hollywood, he insisted on authenticity in the Native languages actors spoke and the dress they wore. Countless times he was filmed or photographed in feathered headdresses and headbands. Displaying feathers incorrectly would have been an insult to his ancestors and culture—and to birds.

Feathers were a conduit to the spirit world. Many early North American cultures that followed the lunar calendar displayed feathers to represent the number of days of the moon. Four feathers indicated the four corners of the Earth, the four winds, the four cardinal directions, or the four elements of the Earth. Plains Indians exhibited feathers in reference to the number of ribs in a bison, a most sacred and life-sustaining animal. They stood for strength, honor, trust, valor, wisdom, fertility, and freedom. A feather's meaning typically related to the traits of the bird it came from.

Eagle feathers almost always corresponded with bravery and the wisdom of the Great Spirit and Creator. The eagle that crossed overhead was a sacred bird. Its spirit existed in the feathers worn straight up or to the side, or displayed with ceremonial objects, and appearing in dreams and stories explaining the origin of life and speaking to the Native reverence for the bald eagle.

NATIVE PEOPLES AND
THE SPIRIT BIRD

In the origin story of the Te'po'ta'ahi, an indigenous people of California, Bald Eagle is the chief of all animals and creator of humankind. When the world was finished and began to support life, Bald Eagle decided it needed people, so it modeled a man from clay and laid him on his back. Man grew to full size but otherwise continued to lie dormant. He "must have a mate," Bald Eagle said, and placed beside man one of its feathers to form into woman. When that happened, Bald Eagle flapped its wings to wake man and then left. When Bald Eagle returned, the great bird smiled and asked whether man and woman had had intercourse. Man said he had been hesitant, so Bald Eagle called Coyote to lie with woman. Coyote was more than happy to oblige but died afterward. Bald Eagle then brought him back to life to learn how well intercourse had worked. Coyote said, "Pretty well, but it nearly kills a man!" Coyote agreed to try again for better results and survived the second time. Bald Eagle looked at man and said, "She is all right now; you and she are to live together," and from this man and woman that Bald Eagle made, the Te'po'ta'ahi people came to be.[6]

"Te'po'ta'ahi" is a name rarely heard. Without ever setting eyes on the Te'po'ta'ahi, a nineteenth-century British ethnologist and linguist renamed them the "Salinan" by varying the name of California's Salinas River, which the Te'po'ta'ahi lived beside. Before American engineers dammed and diverted the river, reducing it to a rivulet that trickles into the sea, steelhead trout and salmon ran it, and Indians and eagles fished it.[7]

Like wildlife, people existing in the days before elaborate engineering projects preferred to live near bodies of water that could provide sustenance. East of the most stalwart of bounding waters, the Mississippi, the Tuskegee lived within the basin of multiple rivers—the Tennessee, Alabama, Coosa, and Tallapoosa—where bald eagles regularly wintered and nested.

Tuskegee territory spread across a luxuriant land created from washed-down sediment of the Appalachian Mountains, eons' worth of decaying plant matter, and rivers in spring flood. The Tuskegee and other area Natives hunted deer and gathered nuts, berries, wild oats, mussels, and crawfish, and they grew corn, beans, squash, sunflowers, and tobacco in the fertile soil. During the colonial era, the British believed that the preference for producing these "earth sweets" rightfully belonged to "civilized people" such as themselves, not to "savages."[8]

The British also believed that Indians had failed to adequately exploit nature's assets before the arrival of the commercial economy from across the great ocean. To the newcomers, a woodland or river with marketable holdings that had been left untouched represented lost opportunity; letting all go to waste was evidence of backward cultures and lazy people. Here was justification enough for Westerners to push Indians off ancestral lands. The woods "abound with fruits and flowers, to which the Indians pay little regard," wrote Henry Timberlake, an eighteenth-century British American military officer, "particularly with several sorts of grapes, which, with proper culture, would probably afford excellent wine." Envy, desire, craving, and perceived possibilities emanate from his words. The forest "admits of no scarcity of timber for every use"; the soil "require[s] little stirring with a hoe to produce whatever is required," and the "meadows or savannahs produce excellent grass; being watered by abundance of fine rivers, and brooks well stored with fish." Beside all this is where Timberlake's countrymen intended to live and where bald eagles did live.[9]

The British were not around long enough, though, to drive the Indians off this good land. Their restless progeny, and sometime enemy, the Americans, took up the matter of expelling Indians immediately after the War of 1812. Having defeated the British at New Orleans and sent them on their way, General Andrew Jackson strong-armed southeastern Indians, including bands that had fought beside him, into ceding twenty-one million acres, about half of Alabama, as debt payment to the US. Then, as president, Jackson wrapped up unfinished business by urging Congress to pass the Indian Removal Act of 1830. Chauvinist lawmakers happily obliged. The new law, along with the cotton gin and the hunger for cotton land, was the impetus for the Trail of Tears, a military-escorted march of some ten thousand eastern Natives to "Indian territory" in the West. Wintering bald eagles were witness to the forced and fatal exodus, one that foreshadowed their own in the next century.

At the outset of that century, the early 1900s, a University of Pennsylvania graduate student in anthropology named Frank G. Speck set out to study the Tuskegee and other eastern Indians. He went to Oklahoma, the bitter physical end of the Trail of Tears.

The federal government was entering its third decade of trying to erase Native cultures from the land. Government bureaucrats and social progressives who called themselves friends of Indians believed that the merciful way to deal with the tattered remnants of those who had survived relocation, war, and disease was to send their young, as with Luther Standing Bear, to government schools to abolish the Indianness in them and bring them, reconstituted, back into conventional society as new Americans. Central to systematic reeducation was dismantling traditional religious practices and beliefs, including the spirit world where the messenger bird dwelled. The architects and agents of this cultural annihilation maintained that they were killing the Indian and saving the person. As with the bald eagle, white Americans formed a paradoxical relationship with Indians, extolling them as noble savages in popular culture while at the same time erasing them from much of the land.

In the midst of this ethnic cleansing, ethnologists like Speck rushed to reservations to record Native languages and stories, just as archaeologists scrambled to rescue artifacts and bones before a developer bulldozed an ancient town or a burial site to raise modern society. What

Speck was doing in Oklahoma became known as "salvage ethnology," preserving a record—language, rituals, images, and stories, similar to the tale of Bald Eagle and the first man and woman—of dying aboriginal civilizations.

Eager to prove his scholarly chops, Speck the graduate student gathered reams of information on the Tuskegee culture. He wrote down eighteen ancestral stories, commodities of Tuskegee identity. One story speaks to the origin of the Earth and how Crawfish and Eagle made the first land when the Earth was only water. In the beginning, when there was "no beast of the earth, no human being," a council of birds met "to know which would be best, to have some land or to have all water." The birds appointed Eagle as chief, and he decided for land. With gamboling legs, Crawfish then descended below and on the fourth morning surfaced with an offering of sea bottom clenched in its claw. After the council of birds formed the offering into a ball, Eagle flew off to transform the ball into land. Initially, the land was small. Still, the first beasts assembled on it, and gradually the seas retreated, exposing more land, connecting all into a vast terra firma. In time, the Great Spirit chose the Tuskegee to be the first people to live on the new land—the rich dry earth that Jackson and white people would one day take away.[10]

Speck's research and that of others, as well as the brittle pages of much earlier journals and memoirs that preserved observations of Native life, offered reminders that there existed a different way of living with and knowing the bald eagle, of honoring it without eliminating it, and abiding its freedom as one's own. The agreements that early North American cultures maintained with nature's nonhuman constituents kept the plant and animal world whole. Casting its gaze over that whole from high above, according to many cultures, was the bald eagle.

IN THE NATIVE LANGUAGES that linguists and ethnologists recorded, lest they disappear, Indians had their own names for *Haliaeetus leucocephalus*. The Lakota called it *pe' sla wanbli*, the Hidatsa *íibadagi*, the Cherokee *awohili*. The Lenape *woapalanne*, the Zuni *bak'oha k'yak'yali*, the Osage *qiiea' pa sa*, the Quapaw *xa-da ska*, the Mohegan *wôpsukuhq*, the Navajo *atsáhqq*, the Menomini *pinashi*, the Ojibwa *migizi*, the Yakama *k'ámamul*, and the Crow *dúuptakoischialeaxe.*

Bald eagles were birds of heaven, soaring at great altitudes. Native peoples and later Americans often thought of them as the highest-flying of the avian species. Yet cranes and mallards soar closer to heaven. Sandhill cranes have been recorded at thirteen thousand feet, and common cranes, natives of Europe and Asia that sometimes wander over to the western US, can reach a stunning thirty-three thousand feet. Cranes hit their peaks when migrating over mountain ranges, staying well above gorges frequented by predacious eagles, which top out at about ten thousand feet. Still, an impressive height.

To the Lakota, wrote Luther Standing Bear, the eagle "flew so high in the sky that he reached the realm of the gods and could look the world over." Seeing one traversing the sky or piercing clouds suggested a messenger delivering prayers to departed ancestors or solemn promises to the Creator. "In an eagle," said the Lakota holy man John Fire (Lame Deer), "there is all the wisdom of the world." He and other holy men would brush feathers over the ill and wounded to keep away evil spirits and sickness. The medicine man's wingbone whistle that woke George Standing Bear from his dream would have served a dual purpose as a sucking tube to draw potential infection and evil spirits out of the young warrior's wound and as an instrument for communicating prayers, asking the Creator to spare his life.[11]

The whistle came from a sacrificed eagle, either golden or bald. Both lived on the prairies. Far to the east and not far from the Atlantic coast, the Lenape saw many fewer goldens, and likely only in the northernmost reaches and in winter months. Bald eagles retained a near-exclusive residency in the territory east of the Mississippi River. The Lenape lived throughout the Delaware River basin, from the Catskills south into current-day southern Pennsylvania. Among the first Natives that English and Dutch settlers traded and fought with, the Lenape were historic rivals of the Iroquois and Susquehannock. Legendary heroes and chiefs in Lenape mythology brandished eagle-feather coupsticks and wore headdresses with a bald eagle tail fixed to the bonnet.

For the Lenape, wearing a tail feather deftly plucked from a living eagle signified courage and portended good fortune. Males coming of age commonly retreated to the hills to seize plumes from the bird that consorted with higher powers. What brought bald eagles to the Catskills was the profusion of fish-filled lakes and streams and rivers, including the 419-mile Delaware, with two branches dissecting the rocky and wooded range.

To Indians, animals were like the pluck in the high-flying eagle and in the braves who wore its feathers; that is to say, they were fundamental. Foremost were the basics: animals were food, clothing, and shelter; they were also weapons, tools, and utensils. Native peoples were creative and exhaustive in the ways they incorporated the physical features of animals, leaving little to waste, which is one reason why many people today think of Indians as the original environmentalists. If nothing else, they were shrewdly efficient and practical. There was a survivalist and energy conservation logic in getting the most out of a hard-won resource, which was the same with the birds they observed.

Just as animals were provisions and utility, just as they were beasts, they were also tribal gods or controllers within the spirit world. The land, the water, and all the earthly things embodied a sacred order for the interior life of the individual and rituals of the community—in song, dance, prayer, storytelling, and rites of passage. Before science and monotheistic faiths came along with their discouraging tendrils, early humans idolized those natural expressions that entered their daily lives—sun, moon, stars, wind, and weather. Gods manifested themselves in these invincible forms, and lesser gods in animal beings. People spoke to animals as if speaking to an elder: with respect. They developed intimate relationships with those they hunted and those they observed hunting. Prey was life that supported life, predator a teacher for perpetuating the life of the observer—each fulfilling its proper and necessary role in a world of cooperative relationships.

Native cultures, for the most part, conceived of a web of existence and perceived death as a partner to regeneration. The power of life cycled across the web and in and out of the spirit world, as in a circle. At some point, early humans looked into the eye of an animal and saw the fire of aliveness. And looking deeper, they established a connection, a kinship through a soul, to all that came to be and continued on Earth. "They drank from the same water as we did," wrote Luther Standing Bear, "and breathed the same air." Humans and animals were "all one in nature." All souls flowed into the others. "The whole universe was endowed with the same breath," according to the Hopi—"rocks, trees, grass, earth, all animals, and men." People "sat or reclined on the ground with a feeling of being close to a mothering power," Luther Standing Bear observed. "It was good for the skin to touch the earth."[12]

On the Native spiritual landscape, the unseen energy that flowed from sun to plant to beast to humans was the cosmic energy that moved through the unbroken circle of souls. The only omnipotence in indigenous worlds was the collective of souls. They constituted the immortal heart of the Great Spirit that resided in the land or the upper strata above. Gradations of authority existed only in a hierarchy of animals within the spirit world. Just as Indians had spiritual leaders and chiefs, so too—as the American essayist David James Duncan puts it—did the "swimming, flying, and four-legged peoples." Certain animal people with outstanding traits possessed godlike powers related to the integrity of the collective.[13]

Aware of and keeping a watch over mortals, animal deities were arbiters of human actions—advising, praising, and censuring them. No animals were better positioned to do that than birds. A character in the 1913 *Sun Dance Opera*, the first Native opera, by Zitkala-Sa, a Yankton Dakota, explains to his daughter, "The eagle, who soars to loftiest heights . . . the wonder-bird who also scans the hearts of man, the great eagle, wisest of all, shall testify." Birds stood apart from land animals, which traveled on foot and along the same worn trails through prairie and wood as humans. Land animals hunted many of the same prey that humans pursued, lay on much the same bedding material, and in some cases, such as a bear when raised on its hind legs, suggested physical attributes that reminded humans of themselves. Birds flew. They attained otherworldly heights, voyaged to remote places, and inhabited tall trees and rock ledges that exceeded human strivings. Land animals often communicated with humans through reprisal—the swipe of a bear claw, for instance—and birds through a screech or cackle. To most Native groups, the hoot of an owl was the sound of bad luck, if not a harbinger of death.[14]

The need for sustenance continued alongside the honoring of animals, as when people hunted birds for food—even eagles. In the 1950s, scientists discovered remains of bald and golden eagles in an Indian midden beside the Columbia River, offering pretty clear evidence that eagles were occasionally on the indigenous menu. The bones were several thousand years old and not necessarily indicative of a continuing dietary trend through the ages; nor is one midden tucked in the northwest corner of Oregon representative of the continent. For many Indian groups of recent centuries, if not most, owls, hawks, and eagles were off the table as food. The Creek of the Southeast believed that traits of the animals they ate passed on to them, which made birds of prey, with their aggressiveness, forbidden fruit. There was the added fear that the spirit of an eaten bird might seek vengeance against those who partook of its flesh. And in some cultures, the eagle, the bird that touched the heavens, was too sacred for consumption.[15]

Like a claw-swiping bear, birds could be physically disapproving of human behavior. There is an old fable of the Lenape, who regarded eagles with the respect accorded an elder, that uses the strength and wisdom of a mature eagle to teach a moral lesson about youthful arrogance. As were all Lenape stories, this one was shared after dark descended, in winter when the ground was frozen, when crawling, slithering, stinging creatures that might object were dormant and unable to hear.

The story begins with a boy who is endeavoring for manhood. He mounts a cliff to pluck an eagle's tail feather to wear in demonstration of his maturity. At first, he repeatedly drives away younger and smaller eagles that come to present their plumage. He wants validation in a feather from a mature bird of proven integrity and respect. When such finally alights, it has come not to make an offering but to seize the boy in its huge talons and fly him to its nest as punishment for rejecting the benevolence of the other birds, sons of the older eagle. "I am the head chief," it tells the boy. "Now you shall stay here and suffer for your greed." The eagle tasks the boy with caring for its four eaglet grandchildren until they grow large enough to fend for themselves, after which they will demonstrate their own maturity by returning the boy to the cliff. The boy soon develops a kinship with and love for his surrogate siblings of feathers. But when they learn to fly and leave for extended periods, he is reduced to staying behind as a mere nestling. He feels rejected, as he once rejected young eagles on the cliff. Finally, they fly him back and leave him with feathers, which, in now recognizing their value, the young Lenape is "glad enough to take."[16]

In a practical sense, birds were regarded as teachers. For thousands of years, Native peoples lived among wise, vibrant avian crowds. In their daily routines, birds taught people about the weather, seasons, hunting, camouflage, and planting and harvesting. Nearly every North American culture revered a bird of some type, named clans after one, or carved one at the top of totems, reflecting the bird's status above other animals. The Apalachee of present-day North Florida played a religious-oriented game that vaguely resembled basketball. Players attempted to toss a small buckskin ball into a nest affixed at the top of a tall goalpost that held a mounted bald eagle facing the setting sun. As well as teachers, birds in Native stories were advisors, heroes, tricksters, suitors, and villains. In the middle of the continent and what ornithologists later named the Central Flyway migration route, the Ojibwa extolled sandhill and whooping cranes, which passed through every spring and fall—thousands in winged formations, their yodeling heard from more than two miles distant. The Ojibwa associated the raucous highfliers with lightning, leadership, and good luck. They were a "bird-harbinger of rain," said Luther Standing Bear. They "foretold wet weather by flying high in the air and coming down whistling all the way." Pueblo tribes had Crane (*ka'lokta-kwe*) clans, and the Creek in the Southeast and Chumash in the Southwest performed ceremonial Crane Dances. The birds themselves danced, in courtship ritual, the author Peter Matthiessen wrote, "elevating bustles, leaping upward, posturing with wings, or moving stiffly in the threat displays that are often included in the dancing." Cranes are otherwise rarely aggressive, and they symbolized harmony, as well as vanity.[17]

Cranes are flocking birds, unlike eagles, which are semi-loners. Cranes are peaceable foragers, eagles bloodletting predators. Cranes seek safety in numbers. Eagles are among those that cranes avoid. Predators of many, eagles are prey of none, except as nestlings or egg embryos, when large birds and four-footed pillagers might visit nests, if they dare risk confrontation with a talon-armed parent. The eagle's enduring nature is one reason why it appeared in Native stories as a creator, even as despairing stories befell indigenous peoples and the indigenous bird while Americans were following their manifest destiny.

PORTENDING THE JACKSONIAN ERA and Trail of Tears that would darken Native life a half century later, a 1778 treaty with the US, forced the Lenape to move west from their native land around the Delaware River. They went first to Ohio, then to Indiana, then Missouri, then Kansas, and finally to the Oklahoma Indian Territory. There they were known as the Delaware, and there, in the late nineteenth and early twentieth centuries, Frank Speck, the University of Pennsylvania anthropologist, wrote about them. One of his early works is his study of the Delaware Big House ceremony, as dictated to him by Chief War Eagle, Witapanóxwe, who wore two eagle feathers. The Big House was an annual eleven-day ceremony that revolved around dances combined with recitations of vision experiences. Before and between each dance, wrote Speck, attendants sweeping the wing feathers of " 'pure' birds, such as the eagle," toward the ground spirited away sickness and witchcraft from the dance paths. In separate peyote ceremonies, participants built a fire with twelve sticks, each representing the dozen tail feathers of the eagle, which in themselves stood for the twelve strata of heaven. Peyote, the sacred medicine, opened pathways to some or all of those strata.[18]

Speck was a dimple-chinned white man who stood out in these cultural environments, looking like the professor that he became after graduate school. In the field he wore a tie, fedora, and tweed coat or wool suit most of the time. Before becoming a product of the Ivy League, Speck grew up in New York and spent summers in rural Massachusetts and Connecticut. Though not quite blue-blooded, the family was positioned well enough to employ live-in help. The nanny and maid was a half-Indian woman named Gussie Giles from South Carolina. All of which raises a thorny question. Most of the early research documents on indigenous cultures were produced by privileged white men like Speck, effectively foreigners in Native worlds. Since the words in these documents were not written in the hand of the subjects themselves, how do we know that we can trust their accuracy, that racial biases common at the time have not tainted them?

Indigenous people have argued that reservation Indians were golden opportunities that gilded the careers of early Anglo anthropologists. The latter often published oral traditions and Native stories, which Luther Standing Bear called "libraries of our people," while rarely giving intellectual credit to the storytellers. Researchers took cultural artifacts without offering compensation, sometimes without permission, like thieves in the night. "We have no secrets left," wrote the Zuni elder Virgil Wyaco directly to his oppressors. "You Whites have stolen them all."[19]

After he became a professor at the University of Pennsylvania, Speck himself took leave from reservations with crates of Indian antiquity bound for museums. Although he shared intellectual credit and encouraged Native graduate students to pursue independent research projects in Indian studies, the professor remained the principal beneficiary of partnerships he had forged with informants.

As one-sided as arrangements may have been between Speck and his informants, he was scrupulous in taking down stories accurately and securing for them a proper repository, even if it was located at the white man's institution. Showing a level of trust in Speck, the Seneca inducted him into their Turtle and Eagle clans (the latter was also known as the Hawk clan). To his credit, he recognized tribal members as more than field subjects. He invited many to join him in research,

Reverse

[Drawing by Benson J. Lossing from the description]

Face p. 12

Pierre du Simitière's rejected rendition of Moses (Benjamin Franklin's choice) for the Great Seal.

WILLIAM BARTON'S SECOND DESIGN

[Traced from the original and reduced one-half]

Face p. 13

William Barton's rejected proposal for the Great Seal.

Charles Thomson's original bald eagle proposal.

CHARLES THOMSON'S DESIGN

[Traced from the original]

The Great Seal of the
United States.

Society of the Cincinnati medal, by Pierre
Charles L'Enfant (1783), which Benjamin
Franklin denounced and then wore.

Fishing. *(Justin Bright)*

Bald eagles were described as lazy thieves for stealing fish from ospreys. This rendering is from *The Burgess Bird Book for Children* (1919), published by the ornithologist Louis Agassiz Fuertes.

An osprey's warning. *(Paul McKenney)*

The earliest known European image of a bald eagle, 1583, by John White or Christopher Kellett. *(© The Trustees of the British Museum)*

John James Audubon's nonexistent Washington's eagle, in reality a juvenile bald eagle. Engraving by Robert Havell, 1827.

Juvenile bald eagle, likely two years out of the nest. *(Ray Fetherman)*

Eaglet, perhaps one to two weeks old, 1934. *(Preston Cook Collection / W. Bryant Tyrrell)*

Most nests hatch two eggs.
(Dave Manke)

Bald eagles often hunt from perch trees near water. *(Paul Armbrust)*

Ice fishing for salmon in Alaska. *(© 2018 Matt Shetzer / Shetzers Photography / AMS Enterprises)*

Bald eagles prefer to eat fish, but they will also consume birds and land animals. *(Jason I. Ransom)*

Bald eagle with a canvasback duck. *(Dave Manke)*

Edward Savage's *Liberty in the Form of the Goddess of Youth*, 1796.

Thomas Crawford's controversial eagle helmet on the *Statue of Freedom* atop the US Capitol dome.

In the nineteenth century, children were told that bald eagles were a threat to society. Nathan L. Silverstein postcard, ca. 1900.

(Preston Cook Collection)

Taxidermy-and-guns postcard. Killing bald eagles in the nineteenth century was a legal and common event.

(Preston Cook Collection)

"A Royal Marauder Comes to a Deserved End," ca. 1930s.
(Preston Cook Collection)

Popular culture, such as
the 1908 silent film *Rescued
from an Eagle's Nest*, perpetuated
the widespread myth that eagles
kidnapped babies.

"The Eagle Catcher," 1900.
Photo taken by Roland W. Reed.

"The Eagle," ca. 1913. Blackfoot men with
coupsticks and feathers.
Photo taken by Roland W. Reed.

Zuni with juvenile bald eagle atop eagle stockade, 1879. Photo taken by John Karl Hillers.

Marching band, ca. 1900. As Americans shot eagles out of the sky, the bald eagle remained a popular symbol. *(Preston Cook Collection)*

NEWARK EAGLES 1939

The Newark Eagles of the Negro National League, 1939.
(Preston Cook Collection)

"Alaska Eagle," ca. 1900. The territory paid a bounty for eagle talons from 1917 to 1952. Photo taken by S. Sexton.

Rosalie Edge, who founded Hawk Mountain Sanctuary and pushed for the passage of federal bald eagle protection, ca. 1940. *(Hawk Mountain Sanctuary)*

Willard Van Name, third from the right, with "hawk-watchers" atop Hawk Mountain, 1949. *(Hawk Mountain Sanctuary)*

A DDT plane spraying for budworms, 1955.

Charles Broley, 1948, one of the first to band eaglets and to link DDT to declining eagle populations. *(Florida State Archives / Joseph Steinmetz)*

Sixty-year-old Doris Mager and George Washington meeting Florida governor Bob Graham after bicycling across the country to raise awareness about threats to bald eagles, 1982. *(Florida State Archives / Donn Dughi)*

Touring raptor education programs in schools reached millions of children during the restoration years in the late twentieth century. *(University of North Texas Libraries)*

Boy fishing in polluted waters. No federal initiative was more important to the bald eagle's late twentieth-century revival than the 1972 Clean Water Act *(Florida State Archives / David E. LaHart).*

The hack tower at Quabbin Reservoir, Massachusetts, 1985. Eaglets lived in hack towers several weeks before release as part of bald eagle restoration in the late twentieth century. *(MassWildlife / Bill Byrne)*

Dianne Davis (née Lefrancois) and Marjorie, being banded and released, 1985, as part of the Massachusetts restoration program. *(MassWildlife / Bill Byrne)*

A juvenile and an adult fight for a fish. An expanding bald eagle population meant more conflicts not only between humans and eagles but between eagles themselves. *(Justin Bright)*

Cameras in bald eagle nesting trees became the most popular wildlife cams in the twenty-first century. US Fish and Wildlife Service, Mountain-Prairie Region, 2004.

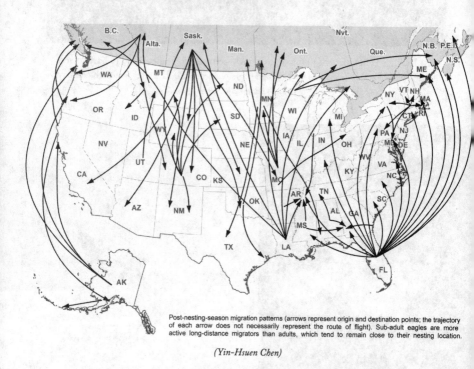

Post-nesting-season migration patterns (arrows represent origin and destination points; the trajectory of each arrow does not necessarily represent the route of flight). Sub-adult eagles are more active long-distance migrators than adults, which tend to remain close to their nesting location.

(Yin-Hsuen Chen)

Bald eagle eyries, the largest raptor nests in North America, were becoming a familiar sight in the twenty-first century.

Processing eagle feathers at the National Eagle Repository, 2010.

Returning eagles to the wild after rehabilitation at a raptor clinic is a time for celebration, as when the bald eagle devotee Preston Cook released this recovered bird in Alaska, 2003. *(Sofia Urata)*

A couple with nesting material. Even early critics of the species admired the fidelity that bald eagles maintain to their partnerships and nests. *(Guiliano De Portu)*

Bald eagles passing safely through a wind site in Wyoming. *(Logan Cyrus for Duke Energy)*

Sarge and a Gulf of Mexico sunset, 2019. *(Linda Weekley)*

and he once engaged an elder, Will West Long, as a collaborator on a book of Cherokee rituals.

The Cherokee were among those driven onto the Trail of Tears. West Long belonged to a band that had eluded removal by taking refuge in the Great Smoky Mountains. He was born in 1870 as Wıli´ westi`. That was around the time when the state of North Carolina affirmed the right of the Cherokee to keep their land and when the federal government officially recognized the Eastern Band of Cherokee Indians (even though the Cherokee had fought on the side of the Confederacy). They were a fortunate few. Wars between other Indians and the US Cavalry were raging out west at the time. Growing up on the Qualla Indian reserve in western North Carolina, officially known as Qualla Boundary, West Long was physically yet not emotionally far from the conflicts, and never from his Cherokee heritage.

Qualla was located in what both Christianized and traditional Indians called "God's country," a softly vertical land of mountains and valleys intense with seasonal colors of mixed hardwoods: birch, maple, poplar, oak, and sourwood. Flowing beneficently through the center of this divine place was the Oconaluftee, a shallow freestone river, trout-filled and mostly flat, quilted with reflections of the sky and trees. West Long's father was a Cherokee Baptist minister who maintained strict tribal customs within the household, and when Wıli´ was a boy, a cousin taught him the words of traditional songs and techniques in carving ceremonial masks—a dexterity he retained to the end of his seventy-seven years.

West Long was a member of the Bird clan and directly involved in rituals and rites that included eagle feathers. He was also the principal keeper of Cherokee sacred ways, a source that both his people and academics turned to. Over the years, he worked with no fewer than nine anthropologists, mostly later in his life.

There is a black-and-white photo of West Long from the 1930s, probably taken by one of those anthropologists, when he would have been closing out his sixties. He is wearing a cotton smock with feather pendants dangling from the sleeves, and a headband securing at the center of his forehead a perfectly erect, white, bald eagle tail feather with a black-dyed tip. Five of the same, fastened at their shafts to a twenty-one-inch horizontal wooden staff, fan out into an Eagle Dance ceremonial wand that he holds high with outstretched hands and arms. West Long's graying hair, combed straight across the top of his head and down the sides to the enlarged ears of an aging man, frames the stereotypical stoicism that whites associated with *the* Indian. West Long may have put on that look for the sake of the picture taker. Tribal members recalled him as being "all smiles and kindness to everybody," tokens of a boyish sense of humor.[20]

There was a constant prick for him, though—an internal struggle in worrying over others exploiting his people for personal gain. "He saw the men for whom he worked prosper," reads his obituary in *American Anthropologist*, "while he remained impoverished." Perhaps in that black-and-white his countenance isn't posed. Perhaps it bears, unconsciously or consciously, an accusatory censure, retiring into the thought that he has stood before the lens one too many times at the insistence of the professors. He was always doing something for them. At the time of his death in 1947, he was helping one who was compiling a Cherokee dictionary, and another who was conducting an ethnobotany study. And there was the book with Professor Speck.[21]

A product of some two decades of interviews, observations, and picture taking, the book, *Cherokee Dance and Drama*, opens to an expansive view of Cherokee songs and performances. At the outset, Speck and his coauthor, Leonard Broom, a UCLA sociologist fond of turtlenecks and tweed, acknowledged that their endeavor would have failed if not for the "cooperation on the part of our collaborator," West Long. The Indian collaborator introduced the two scholars to several traditional dances: Booger, Ballplayer's, Peace Pipe, Pigeon, Green Corn, Running, Friendship, and Eagle.[22]

The first white man to put down in writing his observations of an Eagle Dance was likely the young colonial Virginian named Henry Timberlake, the British American military officer who salivated over the land where the Tuskegee lived before they were sent to Indian Territory. He called the performance an "eagle-tail dance." Each participant used all twelve of the bird's tail feathers attached together to form a wand. Timberlake and a small expedition—a single soldier, an interpreter, and a servant—had canoed down the Tennessee River into Cherokee country past what he later penned on a map as "Enemy Mountains," a province of the Blue Ridge range and the Great West of the day. His impressions of the dance were mixed. He often reverted to cultural norm when portraying those he considered his inferiors, although he was not above siring a child with a Cherokee, she herself in her early teens. In a memoir recording his time with the Cherokee in the winter of 1761 and 1762, he described the dance as a "violent exercise" and its participants "uncouth beyond description." There were maybe four hundred of them, a "mob," many nearly naked, the men wearing only a breechcloth, circling in rows around a blazing fire, their bodies painted "in hideous manner."[23]

Almost two centuries later, with West Long's assistance, Speck and Broom observed a simple handful of participants partaking in "undoubtedly . . . the most spectacular of the Eastern Cherokee dances"—all of them fully clothed.[24]

The Cherokee employed Eagle Dances for multiple purposes, restricting them to the winter in the belief that holding them in other seasons might bring unwanted frost to crops. What Timberlake styled as a "violent exercise" was a peace ceremony, which the circumspect Brit not only scrutinized but sat before as a guest of honor. An ensign in the Second Virginia Regiment, he had volunteered to travel into Cherokee country to complete a peace ritual recognizing the end to recent violent exercises of another sort, between his people and the Indians. Timberlake's was an exceptional task for an officer of the lowest rank. The two sides had been fighting over land, and were destined to fight over it again. White settlers and soldiers had been taking it and would continue to take it, parcel by parcel, deeper and deeper into the mountains—land where eagle feathers were used for ceremony.

Sitting with tribal elders at the ceremony, and following his own culture's protocol, Timberlake was likely attired in dress blues. In front of him, the half-naked dancers circled a fire in a series of tightly orchestrated moves, shaking rattles and waving eagle tails. Others, their eyes the orange of the ceremonial flames, prepared peace pipes.

Dangling from their long stems were "porcupine quills, dyed feathers, deer hairs, and such like gaudy trifles." Passed around, the pipes came at Timberlake as a bucket brigade, "about 170 or 180" of them. Their nearly three-foot stems made for clumsy handling, to say nothing of the steady inhalation of tobacco smoke. It left the white man too "sick" to "stir for several hours" afterward.[25]

The French called the long-stemmed ceremonial pipe a "calumet," or *chalumeau*, which is a flute made from a reed. Ceremonial pipes were common across the continent and sacred among Native peoples, much like communion cups at Catholic mass. "Less honor is paid to the Crowns and scepters of Kings than the savages bestow upon" the calumet, wrote Father Jacques Marquette, a seventeenth-century Jesuit. "There is tyed to it," two French traders noted in 1662 of the calumet of the Ottawa on the upper Mississippi, "the tayle of an eagle painted over with several colours and open like a fan." Dressing the flutelike stem with feathers, from those winged intermediaries between mortals and maker, evoked the sacredness of the pipe, and the rising smoke, like the flying bird, lifted ceremonial utterances heavenward.[26]

Sometimes the Eagle Dance was a spiritual invocation of the opposite to peacemaking. Another eighteenth-century observer watched a performance that was intended to "stimulate in the minds of the young growing people the spirit of war." The dance was also sometimes performed as a "cure" against fatal retribution of the eagle spirit after eagles or eagle feathers had appeared in the dreams of someone, who might have awoken in an anxious sweat. The dreamer had trespassed into an exclusionary realm of a deity and was obligated to give apology, to plead that the subconscious had not been exhibiting hubris or disrespect, had not been peering or snooping. Instead, in its wanderings, the subconscious had become lost and stumbled upon a place of which the humble dreamer was unworthy.[27]

When Speck and Broom visited Qualla Boundary almost two centuries later, an equal number of men and women performed the Eagle Dance in a mock peace ritual. Against the night, the fire and reflective light danced on their own. The performers circled one way and then the other, shadows dropping from their feet, strutting out and back, the entire ensemble moving to song and drumbeat, the dancers variously walking, shuffling and trotting, and stooping and crouching with one knee near the ground. The women bore eagle-feather wands like the one West Long held up for the camera—the shafts made of sourwood, a sacred wood. The men grasped the same in their right hand with gourd rattles in their left, shaking one and waving the other, aloft and then down and aloft again. If a wand touched the ground, apparently suggesting an eagle falling from the sky, the belief was that death might soon visit a dancer or another tribal member.

The performance comprised four movements. In the last, women sometimes held out empty baskets symbolizing a feast offered to all present. "It is an allegory," wrote Speck and Broom, "of feeding the eagle to compensate him for his feathers"—a compensation not extended by whites who took eagles out of their natural place.[28]

This is not to suggest that Indians did not take eagles too. Native peoples "loved to wear his feathers," wrote Luther Standing Bear of the bald eagle, "for he had such great strength and vision." So, Indians acquired them, sometimes by gathering molted feathers and sometimes by plucking them from either a living eagle or a sacrificed one.[29]

❧

You wouldn't want to try plucking one of the Earth's original feather bearers, dinosaurs. Theropods, the ancestors of modern birds and the suborder that includes *Tyrannosaurus rex*, had feathers, although these two-footed, three-toed, hollow-boned, lizard-hipped dinosaurs could not do what most birds do: fly. Despite popular belief to the contrary, no dinosaur could.* Plumes served to camouflage or insulate their cold-blooded bodies, or to show off their good looks. These Jurassic feathered friends were potentially covered from head to toe to tail, even *T. rex*. When a wind kicked up, the mightiest of dinosaurs could look mighty fluffy.

Of all types of coverings from the animal wild, feathers make a superb one. They are light and warm and necessary equipment for fly-

---

* There were flying reptiles that existed alongside but were unrelated to dinosaurs.

ing. They can also be colorful. If we think of beauty in nature as art in nature, feathers are among the most exquisite, the perfect marriage between form and function.

First of all, only the spectrum of colors of fish and insects can compare with that of the estimated ten thousand feathered bird species that bedeck the world. The four-inch, six-gram fiery-throated hummingbird, with tiny scallop-shaped contour feathers of more than a half dozen colors—yellow, fuchsia, violet, orange, green, blue, black, gray, and white, with neon shades in between—is a kinetic rainbow. Wilson's bird-of-paradise, all two ounces of it—flitting around on indigo feet and legs with a matching head of bare skin—has a yellow-feathered nape, blood-red back, emerald-green breast shield, rust-and-black wings, and pitch-black face and chest. As if not already flashy enough, it also sports curious handlebar-shaped tail feathers that would be the envy of the famously mustached Wyatt Earp and Buffalo Bill, not to mention the villainous animated antiheroes Snidely Whiplash and Dick Dastardly.

Yet no evil lurks behind this spiraling allure; just the opposite. Only males have these jaunty tail feathers, and as most of us know, males among many bird species are the vainglorious ones. When playing the suitor, they flaunt their colors to attract a mate, but when engaged in the role of parent, they do the same to distract predators from the nesting partner and her eggs and hatchlings. Color also enables offspring to recognize their parents. Compared with their fathers, their mothers might seem drab in tone, but offspring and mothers rely on the browns and grays common to many as essential camouflage when confined to the nest. This may be the reason why eaglets sprout brown head feathers before multiple molts during adolescence bring in white ones. Their parents—even the mother—don't have such concerns about their own safety. Her signature white head visible above a nest likely serves as a fair warning to hungry interlopers that a bird of prey is in residence.

The basis for coloration is the same across bird species; the difference is in what each is born with, what each eats, and how each processes those food sources. Glossy greens and blues represent a correspondence between light refraction and the microscopic structure of feathers, including underlying pigments. Pigments, both dull and bright, come from either genetically determined melanin or dietary-derived carotenoids. Carotenoids produce the show-offy red, yellow, and orange pigments; melanin the less lustrous black, brown, beige, and gray. However wondrous the entire avian palette seems to humans, seeing the world as we do in a three-dimensional color space, the array of hues appearing in the four-dimensional view of birds, which captures the ultraviolet spectrum, is much more lavish.

The same is true of marine life. In the 1980s, ichthyologists discovered that female guppies harbor a fancy for males with a lot of red spots. A decade later, house finches revealed something similar to researchers: the redder the male wooer, the more successful his seduction, although handsomeness is not necessarily what makes him a good catch. Carotenoids, the source of the red, are acquired from a vegetarian diet of plants and seeds—algae, in the case of guppies. More and brighter spots suggest a skilled forager and better provider to mate and brood with. They also imply that the conspicuous bird is good at evading predators and might be around for the long haul. The white head and tail feathers of bald eagles have no pigments. Their breast and back feathers are a blackish brown, their wing feathers a dark gray and gray-brown, and all colored by melanin. This is not to say that their diet lacks carotenoids. Eagles are not vegetarians, but they eat plenty of herbivorous rodents and fish (the mastodon, remember, stirred up the mouse), and through them absorb carotenoids, which infuse their beaks and feet with yellow as they mature. That yellow potentially plays the same role as red spots in house finches and guppies. Bright yellow equals good health equals alluring mate.[30]

The truth is that scientists are uncertain how eagles choose mates. Females and males look alike in feather dress, at least to the human eye. Perhaps the sharper vision, that four-dimensional color spectrum, of the birds picks up something—a tint or physical trait—that we cannot see. Maybe the crucible is size. Or conceivably, the Wintu of the northern West Coast have it right in believing that the real test is the aerial abilities of the male and how well he uses those yellow toes with talons. Wintu elders tell the story of the female eagle selecting a partner by taking off with sticks, one at a time—each stick bigger than the previous—and dropping them to evaluate a potential male's flying skills. If he fails to grab any of the sticks before they hit the ground, she flies off in search of a new candidate. According to the Wintu, a prospective mother wants a partner who will be able to catch a fledgling that falls from the nest. She may also prefer one who can capture prey in the air or swoop in to retrieve victual that another bird in flight drops—a misfortune that any capable eagle can bring upon another.

These colorful coverings are also ingeniously designed. When you hold a feather, you usually pinch it between your fingers at the base of its stem, or rachis. On eagle feathers, a small V-shaped groove runs along the underside of the stem to add strength against bending. Feathers are formed from beta-keratin, nature's carbon fiber equivalent, the structural proteins that build hair, wool, hooves, beaks, claws, and finger- and toenails. Each grows from a follicle, similar to the hair follicles on the human head. Unlike what happens with a person who goes bald, when a bird loses a feather, the follicle is programmed to produce a replacement. No two follicles and no two feathers are precisely the same, even those in a kindred group—tail or primary wing feathers, for example.

On the bird, each feather laps over another, like clapboard siding on a house or shingles on a roof, shedding away rain and wind. Birds preen in part to keep feathers lapping, and typically they have seven kinds to fuss with. The most commonly known feathers are contour, wing, tail, and down. All told, balds wear seventy-two hundred feathers.

No Lakota, wrote Luther Standing Bear, wore feathers "to 'look nice,' but with awe and appreciation of the wonder of nature." Yet few cultures, even Native, put themselves above pride in appearance. The people of *Tanam Unangaa*, the indigenous name for Alaska's Aleutian Islands, a people of kayaks and fishing and waterproof parkas, called *kameikas*, made those parkas of seal or bird skin and insulated them with feathers. Those for children they lined with the downy skin of young eagles. This was a practical use of feathers, but on rare warm and dry occasions, they wore the parkas with the feather side out to fluff and air them, with the added benefit of changing one's appearance, bringing a little of nature's attractiveness to oneself. Other Natives in cold-weather regions wore plumed outer garments, such as shawls, that were regarded as effects of beauty. The Aleut also pierced the extremity of their ears with quills, and their noses with the shaft of an eagle feather, from which they might dangle small coral, decorative strings, amber, or other feathers. They wove feathers of puffins in baskets as decorative, colorful accents.[31]

When an earnest young ensign of the colonial forces and, centuries later, an erudite professor of the Ivy league, visited an Indian village, each entered a place suffused with representations of the nonhuman world, in wood, stone, shells, furs, skins, and feathers. The settings practically quivered. Feathers hung not only from ceremonial pipes (sometimes with the scalps of birds), but from tomahawks, drums, gourds, spirit sticks, dream catchers, ceremonial shields, lances, satchels, clothing, and hair. Girls and women were sometimes awarded plumes of an eagle—the small, white, billowy feathers beneath the tail—in recognition of reaching a milestone in age or exhibiting a deed of valor. White-feathered heads, frequently along with feet, were mounted on the end of staffs that tribal leaders embraced during ceremonies. Wing feathers fastened to the butt of a staff communicated prayers to the deities, and an oscillating feather often meant the Creator was speaking to the people. Other prayer feathers lay in shrines and floated on water. Feathers were the fletching on arrows and the feature of headdresses. As part of their tribal identity, the Eastern Band of Cherokee Indians adopted, as did Will West Long, a single white bald eagle tail feather with a black-painted tip. Once horses became part of Native culture, Indians attached feathers to their manes and tails.

If you include the golden and bald together, eagles were likely the most popular animals identifying clans, including those of the Caddo, Osage, Ojibwa, Haida, Kwakiutl, Tsimshian, Tlingit, and Chippewa. Some Indians artfully trimmed their hair to resemble their clan animal, with head, wings, and tail in the case of the eagle, looking something like Liberty's headwear on the US Capitol dome. The Zuni had

Turkey, Crane, Grouse, and Eagle clans, yet among all, eagle feathers were paramount to identity and ritual. Medicine men used wings as fans to brush away bad spirits wherever they might exist, within a place or a person. When the Zuni Native Virgil Wyaco returned from France in 1945 fresh from victory over the Germans, before he crossed the Zuni River to home, a medicine man blessed him with a dousing of sacred cornmeal and brushed him down with an eagle wing, "taking away all the evil I might have brought with me."[32]

During their Snake Dance, the Hopi used an eagle wing to stroke the back and rattler of the living reptile to keep it from coiling up in a ready position to strike. At birth, children of Pueblo peoples were given kachina dolls, which represent spirit beings, to teach them how to identify kachinas, of which there are more than two hundred among the Hopi. Feathers were significant to the doll's dress. As breath announces the beginning of life, fluttering feathers, usually the down or smaller ones of the breast of eagles, imitated breathing. The dolls bore "these feathers because the eagle is strong and wise and kind," wrote the ethnologist Ruth Bunzel. "The eagle feathers must always come first."[33]

RITUAL CREATED DEMAND. AMONG the bald eagle's plumed inventory, Natives desired down and tail feathers the most. Lacking pigment, these feathers were ideal for painting or dying. Tail feathers measure approximately eleven inches to their rounded end. An eagle has a dozen. The three hundred dancers that Timberlake encountered with their eagle tails translated into thirty-six hundred feathers. An adult eagle molts approximately once a year, losing only one or two feathers at a time so that it never grounds itself. Gathering feathers from below a nest, in a field, or along a walking path was not sufficient to meet the Indians' demand. Or how most feathers came into their possession.

Some Natives captured and plucked eagles, some domesticated them like livestock for routine feather harvesting, and some killed eagles for the desired gifts. The first Westerners to record domesticated eagles in the New World were the foot soldiers of Francisco Vázquez de Coronado's sixteenth-century search for the legendary Seven Cities of Gold in present-day New Mexico. One among the hopeful treasure hunters was Pedro de Castañeda, who documented aspects of Pueblo life in the region. Outside of trinkets, he saw no gold—no one did—but he took special note of "tame eagles, which the chiefs esteemed to be something fine." The only other domesticated animals were dogs, at best cherished as pets but not idolized. When, centuries later, white men and women descended from American universities and museums, to see, study, and take—not unlike invading conquistadors (more than ten thousand Zuni objects alone)—they found eagles just as the Spanish had, housed in stockades and tethered to perches.[34]

Probably the earliest photograph of an eagle stockade was taken in 1879 by John Karl Hillers, the first official photographer of the Bureau of American Ethnology. A former New York City policeman, Hillers had been trained to shoot bad guys, not pictures, so he learned the latter on the job. An immigrant from Hannover, Germany, he had known mainly city life and could not help being enthralled by the indigenous essence and sprawling geography of the West. Its big-sky light is evident in his photographs, including the one of the stockade. He took it when the sun was low but luminous, sharpening the features of the stockade and a Zuni standing next to it—a young man or teenager. His clothes are a hybrid of Western and traditional: a smock and shin-length trousers made of some type of textiled fabric, and moccasin boots, their soft leather sagging at the ankles. He's also wearing beaded necklaces and a loop earring. The shadow cast on the wall behind him suggests an unseen feather at the back of his plaited cloth headband.

Hillers likely posed him for the moment. His legs are crossed at the ankles, right hand on hip, and left elbow on top of the stockade's side wall, a misshapen form made of mud and rising askew. Wooden poles standing vertically and set apart constitute two other sides, and a square of canvas stretched over a pole frame covers the stockade. It seems like anything but a cozy cage. An eagle is perched on top. It and the posing Indian look away from the lens toward the light. The eagle is a young bald, as handsome as the young Zuni, and wearing all its feathers.

In Zuni culture, taking charge of birds and feathers was a sacred responsibility reserved for males. Feathers were collected during molting, like eggs of laying chickens, and at times plucked between molts and the bare skin rubbed with a salve to quicken healthy regrowth. The Vandyke-bearded Hillers was part of a research party led by a handlebar-mustached James Stevenson, a colonel with the US Geological Survey. Anthropologists saddled up their expeditions into the West much like the early campaigns of exploration and conquest, with horses, tents, bedrolls, and food supplies, sometimes chow wagons and a cook as well. Soldiers were not usually part of the detachments, but rifles and pistols were, mainly as protection against rattlesnakes and territorial mountain lions. A gun toter, Stevenson was a self-taught anthropologist whom the Pueblo despised for his sticky-fingered thievery. They could have reasonably perceived his behavior as having karmically fated his death by Rocky Mountain fever in 1888. Among the Zuni, he claimed to have seen some seventy-five caged birds.

Pueblo territory was ripe for bald and golden eagles. The Rio Grande and Rio Chama traveled ancient serpentine courses through incised valleys and gorges, where the rare rain brought out the sweet, homey scent of the creosote bush. Surrounding sandstone hills, in hues of amber and whiskey, supported a thin, low-lying greenery of pine and piñon. Scattered among them, prickly pear bloomed seasonally with flowers in saffron, violet, and red; the flowers were cup-size vessels of pollen and besotted bees. This was a land of rabbits, weasels, prairie dogs, lizards, turtles, and diamondback rattlers; and a water of sturgeons, eels, trout, bass, bluegill, crappie, and walleye. Food for the winged plenty.

The Pueblo captured eagles (*k'yak'yali*) in two ways. The simplest was to raid a nest for an unfledged eaglet wearing juvenile feathers, the plumage that grows in after the natal down. According to another bearded white man—J. Walter Fewkes, an ethnologist who came calling from the Bureau of American Ethnology—clans of the Hopi assumed "property-rights" claims to nests in the wild. One clan's nest was another's prohibition. To ensure the permanence of the species, a ritualistic ceremony or offering preceded the taking of an eaglet, never more than one per nest, and only if the brood numbered two or more. At the winter solstice, the Hopi offered an eagle-egg prayer to foster a season of many hatchlings. Unlike the Zuni, the Hopi did not keep eaglets to grow old and bear feathers in stockades. Upon returning to the pueblo with a capture, the bird wranglers washed and sprinkled its head with ceremonial cornmeal before squeezing its sternum and the life out of it, then removed its feathers. Amid prayers, the dead were buried in a special eagle cemetery, watched over by kachina dolls.[35]

Back "in ancient times," as Fewkes put it, before the Hopi took possession of nests, they hunted adult eagles. Again, ritual metered the event, which took place in a small stone hut on a distant mesa or crag. The hunter baited his trap with dead rabbits tied to the wooden beams of the "eagle-hunt-house." He then hid inside and chanted ceremonial songs that called for a feather bearer, humbly. When one alighted and sank its talons or beak into the fleshy bait, the hunter reached through from his stony cocoon, seized the bird by the legs, wedged it free of the rabbit, and pulled the big bird inside. After capturing multiple specimens, the hunter attached a prayer stick to one that would be released, to deliver his people's solemn gratitude to the eagle spirit.[36]

When the Spanish came to the high desert, Pueblo were pursuing eagles with one major variation. Rather than raise a hut of stone, the eagle catcher dug a pit and then lowered himself inside and under a blind made of brush. Being cramped in the dark and enveloped in the earth-heavy air could make the wait long. Using a bowl of water as a mirror, the hunter could watch for activity above and outside around the bait, usually a rabbit.

The Lakota, who caught eagles in the same fashion as the Pueblo, believed that the spirit of a slain eagle would shadow the hunter and his people with misfortune. To reduce the risk of reprisal, the Lakota appointed a designated eagle catcher and two assistants. Their task "was no playing matter," wrote Luther Standing Bear, occurring only after a purification ritual lasting days, three or four devoted to the hunter's fasting in a sweat hut made of willow branches, its entry facing east in "reverence for the sun." Properly cleansed, the eagle catcher joined his assistants, who had by then dug an elongated hole, using bison bones near a nest or common roost. After he climbed down in, the assistants covered the hole and the hunter's muted stillness with grass and shrubs, baiting the outside with raw meat, fish, venison, or rabbit—the smellier the better to attract the bird, they believed. Descending to the bait, the bird that had no natural predator remained cautious, "turning his head this way and that, looking and listening," Standing Bear wrote. "His ears were keen, just as his eyes." When the bird's vigilance failed against a snare or the hunter's bare hands reaching its feet through the grassy cover, death came quickly. Its captor rung its neck. Once feathers were plucked, the hunter or an assistant painted the bird's bare head red and placed a piece of meat in its mouth, all before an altar showing the four cardinal directions and adorned with sweetly aromatic pouches of sacred tobacco. The trio then retired the sacrificed eagle to the place where it had been caught, leaving its naked body on a white buckskin "consigned to the Great Mystery who takes care of all things."[37]

One might think that a man holding fast to a live eagle's legs would suffer bloody hands and arms from resisting talons. Standing Bear said that when a hunter gripped the legs "softly and gently," the eagle would remain quiet, as if entranced. "Perhaps he was held by curiosity or perhaps by fear." Robert L. Hall, a Harvard-educated Native scholar and author of an eloquent book with an eloquent title, *An Archaeology of the Soul*, offered a qualifier: "There is little defense against the beak of the eagle or the talons if there were some misstep at the moment."[38]

Avoiding potentially insulting lapses and mistakes was key, particularly with rituals. The habits in spiritual mind and practice varied little among Native groups that hunted eagles. For one, when execution was prescribed, the fear of the eagle spirit's anger weighed heavily. "They believe in rewards & punishments" and in "infamy and misery" in a "future state of existence," wrote William Bartram of southeastern Indians, "just in the manner we do." Among those he observed during travels through the region in the 1770s were Will West Long's ancestors. The Cherokee took a bird only between growing seasons, in the event that their reverence for the sacrificed life fell short. To be an eagle catcher, one had to be a member of the *Ani Tsiskwa*, the Bird clan, the clan of West Long. The task began with the hunter going alone into the hills to fast and pray for four days. Afterward, he killed a deer, set its enticing carcass out, and from a hiding place sang a song from the Eagle Dance. When the predator as prey landed, the hunter used a bow and arrow or blowgun to slay it. Without touching the dead, he prayed, asking for forgiveness, and pushed seven hand-length stakes of sourwood into the ground around the body. He then returned to the village to retrieve helpers, who had also been fasting. Together, they removed the feathers, leaving the body within the stakes or wrapping it in the skin of the deer that had been used as bait. The soul of the sacrificed could then retire to the spirit world and back into the form of the noblest of birds.[39]

The rituals that preceded and followed the killing of an eagle were an essential expression of humility that enabled the spirit to fly away and leave the hunter and his people unharmed. "All these things you must do," the elder Father Bear explained in a Hidatsa story. When clenching and stilling the eagle's feet, taloned feet used to clench the life of another animal to sustain its own, wrote Robert Hall, the hunter in his "grave-like" pit was "dramatizing the sky-earth union," keeping the circle of the essential oneness unbroken.[40]

IT'S EASY TO ROMANTICIZE Native relationships with nature, and to draw on clichés about harmony and aboriginal environmentalism. Scholars debate whether Indians were "ecological." Did their animism translate into ecological science? Inquiries of the sort too often collapse cultural diversity and historical change. The concept of the ecological Indian that many scholars and romanticists preach is flawed. "Ecological" is a social and scholarly construction that originates outside Native cultures and the past.

In speaking on their own behalf, through stories and rituals, and directly to others, Natives regarded themselves as spiritual people and the world as interconnected by the souls of all things. One did not break that sacred circle. No two indigenous cultures were the same, but shared among all was the place of nature in spiritual life and an emphasis on the integrity of the whole. Nature, in turn—how it worked and persisted and meshed together, how its constituents related to the rest—affirmed spiritual beliefs.

All the same, for both human people and animal people, not all was equal. Across all groups of the early North American cultural landscape, there existed no expectation or covenant that a human be sacrificed for an animal. Natives believed all life was eternal, existing for itself, yes, but also to serve humanity's priorities, beliefs, and practices. In the case of the bald eagle, bird and people were at the top of the food chain together, but not quite side by side. Bald eagles elicited the reverence of Indians, yet in worshipping eagles, whether they killed or stockaded them, Indians exercised power over them.

That said, the power difference was not biologically fatal. Indians may have run bison off ledges, but they did not undermine the prairie ecosystem. They may have cleared swaths of land that stole animal habitat and may have piled up pyramid-size oyster-shell mounds up and down coastlines, but they likely did not drive any species to extinction.* They did not reverse the evolutionary process. As Standing Bear said of indigenous lands, "The birds that flew in the air came to rest upon the earth and it was the final abiding place of all things that lived and grew."[41]

Eventually, the invading Euramerican culture changed that abiding place—that is, the natural environments of Native customs and traditions. Even if Indians had escaped European diseases and conflict, the land without the same fruits and animals could not support Native life as before. Feathers of the spirit bird, for one, were harder to come by. As salvage ethnology and armed ornithology suggested, as range wars and predator control indicated, as the experiences of Luther Standing Bear and every slain bald eagle in the United States affirmed, eagle and Indian were both pushed toward unpleasant ends. Anthropology

---

* Humans may have contributed to the extinction of some megafauna species during the Pleistocene epoch, but scientists remain unsure about this assumption.

preserved artifacts but not Indians; the Great Seal of the United States preserved a symbol but not a bird.

These courses would not sustain themselves. A thing lost or on the verge of being lost often acquires a value it did not previously inspire. Concern, even contrition, replaces indifference and callousness. That happened after one too many acres of Indian lands had been taken and when Native populations had shrunk to despairing sums and the language of feathers lay in jeopardy. One response to these developments was the Indian Reorganization Act of 1934, which, with the objective to rectify the past, strengthened tribal sovereignty and put an end to cultural assimilation.

Again the paths of eagles and Indians ran parallel. In Standing Bear's lifetime, when Old Abe was still a mounted display in a glass box, the American way with the predator bird began to evolve toward amity. By the 1890s, Americans had filled out the continent with their farms and cities, and laced it with more than 210,000 miles of railroad tracks, nearly enough to reach the moon. Their frontier no longer existed, and the nation's compelling natural heritage was passing with it. The presence of bald eagles, the living picture of that heritage, was fading too. Americans noticed what was happening, took a moment for self-reflection, and determined that the integrity of their national symbol could not stand without the living bird behind it. The indigenous species that indigenous people called the "spirit bird" spoke.

The Great Eyrie, 1890–1925. (Francis Herrick)

Part Three

NEW
SCIENCE
AND NEW
ATTITUDES

SIX

Eagledom

ALONG THE SOUTHERN RIM OF LAKE ERIE, THERE ONCE existed a natural eagledom ruled by a feathered royal couple. Nearby was the village of Vermilion, Ohio, a tidy, charming leisure destination with a lakefront, riverfront, and lagoon front. The eagledom incorporated a modest-size copse of hardwood—ash, elm, hickory, oak, sycamore, honey locust—that the restless culture beyond had allowed to exist undisturbed. A local newspaper called the eagledom "almost a sacred institution."[1]

Then, in the winter of 1922, a work crew of men entered the sylvan sanctum, not long after the eagles had returned to their perpetual nest in an old shagbark hickory. The crew arrived driving across the frozen ground a small team of horses that drew a wagon loaded with tools and building supplies. The men intended to erect their own nest a dozen yards away from that of the feathered residents. When the work began each morning, the male eagle alighted on a topmost branch of the old hickory to sound a *kark-kark-kark* alarm, while turning his don't-tread-on-me stare on the intruders.

Like any monarch, eagles don't give up their domains easily, or their nests, which can represent years of work. They will abide others in their midst that don't pose a threat and will chase away those that do. Nesting eagles insist that other nesting eagles maintain a wide berth between their domiciles. How the Vermilion pair perceived the work crew is uncertain. Ultimately, the disturbance settled, and the wood remained the sovereigns' realm and their nest the conspicuous castle at its center.

As reported by local old-timers, the eagle nest had occupied the same hickory since 1890—thirty-two years, which is the far end of the life span of wild eagles, if they can avoid gunfire. If a nest endures after the birds' death or some other circumstance has prompted its vacancy, another couple looking for a home will often take it over. Generations had commanded the Vermilion wood from various nests at least as far back as 1830. That's how long whites had been in the area and had known of nesting eagles. But eagles were probably around long before then, hearing the voices and gunfire of the Huron, Ottawa, and Chippewa who were hunting partners with French fur traders, and before that, listening to the sounds of an eagledom free of human incursions. Locals believed that three successive nests predated the one built in 1890. Comparing it to a castle, as they did, was no glib analogy. The nest grew so massive from the eagles laboring over its annual expansion that many people believed no other of its size existed in the country. They called it the "Great Eyrie."[2]

The term "eyrie"—pronounced like either "airy" or "eerie"—dates to medieval Latin and refers to the nest of a bird of prey. In American English the word is spelled "aerie." The British spell it "eyrie" and prefer to use it in place of "nest" when speaking of raptors. Although an American, the man responsible for sending the horses and noisy workers into the eagledom favored the British standard. His name was Francis Hobart Herrick, and he was born among the curling hills of Vermont when the eagles' old hickory was already old, approaching its fourth century of existence.

Herrick was a professor and founder of the biology department at Western Reserve University (now Case Western Reserve University) in Cleveland, a forty-five-minute train ride east of Vermilion. The author of a two-volume biography of John James Audubon and a groundbreaking book on nesting behavior, *The Home Life of Wild Birds*, Herrick was a highly respected and innovative scientist. The sawing and hammering in the Vermilion wood was the consequence of one of his innovations, a wooden structure in a tree on which he would pitch a tent. Herrick got the idea for it from another eyrie he had seen on a trip twenty-three years earlier, in the fall of 1899. He'd been aboard the

Lake Shore and Michigan Southern Railway, crossing Pennsylvania's little toe of Lake Erie waterfront. Out his window, a stick nest in a tree caught his eye. It had the unmistakable lofty presence of an eagle eyrie. In June of the next year, he got off the train to investigate the nest. When he asked around, townsfolk told him that a January storm had blown down the sycamore that had hosted the nest for fifteen years.

Two decades would expire before Herrick put his idea into motion. In the meantime, he published the Audubon biography and bird-nesting book, opened the department's lab to female students, taught thousands of other young people, won professional awards, started collecting Delft china, and stopped curling the ends of his mustache. In all those years, he never lost the enthusiasm for his idea.

What Herrick had in mind was to study the domestic life of *Haliaeetus leucocephalus* up close. To do that, he needed to enter their homes. But balds almost exclusively build their nests near the tops of tall trees, and from the ground, Herrick's view was hardly better than a blind earthworm's. If he climbed into the canopy, he would anger the adults and frighten the young, distracting all from normal domestic behavior. He thought then to build a structure close by in a neighboring tree that would put him up at eye level with the domestic action. On top of that structure, he would add a platform where he'd pitch a small tent, about the size of a "sentry box," as one observer put it. From inside the tent, he would peer stealthily out through a peephole and across the deciduous canopy to see eggs laid and hatched, white fuzzy babies fed and fattened, eaglets growing new feathers and stumbling about, juveniles testing their wings and leaving the nest, and the adults shuttering the house in autumn until the next season.[3]

Institutionally trained, Herrick personified a new generation of ornithologists. He earned a doctorate at Johns Hopkins University in 1888, when PhDs in the US were still a rare commodity and when natural science was evolving beyond mainly identifying and classifying species.

He was also among the new breed of scientists who, around the turn of the twentieth century, rejected guns as tools of the trade. Some scientists who ventured into the wildest of places still armed themselves for protection, but Herrick went into the field bearing nothing more lethal than a box camera and shooting nothing more than pictures. Published in 1902, *The Home Life of Wild Birds* introduced the pioneering methodology with 136 photographs and advice on how to get that ideal shot a mere "arm's length" away. Herrick had not yet pitched a tent on a structure in a tree. That finally happened in 1922, in the eagledom in Vermilion. Through a hole from inside the tent on the platform in the tree, a hardy elm, he began to shoot away and rarely ruffled a feather.[4]

Whether the Great Eyrie was viewed and photographed up close or from a distance, the sheer mass of this structure built by birds boggled the mind. Cone-shaped and spiky, the nest looked like a fallen meteor that had wedged itself between vertical tree branches. It measured twelve feet tall and eight and a half across its circular top, amounting to some fifty square feet of living space, and it weighed approximately two tons. It was larger than the Pennsylvania eyrie that had inspired Herrick's elevated platform, larger than any of the twenty-five eyries the professor measured in the southern Great Lakes region and Canada, larger than those of similar age. The others measured, on average, approximately six feet in height and more or less the same across.

The Vermilion eagles that built the Great Eyrie had longevity on their side, but they had also been tireless renovators. Over the nest's life span, various eagle couples had carted a literal ton of sticks back to the hickory to build and continually add onto the royal pair's home. Herrick referred to an eagle's nest as a "composite structure." It was, he said, a permanent dwelling place, unlike the nests of most birds, "being built upon each year."[5]

In arguably his most alluring photograph of the Great Eyrie, one of the adults is feeding three eaglets, little heads peeping above the bowl. The four look like a family out on the sundeck of their penthouse. The nest is stupefyingly large, seven times taller than the adult. The woven-stick architecture is so dense that it could probably stop a cannonball. It's a wonder that the old hickory kept it aloft. The view from the castle was the best in the eagledom, with a 360-degree panorama of the eagle's domain and beyond to the quilted lake. The top branches were favorite lookout perches for the adults.

Herrick was able to estimate the weight of the Great Eyrie and measure its size after a snarling squall off the lake sent the hickory and nest crashing to the ground in 1925, taking eggs with it. The eyrie was in its thirty-sixth season. It had been a standard of life and persistence in the wood. For generations, the Great Eyrie had been a seasonal meeting place, feeding place, home place, birthplace, nurturing place, flight-training place, and historic place, now gone. If animals experience emotions equated with love and anxiety, one might assume that the Vermilion eagles experienced sadness and frustration from losing a longtime home.

If they did, they channeled those feelings into a search for a new nesting tree. Starting immediately, they scouted the gray wood, as a human couple would shop for a home lot, alighting on upper branches of the tallest trees, apparently checking out each for its visual vantage point and fitness for a nest. Eagles are not particular about whether the host tree is dead or alive. The most important provisions are its location and the perceived strength of the supporting branches that will cradle the nest. The branches are like the ribs of a wooden ship. They brace the overall structure, and ultimately, their arrangement and trajectory determine the shape of the nest. Like the Great Eyrie, some nests resemble a cone, others a saucer, bowl, or wineglass.

After ten days of searching, the homeless eagles settled on a "venerable oak," three and a half feet in diameter and 330 feet south of their old nest. On the first day of building, both birds gathered sticks ranging in diameter from little-finger to fat-cigar size, some half as long or more as their wingspans, picking them off the ground or breaking them off trees. On the second day, one eagle collected and delivered sticks in its talons or beak one at a time to its partner in the tree. The partner then interlaced each with the others to create a foundation and walls. By day four, a nest, complete with a bed of dry grass and straw, was ready for occupancy.

A bald eagle's eyrie is pretty much a solid wooden mass, with grass and leaves and pine needles woven in like mortar between bricks, which is one reason why eyries weigh so much. Unlike a common songbird's nest, which has a deep bowl, the bald's nest across the top has only a shallow dip in the very middle where the eagles place the bedding material. This is the nest itself, on top and at the center of a large wooden edifice. The sticks and sprigs and twigs encircling the bedding are both walls and floor space where the parents can sit or stand and tend to the nestlings and where older eaglets can romp.

The next February, when a chill hung in the wood, workmen came to build a new observation platform near the new nest. Herrick had perceived the fallen Great Eyrie and "tempestuous character" of the lakeside weather as a warning to construct a free-standing steel tower mounted in poured concrete instead of a wooden structure bound to a tree of uncertain durability. Already in residence, the eagles had begun preparing the new eyrie for its inaugural season. The nesting, or breeding, period for bald eagles generally runs nine to ten months, and the first months are devoted to gathering materials and enlarging the eyrie. The Vermilion couple put the refurbishing on hold when the steel tower started going up. Herrick worried that the construction would encourage the sovereigns to abdicate their domain and relocate to parts elsewhere.[6]

Ultimately, they stuck it out. As soon as the workers completed the tower and left with their horses, tools, and busyness, the eagles resumed their annual domestic rituals, which included laying eggs and tending to two eaglets, the most common number for eagle broods. Herrick described the *Haliaeetus* eyrie as a "gymnasium and practice field." About two weeks before the young fledged, they began to spring straight up and down, like gymnasts on a floor mat, to strengthen their muscles and grow accustomed to using their wings. The first few times, the wings unfolded awkwardly and heavily, flapped clumsily, and lifted meekly. There was less air lift than bouncing, and the broad and sturdy nest felt none of it.[7]

The year after the Great Eyrie went down, misfortune shattered the domestic tranquility of the eagledom once again. Beetles had invaded the support branches of the venerable oak, weakening them like wood rot from the inside out. When intense weather inevitably broke in mid-May, the tree's branches did too, and the year-old nest plummeted to the ground, killing all on board—three eaglets grown to half their full size.

The adult eagles waited until the next nesting season to build again, but build they did. They were nothing if not persistent—an impulse that has clung to their species since eagles consorted with mastodons.

Wild animals are versed at rebounding from a storm, fire, or flood—all of which are natural fallouts—responding to them with gestures toward vigorous new creations, as in a new eyrie. Animals have a tougher time coming back from the assaults of humankind, in part because the aftereffects—poison or invasion—often linger and slam the door to creation. Still, as long as animals are not polluted or shot out of a healthy habitat, they can generally tolerate the presence of the human species. *It's humans who take less kindly to coexistence.*

PULLING BACK
FROM EXTINCTION—
THE FIRST TIME

IT WAS INEVITABLE. THE VERMILION EAGLEDOM WAS challenged by the bald's sole predator. On Thanksgiving Day 1924, not long before the storm took the Great Eyrie, a local by the name of John Bovinsky shot the male. The female did as she had done some years before when someone shot her previous mate, and did five years later when the tragedy repeated itself once again: she answered that evolutionary call to species survival and flew off to find a new partner. Eleven weeks after Bovinsky's violence, she returned to the Vermilion eagledom with a new beau. He was, as they say, age appropriate, suggesting that he, too, had lost a mate at some point, although he did well in joining with his latest one. If he had been a *Homo sapiens* male, he wouldn't have believed his luck in landing in a conjugal partnership that came with a castle.

The good fortune didn't hold, however. A month later came the March storm that toppled the Great Eyrie's old hickory. After quickly rebuilding, the couple hatched and raised two offspring. Yet finishing out the season with a fully successful comeback turned uncertain. In late summer, farmers from the next township charged the nesting couple with a capital offense: livestock theft. Turkeys, chickens, and lambs, they claimed, had disappeared, and so, too, should the eagles. The headline of a front-page article in the newspaper read, "Eagles May Pay Death Penalty for Raids on Lambs, Fowl."[8]

It was a familiar story that might have come with a familiar execution to punctuate it—with the duties of judge, jury, and hangman carried out by whoever got to a gun first. But not this time. Herrick believed that executing the eagles would be tantamount to a "lasting disgrace." He wasn't alone. "Friends of our rapidly vanishing wild-life," he wrote, "became deeply concerned over the fate of these eagles, and offers of cooperation and support in seeing full justice done in the case came from many parts of the state and nation." The Izaak Walton League, one of the oldest existing conservation groups in the country, alerted all its chapters nationwide and proposed creating an indemnification fund to reimburse the farmers for their losses. An Ohio congressman made a donation.[9]

All along, Herrick questioned the eagles' alleged guilt. With the cooperation of the Ohio fish and game commission, he conducted a forensic examination of the fallen Great Eyrie, looking for clues into the eagles' diet. Balds, fortunately, were "like primitive man . . . too indifferent to sweep house." In dietary fare, the Great Eyrie revealed no mammals exceeding the size of rabbits. Herrick found feathers and bones of wild birds and, indeed, chickens, yet only a few of the latter. The unkempt house showed no evidence of larger fowl. Feasts of fish had mostly sustained the home place. The "contents of their eyrie," Herrick wrote, "spoke with authority and told the truth." The eagles were cleared of major wrongdoing and spared from punishment.[10]

What of punishment for John Bovinsky, the perpetrator of the Thanksgiving Day murder? As late as 1930, *Nature Magazine* noted that "it is rare that a killer of our eagle is punished in any manner." Yet not on this occasion. A judge fined Bovinsky $25 (equivalent to $400 today) for shooting a bald eagle in violation of state law—an utterly uncommon action taken on behalf of a bird of prey. In a similarly rare instance for journalism, the newspaper ridiculed Bovinsky as an "igno-rant" man whose crass, pointless act cost his large family dearly and upset the good people of Vermilion deeply.[11]

Endearing themselves to Lake Erie residents did not alone save the eagles and their eagledom. In 1934, Herrick published an impassioned plea to stop the mindless slaughter of their species, writing that a "symbol or an emblem represents an idea, which if commendable should be among our country's most treasured possessions. A symbol speaks without words, but if living it speaks with greater power and authority than when dead."[12]

Asserting a shrewd awareness of the bird of paradox, the appeal was more than an appeal; it was a set of shears clipping the wings of the parallel universes. The man working the shears had become a crusader. How could he not? Eagles had opened their homes to him and revealed themselves not as rank cowards and thieves but as industrious, family-oriented, and steadfast avian paragons, representing a species entirely worthy of the symbol it inspired.

The raptor was attracting other new friends too. For the most part, they were city people of a professional class who lived in the East: scientists, writers, photographers, birders, men and women, many of them from the higher echelons of society. They rode the crest of the conservation movement that had started to build in the late nineteenth century. Attitudes were changing, campaigns were waged, and laws were passed, including one of the bald eagle's very own. Like Herrick, people were beginning to see in *Haliaeetus leucocephalus* the behavior and intelligence of a respectable rather than a despicable raptor. Going up against entrenched attitudes, harbored in both expected and unexpected domains, this newly enlightened set nudged history in a new direction and invalidated the separation between symbol and species.

AMONG THE BALD EAGLE'S best friends was the esteemed nature essayist John Burroughs. His eighty-three-year life stretched from the eagle silver dollar of the 1830s to the latter years of the Great Eyrie, in the 1920s. Raised at the literary knee of Walt Whitman, with whom he shared a devoted friendship, Burroughs published more than twenty books, including a biography of Audubon and several volumes on birds. Although he was never directly involved in the conservation movement, those who were admired his stirring reflections on nature and quoted him often. Rare were individuals who humbled themselves to other living things as did Burroughs, most particularly the bald eagle. The white-bearded sage did most of his writing in a secluded hillside cabin, named Slabsides, in the Hudson River valley, where he composed one of the most poignant passages ever dedicated to the avian species:

> Many times during the season I have in my solitude a visit from a bald eagle. . . . The days on which I see him are not quite the same as the other days. I think my thoughts soar a little higher all the rest of the morning: I have had a visit from a messenger of Jove. The lift or range of those great wings has passed into my thought. . . . I want my interest and sympathy to go with him in his continental voyaging up and down, and in his long, elevated flights to and from his eyrie upon the remote, solitary cliffs. He draws great lines across the sky; he sees the forests like a carpet beneath him, he sees the hills and valleys as folds and wrinkles in a many-colored tapestry; he sees the river as a silver belt connecting remote horizons. . . . Dignity, elevation, repose, are his. I would have my thoughts take as wide a sweep. I would be as far removed from the petty cares and turmoils of this noisy and blustering world.[13]

The original passage is significantly longer and written somewhat as an elegy to the bird that lifted his thoughts. "Twenty years ago," Burroughs wrote around the turn of the twentieth century, "I used to see a dozen or more [bald eagles] along the river in the spring when the ice was breaking up, where I now see only one or two, or none at all."[14]

In 1923, two years after Burroughs's death, the nature photographers William Finley and Irene Finley put the predicament of the bald eagle community in a larger context. "Of the millions of people who daily see our national emblem on the coins and arms of our country," they wrote in *Nature Magazine* in an article titled "A War against American Eagles," "it is safe to say that a very large portion have never seen an American eagle in the sky."[15]

It was undeniably safe, if also tragic, to say. By the 1930s, nesting bald eagles had gone missing in a dozen states, most of those by 1900.

In the last decades of Burroughs's life, newspapers were commenting on the birds' inconspicuousness. York, Pennsylvania: "The bald eagles, while not an extinct bird, are very rare." Lafayette, Indiana: "It is the first eagle shot in this county for over twenty years." Fall River, Massachusetts: "The American eagle, or bald eagle, is seldom 'round now."[16]

Indeed, the last active eagle nest in Massachusetts went dormant in 1905. Two years later, and two years too late, the state became one of the first in the country to outlaw the hunting of eagles. A year after that, the staff ornithologist for the Massachusetts State Board of Agriculture followed national custom and listed the bald eagle among "Feathered Enemies." Then, giving perfunctory deference to the law, he clarified that the white-headed raptor was "growing rare" and its threat "need not be reckoned" with.[17]

So paltry was this living vision of America in the East and parts of the Midwest that many people believed balds to be primarily Rocky Mountain birds. Yet even in the West, the frequency of their appearance had fallen away. Not only could this family-oriented species not live in peace in its own homeland, but opportunities were hardly better above the forty-ninth parallel in Canada, where many eagles from the Lower Forty-Eight had historically migrated between nesting seasons. Canadians stubbornly bought into the myth that "bald-headed" eagles were "known to carry away babies," as the Vancouver *Province* maintained, while also peddling the companion fiction that the kidnapping raptors' "lifting power is about twice their own weight." Canada still had amplitudes of humanless tracts where natural catastrophes were the eagle's only risk. Yet outside those lonesome places, Canadian Mounties weren't chasing down eagle killers.[18]

Having always pursued their endeavors free of objection, gunmen for the most part didn't see a backlash coming. An expanding contingent of Americans had begun agonizing over what eagles had been witnessing and experiencing for as long: the loss of wilderness, those natural elements and places that had made America the envy of European countries.

In a familiar pastoral scene, farms and ranches across the land displaced trees and wildlife habitat. Driven first by shipbuilding and then by the housing market, timber outfits routed wild living places wherever they marched out battalions of axes and saws. Erasing the hill-and-dale bucolic milieu that had inspired indigenous schools of art, they left behind a horizon-deep desolation of naked stumps and charred logs, with washways of silt and soil fouling lakes and rivers, the bodegas of bald eagle territory. New England had lost sixty to eighty percent of its forests as early as the 1830s. Hardwoods in the Great Lakes region fell faster than bald eagles. Axes and saws assembled a full assault against the Northwest too. By the mid-twentieth century, ninety-two million acres of southern longleaf pines, trees seemingly designed by nature for the eyrie, would be well on their way to a near complete wipeout. Mining operations from Florida to Maine to California were gouging mountains, hills, flatland, and underland. Fewer and fewer rivers, clear or silty, ran free. Everywhere, dams industrialized the landscape. In little New Hampshire, people had built well over a thousand dams and would not stop before the tally reached four thousand. Towns from ocean to ocean grew into sooty, sun-blocked, hard-to-breathe cities where imbibing drinking-water risked life and gastrointestinal tract. Their economies—the nation's economy—drew heavily on raiding nature's stores, taking away significant parts of the country's natural heritage, what Jefferson said was America.

By the late nineteenth century, more and more people were waking up to the reality that the living eagle's absence was indicative of a larger and duplicitous world. The spiraling decline of America's winged sovereign implied more than the loss of an animal or an official representative; it implied the loss of something fundamental to the kind of country the United States was becoming.

There was an eerie possibility that the bald eagle could go the way of the passenger pigeon, great auk, Labrador duck, heath hen, and Carolina parakeet, all of them extinct by 1918. They joined the many other lost species that had become museum pieces, some of them not even on display but behind a locked, "Staff Only" door off the exhibit hall, reduced to tagged corpses laid out in broad, shallow drawers of collection cabinets—avian morgues. If a taxidermist's rendition of *Haliaeetus leucocephalus* became the only way children, grandchildren, and great-grandchildren would see the real thing, though really an artificial real thing—mannequin stiff, beak and legs painted, feathers faded, glass eyes inserted, the fire of life gone—what would that say about America and national self-respect? What cultural value or noble gesture of humankind would a mounted relic impart?

Americans at the time were bringing the bison back from extinction and learning that such feats could be done. The principal force behind that rescue was the American Bison Society, founded in 1905, the year John Burroughs published his essay on the bald eagle. One of the society's cofounders was William Temple Hornaday, director of the New York Zoological Park. Few were as influential in animal studies as Hornaday, a man of patriarchal mien and a bone-deep confidence in himself as a scientist and scholar. His initial recommendation for dealing with the imminent extinction of species was essentially the same as that of salvage anthropologists. "*Now* is the time to collect" species, he wrote in 1891. "A little later it will cost a great deal more, and the collector will get a great deal less." A sportsman who traveled widely around the globe, Hornaday started noticing a significant decline in wild animal populations in many parts. Seeing a new light, he published *Our Vanishing Wild Life: Its Extermination and Preservation* in 1913, the year before the last passenger pigeon, Martha, died. His most famous of nearly thirty books, *Vanishing* was a four-hundred-page passionate plea—an undisguised demand really—to stop the ruin and preserve America's natural heritage. Among the near-departed that got Hornaday's attention were bald eagles, "threatened with extermination," he noted, in Maine, Missouri, and Wisconsin. He could have named several other states where the aura of the species was turning ghostly. A few years later he wrote in the *American Review of Reviews*, a leading magazine of the reformist temper, "Civilization is against the eagle."[19]

Unlike some birds, bald eagles were not immediately targeted for rescue. Initially, the plighted raptors benefited tangentially from the conservation movement, a product of progressive reform of the late nineteenth century. The challenges of population growth, urban life, and dying agrarian traditions seemed to precipitate every kind of reform agenda, from cleaning up politics to busting business trusts to shutting down saloons, and so much more.

The history of conservation as policy and practice is older than the movement. It is older than Peter Kalm, a bewigged, portly Swedish explorer and naturalist who said of America in 1748, "Since the arrival of great crowds of Europeans, things are greatly changed: the country is well peopled, and the woods are cut down; the people increasing in this country, they have by hunting and shooting in part extirpated the birds, in part scared them away."[20]

Conservation emerged as belated reactions to the destruction that Kalm highlighted. Settlers from the British Isles, having never seen or had free access to woodlands the likes of those in America, and giving little thought to the need for restraint, consumed untold board feet of wood for shelter, hearth, stove, and fences. The English had hardly gotten to know America when, in 1626, six years after securing a foothold at Plymouth, they adopted an ordinance limiting the cutting and sale of timber, largely to conserve resources for enlarging the British naval fleet, an enterprise that leveled immense sweeps of woodlands anyway.

People devoured wild food too. A number of counties in the colony of New York, forty years before Kalm toured it, adopted hunting seasons for grouse, quail, turkey, and the doomed heath hen, all regarded as game birds. Two years later, Massachusetts outlawed using camouflaged boats to pursue waterfowl. In New York, Massachusetts, and elsewhere, the effort to protect was motivated by a desire to ensure that hunters would have a continuing stream of game birds to bag—some for cooking, others for boasting. Enforcement, nevertheless, lacked fortitude and consistency, and measures to conserve were largely ineffective. Safeguarding trees and birds, consequently, remained the principal concerns when conservation energized a movement nearly a century and a half later.

~

CONSERVATION WAS ABOUT THE future—having heat, houses, food, and sport down the line. It hinged on little pleasures in life too. In 1818, the year that John Trumbull completed his *Declaration of*

Independence, Massachusetts became the first state to set up legal safe-guards for a couple of nongame birds: robins and larks. People attached a value to the two species; they decked the air with lovely song—the lark's gay and cheery, the robin's a staccato laugh and chirr. They also ate insects. Other states followed with protections for game and non-game birds, and in 1869, Michigan and Pennsylvania implemented restrictions in a vain attempt to save the dying-out passenger pigeon.

The organized efforts to preserve nongame birds, one could argue, began at bird feeders. After the Civil War, these outdoor accesso-ries became the vogue among the middle and upper classes. People whose lives afforded moments of leisure wanted to be able to iden-tify wild songbirds—those vibrant, melodious, pleasure-giving creatures—feeding out their windows or off their porch. Initially, they had to rely on ornithologists, whose texts were erudite and bulky. Seeing the need for an alternative, a budding nature writer named Florence Merriam published a field guide in 1890 that was light and compact enough to carry when birding, and it employed an agile key system to simplify identification. She titled the book *Birds through an Opera Glass*, a title that resonated with the social class of bird-watching readers. There were legions. The book was a phenomenal success. It inspired a new genre and scores of other birding guides, including chil-dren's versions. Merriam went on to write several specialized guides, among which was a western travelogue with a frisky title: *A-birding on a Bronco*.[21]

It was not raptors but innocent birds with no evident kleptomaniacal streak in them, those that did no perceived harm to human welfare but indeed contributed to the good life, that invigorated bird conservation. They were blameless and appreciated species that, too often, had to contend with the commercial hunters, trophy hunters, pothunters, and just-for-the-hell-of-it target shooters. No less notorious were boys with mail-order pellet rifles slinking like house cats out to bird feeders and nesting trees. "Don't shoot the little birds!" exclaimed the first line of a popular poem of the 1860s.[22]

Trigger-happy youth were a plague, but nothing really opened eyes so wide as the open season on wading birds and their feathers for ladies' hat fashion in the decades enveloping the turn of the twentieth century. "Millinery murder" was the term bird defenders used. Florida and New York were the first states to try to shut down the season—Florida because that's where the greatest portion of wading birds lived in the wild and died by the gun, New York because that's where the millinery industry transformed bird wear into women's wear. Florida didn't have game wardens to arrest lawbreakers, though, and for need of enforcement, state laws weren't much more than sheer window dressing. Making up for that deficiency, the American Ornithologists' Union (AOU) and the National Association of Audubon Societies hired private wardens deputized by county sheriffs.

The AOU was organized in 1883 to promote the professional study of avifauna. Not long after, the professionals began considering the benefits of abandoning the gun for the camera. Theodore Roosevelt, who was born in the same year as Herrick, 1858, and could hold his own with both marksmen and ornithologists, reflected on this shift, writing that when young, he had fallen "into the usual fashion" of a birdman with a gun and "thereby" committed "an entirely needless butchery of our ordinary birds. I am happy to say there has been a great change for the better since then." Not all AOU founders countenanced the modern age of ornithology with a willingness to shoot pictures instead of bullets. To Charles B. Cory, a vainglorious, old-fashioned specimen-collecting naturalist from Boston, the best use of the camera was to take pictures of him posing with his trophies—cat, bear, or bird. "I do not protect birds," he once snarled. "I kill them." Despite lingering traditionalists, the AOU came to represent the new generation of ornithologists to which Herrick belonged. By then, part of its mission, or one conceived by some of its members, was to educate the general public about the foreboding plight of North American birds.[23]

One of those members was George Bird Grinnell, the much-admired editor of the late nineteenth century's most popular hunting-and-fishing magazine, *Forest and Stream*. He would never get fully on board with saving the bald eagle, and he would popularize the reactionary notion to "conserve the collector," which translated into outdated ornithology that relied on the impudence of the gun. Still, his contributions to the early conservation movement were pivotal. Grinnell spent crucial formative years growing up in Audubon Park, the upscale residential development occupying the former farm of John James and Lucy Audubon. In that hallowed setting, he devoured John James's *Ornithological Biography* and sat as a student of Lucy in his primary years.

Along with an appropriate adolescent setting, George Bird Grinnell had the appropriate middle name for the passion he developed for the avian species. He hunted wildfowl but used the pages of *Forest and Stream* to promote preserving their populations and habitat. To be sure, he turned the magazine into a platform for the protection of game and nongame birds alike, perceiving no contradiction between shooting and saving. "The sportsman," Grinnell observed accurately, "through his clubs and journals, has secured many of the best bird protection laws now in force." The outdoor activities of sporting men and women exposed them to the marvels of nature and the sublimity at the heart of national identity, and turned many into vociferous advocates of protection. Some even put down their guns and rods.[24]

Grinnell, though, wanted to communicate the imperative of saving birds beyond the hook-and-bullet crowd. In the February 11, 1886, issue of *Forest and Stream*, at the height of the plume-bird pogrom, he announced the formation of a new organization. He was its founder, and the name was a foregone conclusion: the Audubon Society. Anyone of any age could join. Membership was free. Grinnell required only that members adhere to the society's rules: kill no bird except those for food, wear no feathers for dress or ornamentation, and remove no eggs from wild nests.

Taking eggs from wild nests had a long tradition. Every spring, for centuries, people in Europe and then America engaged in two egg hunts. One was the scramble for the hand-colored and -decorated prizes of the Easter Day quest. The other was the search for the natural-colored baby-blue, white, brown, aqua, and speckled jewels from bird nests. There were the hobbyist and scientist collectors who coveted eggs from every species possible. The bird on the Great Seal qualified for no exemption. On the subject of the scientists' methodology, John Burroughs had choice words to share:

I once heard a collector get up in a scientific body and tell how many eggs of the bald eagle he had clutched that season, how many from this nest, how many from that, and how one of the eagles had deported itself after he had killed its mate. I felt ashamed for him. He had only proved himself a superior human weasel. . . . what would it profit me could I find and plunder my eagle's nest, or strip his skin from his dead carcass? Should I know him better? I do not want to know him that way. I want rather to feel the inspiration of his presence and noble bearing.[25]

More common were those who took bird eggs of all kinds to fry up with pork bacon and potatoes for a stick-to-the-ribs meal. Their "egging," as it was called, often turned greedy. When word spread that the mother birds had laid, an entire village might turn out for a roundup, leaving no nest unturned or egg behind, filling buckets, baskets, washtubs, and flour sacks with more than all the greasy iron skillets could fry up—the excess (the gone-bad eggs) put out for the dogs and hogs. Or for the kids, who would wind up like fastballer Pud Galvin on his way to 365 career wins, throwing strikes for the Buffalo Bisons against the side of a barn, or into a peach basket set up like Buck Ewing catching behind home plate, a .303 lifetime slugger who could smash the yolk out of the ball.[26]

The game couldn't last, though—not when you were dismantling next season's team, undoing what nature did to protect the future. In the interest of avian posterity, people flooded Grinnell's *Forest and Stream* offices with Audubon membership applications. Within a year, he had just shy of forty thousand members, a number that could have knocked him over with a feather. Local branches organized around the country, and after Grinnell began putting out a new monthly publication, *Audubon Magazine*, another ten thousand members joined. It all got to be too much for the *Forest and Stream* budget and staff. After three years and with a heavy heart, Grinnell shut down the society.

Grinnell's Audubon venture was anything but a failure or a fizzle. It was the Big Bang of bird conservation. It had tapped into a spirit that could not be quashed. Down south, on a salt-dome island on the Louisiana coast in 1897, Edward Avery McIlhenny, a sportsman and son of the inventor of Tabasco sauce, established the first ever sanctuary for plume birds (which still exists today). Up north, influential citizens in Boston, home to hatmakers and feather agents, matched McIlhenny's efforts that year by persuading the state legislature to outlaw trade in feathers from wild birds. The principal lobbying force behind the law was the Massachusetts Audubon Society. The year before, Harriet Hemenway and Mina Hall, cousins who belonged to a well-to-do family, turned their elite sense of noblesse oblige to birds, and dipped into the social register to convince friends to organize. Mass Audubon grew fast and large. Nine hundred women dominated its membership.

Women elsewhere took inspiration from the Boston initiative. By 1900, twenty-two states had Audubon societies, and women were their movers and shakers.

That's the way it was. Women could not vote, but they could agitate. They were middle- and upper-class stay-at-home women mostly, but not exclusively, and not only white, yet generally women of means and free time and with a sense of duty. They turned their garden clubs and poetry clubs into political machines that got things done by campaigning for elected-office seekers, clamoring for the ballot, and agitating for new laws. Membership of one club was often duplicated in another, including in Audubon groups. Women Auduboners lectured other women about wearing feather fashions, sometimes calling out one's vanity on the sidewalk or the steps of the opera house. They used their social standing, their money, and the local newspaper's social page to establish bird sanctuaries in their towns and states—thousands across the country. In 1903, the female-powered Florida Audubon, along with the AOU and the League of American Sportsmen, encouraged President Theodore Roosevelt to create with the stroke of an executive order the first federal wildlife refuge, one for birds: Pelican Island on Florida's east coast.

Outdoors among songs, calls, and wings, Roosevelt was like a kid wanting to be the first person to identify all the birds in the bush. He knew and respected Grinnell, as he also did Frank Chapman, an ornithologist at the American Museum of Natural History who edited and launched, in 1899, *Bird-Lore*. Its cover identified the magazine as the "Official Organ of the Audubon Societies" (it would eventually be taken over by the national organization and renamed *Audubon*).[27]

Bird-Lore called 1900 a "red-letter year in the annals of American Ornithology." Seven new state Audubon societies had come into the fold. In December, Chapman initiated the first Christmas Bird Count, an annual nationwide inventory taken by volunteers that became an Audubon Society ritual. That year the state Audubons and the AOU—the birders and the bird scientists—were instrumental in getting Congress to pass the Lacey Act. The law authorized federal prosecution of anyone trading in wild plants and animals in violation of state laws. The namesake and sponsor of the bill was Representative John Lacey of Iowa. Iowa is on the Mississippi Flyway, and Lacey was a passionate birder. *Bird-Lore* said he acted "not alone as a representative of his constituents, but as a representative of the birds." Before his legislation, the cloak of federal protection had never covered their kind.[28]

A main objective of the bill was to reinforce the so-called model laws that states had adopted at the urging of the Bird Protection Committee of the AOU. The committee had drawn up template legislation that each state could revise to fit its particular circumstances. With the AOU and Audubon chapters combining their lobbying efforts, forty-three states ultimately adopted some form of the law. In 1905, Audubon societies in the individual states, with the exception of Massachusetts, decided to join forces in a new confederation, which they called the National Association of Audubon Societies (NAAS).

The state societies did not sleep. There were bird-watching outings to lead, bird defender clubs to organize, sanctuaries to establish and tend to, and a public to educate. The state groups also schooled elected officials. Generally resourceful and well-connected people, birders proved effective lobbyists. National Audubon exhibited the same energy and ability to bring pressure to bear in the halls and chambers of lawmaking. An early significant test of Audubon's influence in Washington came when it campaigned for the Weeks-McLean Migratory Bird Act of 1913. After the act went into effect, migratory and insectivorous birds flew skies a bit more securely. The bill placed them under the "custody and protection" of the federal government and empowered it to establish nationwide hunting seasons and to prohibit the sale of feathers used in women's fashion. The new law was unprecedented, but it was dead within two years, struck down as unconstitutional by two federal district courts.[29]

Bird defenders remained persistent nevertheless, and Congress remedied the matter with the Migratory Bird Treaty Act, which went into effect on August 1, 1918, the year of the death of "Incas," the last Carolina parakeet. The new legislation codified an agreement between the US and Canada and, surviving legal challenges, reinforced US state and Canadian laws by prohibiting the interstate and international transportation and sale of birds, extending federal protection to migratory birds and their nests, eggs, and future.

Around the time that the outlook for bird life began improving, ornithologists finally agreed on a scientific name for the bald eagle: *Haliaeetus leucocephalus*. No more *Falco*, *Vultur*, or *Pygargue*. The fixed name came in 1898, accompanied by no fanfare and giving no particular comfort to the species. *Haliaeetus leucocephalus* and other birds of prey were not granted the same consideration as birds identified for protection. Raptors were excluded from the Migratory Bird Treaty Act. Furthermore, a few out-of-touch ornithologists insisted that *H. leucocephalus* was a nonmigratory species—an erroneous claim that influenced the thinking in government agencies and Congress. America's bird was not a beneficial insectivore or a premier provider of hat feathers either. It was not a marketable good traded across the Canadian border, although it was killed on both sides of it. And it was not a game bird to be protected for the sporting crowd. It was a predator, an *animalis non grata*, an enemy of civilization.

Many bird defenders carried the same prejudices. In one of her guides for beginning birders, the author Florence Bailey did the eagle no favor by maintaining that this otherwise "splendid bird," when failing to obtain its natural food, "carries off sheep and other domestic animals." When Lucy Audubon died, she left to her former pupil, George Bird Grinnell, her husband's dramatic oil-on-canvas *The Eagle and the*

Lamb, which depicted the raptor committing its ultimate crime against humanity. The painting accurately reflected Grinnell's own views.[30]

Alaska also, and significantly, complicated matters for bald eagles. Their ubiquity in the territory challenged notions of extinction and the need for their security.

⌒

INCORPORATED AS A US territory in 1912, Alaska was a generally tough place to live for people coming up from the Lower Forty-Eight in the early days. It was cold and faraway, ideally suited for rugged individualists and recluses. A person typically felt the loneliness of oneself standing in its lost-in-the-wilderness vastness. Put the fourteen Atlantic seaboard states next to Alaska, and they'd be morsels that its mammoth geography could swallow whole, along with a few more— Kentucky, Tennessee, and Alabama.

Alaska was big, and it was biologically and geologically replete. Glaciers and glacier-cut gorges and fjords trimmed and scarred its planetary expanse and erased the modern-calendar meaning of time. Constitutionally conjoined with the rhythms of nature, the nominal father of American conservation, John Muir, who made several curiosity-driven trips to Alaska, wrote, "How truly wild it is, and how joyously one's heart responds to the welcome it gives, its waters and mountains shining and glowing." Where not frozen, water was restless, tumbling in flumes and sheets down mountainsides and storm-tossed against the rocky coast. Far and near, mountains were always in sight, and rock rubble always at foot. As present as both, deep-shadowed woodlands with scents of the living and recycling dead were both solitude and fright. Large and small creatures foraged in them, as they did out on the tundra and grasslands. Alaska was a province of antlered ungulates (elk, caribou, moose, and black-tailed deer) and antlerless ungulates (Arctic-hoofed musk ox, even-toed bison, mountain goats, and Dall sheep). Wolves and grizzlies moved under the midnight sun—as did birds of prey.[31]

Charles Keeler, an ornithologist and poet, who was part of an 1899 Alaskan coastal expedition, which Grinnell and Burroughs joined, wrote that the bald eagle "makes a striking picture" that "seems pecu-liarly in keeping with the grandeur of the scenery and the solitude of these wave-washed shores." Millennia before early *Homo sapiens* crossed the Bering Land Bridge, bald eagles experienced Alaska's steadfast generosity in its rivers and seas. From the fish eater's perspec-tive, nothing was more profuse and clockwork dependable than Pacific herring and salmon. Every spring and summer spawning run was for bald eagles the equivalent of a Klondike gold rush, except the runs were perennial, and each delivered a mother lode to every prospecting eagle. Muir described the Alaskan raptor as "heavy-looking and over-fed, gazing stupidly like gorged vultures."[32]

Eventually, their pescatarian proclivities got them into trouble after people with commercial proclivities began showing up in Alaska. When the US purchased the territory from the Russians in 1867—with gold in the hills, timber in the woods, furbearers on the land, blubbery and pelted mammals in the sea, and fish in salt- and freshwaters—the corresponding extractive industries already had the run of the place. Bald eagles soon would not. As they were to the livestock farmer down in the Lower Forty-Eight, they were to the salmon and herring fisher, fox-fur trader, and sheepherder in Alaska. "There is no agency more destructive to the fish and game of this country," an Alaskan wrote in a letter to the *New York Times* in 1920, "than the American eagle."[33]

Members of the territorial legislature held that same view. If a leg-islator wasn't himself a fisher or trapper, his brother or father or uncle or mother or aunt or sister was. In Alaska, you made your living off the terrestrial and aquatic fulsomeness, or you provided services or sold goods to those who did. The wild essence of the land was everything in Alaska: resource, commodity, capital, asset, equity, sustenance. The people of the territory pretty much operated in a trade-and-barter "*econ-omy*" tied to the most marvelous of "*ecol*ogies." The two words derive from the ancient Greek *oikos* (that's where "eco" comes from), which referred to home, family, and property. Lawmakers stated their prior-ity for property when they put a 50-cent bounty on bald eagles, those alleged fly-by-day scoundrels that picked the pockets of hardworking Alaskans. After handing in a set of talons to territorial authorities, hunters could collect enough to restring their snowshoes to go out for more talons.

That was 1917, the year that the US mobilized for war against the Kaiser. Food shortages are common during war, and a high priority is understandably given to safeguarding limited supplies against waste, mismanagement, and theft. Where there was no shortage of food was the sea. Fishers were making more money than they had hardly ever seen, and they were fanatically protective of their new riches, giving priority to eliminating their avian competition. The majority of states funded wildlife bounties as part of predator control programs, but only Alaska enacted legislation that put a price on the heads of bald eagles.*

At the same time, the emblematic bird helped win the war. The US conscripted its image for uniforms, documents, shipping crates, and submarine chasers. It had been on the army's regimental flag since 1797, and on its medal of honor since the 1862 inception of that distinction. Alaska was entering its second year of paying bounties when the Federal Reserve issued a banknote with, on the back, a most impressive bald eagle in flight and carrying an American flag in its talons. For two sets of those talons, Alaska would pay a bounty hunter the value of that banknote.

The war years marked a low point in America's relationship with its chosen bird. It was the low point, yet a turning point. It was a nadir that became a catalyst for better days ahead for bald eagles and most birds of prey throughout the US.

⁀

THE ALASKAN BOUNTY OPENED up another theater of action— this one stateside—that exposed the deplorable paradox in seeing the replica bird almost everywhere and the real bird almost nowhere. Alaska, the last frontier, was also the last bastion of bald eagles, yet the territory was on its way to replicating what had happened to America's bird in the rest of the country. While highlighting the paradox, William and Irene Finley wrote their 1923 "War against American

---

* Vinalhaven, Maine, instituted a 20-cent bounty in 1806.

Eagles" article in *Nature Magazine* to denounce Alaska's sponsorship of eradication. In the six years since the bounty's inception, they noted, the territory had paid to kill eighteen thousand bald eagles. That total apparently wasn't large enough for Alaskan policymakers. As if to spite the Finleys and others sympathetic to the salmon gourmand, the legislature doubled the bounty, to $1, for every set of talons.[34]

The response of eagle advocates incited hostilities within the conservation community that would endure for decades. Eagle advocates wanted two resolutions: federal protection and an end to bounties. The number of advocates was small, but their voices were loud. They had to be. They were taking on not only Alaska, agricultural and commercial-fishing interests, and a history of misconceptions, but sportsmen and some of their own kind. Members of the outdoor sporting set had been fairly dependable and crucial allies to conservation, and tended to be well organized, funded, and connected. When a conservation proposal was likely to benefit their pastime, they got behind it. When a proposal threatened their pastime, they went after it. Bald eagles ostensibly threatened that pastime by taking fish from the sportsmen's fishing holes and waterfowl from their hunting grounds. Despite the diminished eagle population, an old scene that played out in the field continued to repeat itself: a hunter shot a duck, which fell dead into the water, which an attentive bald swooped down and carried away, and the fist-shaking hunter then swore to prevent another performance of the sort by taking deadly aim.

Much more disappointing—excruciating even—was the initial rejection of protection by three important institutions. One was the Biological Survey. The bureau had always given a high priority to predator control, so its opposition was not altogether a surprise, except that it was packed with staff biologists who might have been, but generally weren't, more enlightened about eagles. The AOU was another letdown. The head of its Bird Protection Committee, Albert K. Fisher, wouldn't sign off on eagle protection. He was an ornithologist with the Biological Survey and the author of an 1893 Department of Agriculture bulletin titled *The Hawks and Owls of the United States in Their Relation to Agriculture*, which tried to set the record straight on the economic benefits of birds of prey. They were among the farmer's "best friends," Fisher wrote. He called bounties a "folly," although he wasn't thinking of bald eagles. While undertaking research for the bulletin, he saw them on the Oregon coast satiating their appetite for salmon, and in Alaska, after the book's publication, he encountered masses of eagles doing the same. Where they proved "injurious" to economic pursuits, he wrote, they "should not be allowed to become numerous."[35]

The third and greatest disappointment was the leadership of the National Association of Audubon Societies. Its snubbing was not altogether unexpected. Anyone who knew anything about its president, T. Gilbert Pearson, knew that he seemed more at home, and welcomed, among bird-shooters than among bird-watchers.

In photographs, Pearson has straight, horizontal eyebrows, expressionless eyes, a round face, and tulip-shaped lips edging a small mouth, also expressionless. He wears starchy white shirt collars to the top of his neck and perfectly set apron- and bowties. He grew up in Archer, Florida, a rural place thick with longleaf pines and alive with resident pileated woodpeckers, Florida scrub jays, and nesting bald eagles. One would think he'd be partial to balds, but he didn't seem to like them any more than had his organization's namesake. In *Bird-Lore*, Pearson called them a species of "no inconvenient scruples," and he could never let go of the notion of the larcenous raptor. For the people who wanted equal protection under the law for eagles and other birds of prey, he had a name: "extremists."[36]

The so-called extremists didn't care for Pearson any more than he did them. It wasn't just his chumminess with bird-shooters and their ammo suppliers; it was his misunderstanding of avian facts and his authoritative attitude about what was best for birds and humans. He was both imperious and insecure. You could see it in his spells of heightened defensiveness, and in the language he used to decry those who disagreed with him, which many did often.

In 1925, when Pearson refused to speak out against a sportsmen-led hawk eradication campaign, birds-of-prey supporters decided he needed a lesson in the will of the people. One who taught it was Waldron DeWitt Miller, a curator at the American Museum of Natural History in New York. At eighteen years old, Miller was a founding member of the New Jersey Audubon Society, and he remained always a faithful reader of Burroughs. With raptors increasingly revealing their home life and true dietary preferences to science, they had begun to win over people like Miller, for whom identifying and classifying species was only the beginning of animal studies. Miller's fervor went to a deeper moral level too. As he comprehended the world, all species had the right to exist free of humans deciding their fate. Categorizing animals as good or bad was an exercise in arrogance that had resulted in one catastrophe after another.

So, Miller spoke out. He believed that the full life of birds was unknowable, and he questioned conventional belief that they existed for the benefit of humankind. He once shared his thoughts directly to the corporate face of the DuPont powder company about the absurdity of a two-month-long international "shoot-the-crow" contest it sponsored with $2,500 in prize money. The ornithologist in the late Alexander Wilson would have approved of Miller's views, and the poet in Wilson would have appreciated his personal flair. Miller was an opera-loving, dungaree-wearing, motorcycle-riding practitioner of modern science who was known for racing off on birding adventures alone or with a friend or colleague on the back or in a sidecar, equipped for the field with only binoculars and a good ear.[37]

Responding to Pearson's disdain for birds of prey, Miller wrote a circular titled *Save the Birds!* He then recruited several prominent naturalists, nature writers, and artists to endorse it and sent it out to some two thousand recipients. The circular encouraged readers to petition Audubon to stand behind birds of prey. Miller had previously launched a letter-writing campaign, and by the time the circular went out, the mail was already piling up at 700 Broadway, Audubon's New York headquarters. Pearson said that the extremists were motivated by a "false sentimentalism" and out to ruin his and the organization's reputation.[38]

Eventually, the supposed sentimentalists forced the issue on Alaska's eagle bounty, mainly through numerous magazine and journal articles like the Finleys'. Feeling the pressure, the Audubon board instructed its president to go on record requesting a suspension of the bounty until a thorough study of the bald eagle's dietary impact could be completed. It also encouraged Pearson to visit Alaska to personally ferret out the truth about the dining behavior of its chief raptor.

He went in 1927, traveling, by his calculations, five thousand miles by rail and steamboat across the territory, talking to people on both sides of the issue. By then, Alaska had paid bounties on some forty thousand sets of talons. That was part of the truth. When he returned to New York, Pearson brought back another truth. Americans, he said, could not fathom the large number of bald eagles in the Alaskan backcountry. Despite the removal of more of the species than likely existed in all forty-eight states, he believed the population could handle it. Instead of recommending a repeal of the bounty, he proposed to launch an education campaign in Alaska emphasizing the value of birds. He had a packet of material on birdlife made up and sent to Alaska's schoolchildren that included a circular that he, the head of Audubon himself, had written. He made no mention of eagles or birds of prey. Around the time the packets went out, Audubon's magazine, *Bird-Lore*, quoted an Alaskan whom Pearson had met: "What good is an Eagle anyway? . . . you cannot eat an Eagle; it does not sing; and so far as I have ever heard it does not eat weed seeds or insects or perform any other special service for man." *For man*—that's how the fishing raptor's fate and value were decided.[39]

~

Inside New York's American Museum of Natural History, through its "castle" entrance of Romanesque Revival design on 77th Street, Willard Van Name preferred that the fate of eagles be left not to humankind but to themselves. Van Name was gathering information that he hoped would compel Audubon to change its position on Alaska's most controversial birds. One of Miller's compatriots in bird-of-prey conservation, Van Name lacked Miller's panache. Outside of work, he mostly kept to his own counsel. The little we know about him comes mainly from his professional work and colleagues. One coworker described him as a "monastic bachelor," which might have been an allusion to both his obsession with work and, most especially, to his sexuality. Some colleagues accused him of being irritable, irascible, and impatient. Yet all knew that he was brilliant too.[40]

Professionally, Van Name started out studying woodlands and trying to save sylvan habitats from timber harvesters, which, along with avian life, remained a central concern of organized conservation in the early twentieth century. Once he turned his eyes and ears (he could discern birdsong as sharply as Miller could) to birds, his attention fell as much on their welfare as on their biology. Said the same colleague, Van Name "devoted most of his salary and all income from a modest inheritance to protection of wildlife and preservation of scenic forests." Like Miller, Van Name was a progressive ornithologist who argued, "More is being discovered about birds to-day with field glasses and cameras than with gunpowder and shot."[41]

The facts that he had been compiling led him, Miller, and Davis Quinn, one of their colleagues, to write a sixteen-page pamphlet. They gave it a mince-no-words title: *A Crisis in Conservation: Serious Danger of Extinction of Many North American Birds*. Four words in that title—"crisis," "serious," "danger," "extinction"—set the tone of the pamphlet's unabashed indictment of organizations that presumed to be the leading authorities on bird protection: Audubon and the Biological Survey.

Both, the pamphlet alleged, were controlled by "sportsmen of leisure," whose self-interest was a "menace." The pamphlet refrained from disclosing the organizations' names, but readers didn't have to work hard to guess the identity of the one that had "thousands of members" and "$200,000 or more to spend every year," and that published *Bird-Lore*. After devoting several pages to discussing birds "in need of protection," the pamphlet moved on to the Alaskan bounty and the bald eagle. Alaska had been underwriting executions of the fisher bird for two years before *Bird-Lore* finally, in 1919, raised the subject in its pages. Then, any mention of it vanished for the next six years, a period during which more than ten thousand bald eagles went to bounty hunters. In 1927 the magazine brought the subject up again, yet failed, according to *A Crisis in Conservation*, to unequivocally take the side of the eloquent raptor against supporters of its extermination. *Bird-Lore*'s articles were instead "largely anti-eagle propaganda," giving comfort to the "idea that the eagle is pretty destructive of game and fish and that it is not in danger of extinction." The central objective of Audubon's representative publication had been to "save the faces of 'bird protectionists' who for ten years have done nothing to put an end" to Alaska's unjust campaign against an innocent and worthy animal.[42]

After acquiring a list of Audubon members, the pamphleteers sent each a copy. To say that it created a stir is an understatement. With close ties to Audubon through its board members, Van Name's employer, the American Museum of Natural History, put a gag order on him. All future works of his required prior approval of his bosses before publication. Van Name had done something like this before in an article for *Science* that attacked the ornithological establishment for buddying up to eggers, hobbyists, and commercial hunters. No gag order was required for Van Name's coauthor Miller. He wouldn't be stepping out of line again. He was dead, killed on his motorcycle in a head-on collision with a motor bus before he could realize the impact of his activism.

As for that scandalous publication, Audubon decided that the best strategy would be to ignore it. Grinnell, whom Van Name had targeted in his *Science* article, additionally recommended that the museum's disloyal curator be left "severely alone."[43]

Rosalie Edge, a previously quiet member of Audubon, didn't ignore the pamphlet. All that Van Name and Miller gave to saving raptors, the bald eagle in particular, was matched by Edge. Although she was not a trained ornithologist, she surely knew and understood more about birds of prey than some who were ornithologists, some who haughtily placed after their name three letters—PhD—some who believed that the best service women could provide to the avian race was filling bird feeders and suet cages. Edge had been a suffragist and knew where women stood in patriarchal society, to say nothing of professional domains. She wasn't afraid of taking on men—and most famously Pearson—who gave life support to pernicious myths about bald eagles. One newspaper said, Edge makes "monkeys out of men of the woods." (It's easy to imagine her thinking, *My stars, what an insult to monkeys!*) In a letter to the editor of the *Brooklyn Eagle* in 1915, she did what hardly any PhD had done without equivocation: she blasted a story of an eagle dragging off a child as an "impossible tale of attack," utterly "offensive" and "preposterous." The *Eagle* called her the "Paul Bunyan (woman) of conservation." An Oregon newspaper described her as a small person with a "giant's strength," possessing "considerable wealth, an indomitable will, a fanatical devotion to a cause, high social position, and a prodigious oratorical gift."[44]

Born in 1877, she had grown up in Manhattan knowing privilege, once wore feather fashions as would a woman of her social class, and then discovered the invigorating outdoors and the marvelous world of wild birds. A boundary pusher, she was a compact force of perpetual energy, who in the field wore skirts and sporty hats that she might just as well wear to a luncheon or Audubon meeting.

As an Audubon member, she received a copy of *A Crisis in Conservation*, which came to her in Paris while on vacation with her children the summer before the stock market crash. After returning to New York, she met the fifty-seven-year-old Van Name on birding outings in Central Park. She appreciated his expertise and his "distrust of mankind" in living peacefully with animal life. A friendship blossomed, and a partnership in ideas and strategies quickly began to take shape. With Van Name's encouragement, Edge decided to attend Audubon's annual meeting that fall, hosted as usual by the American Museum of Natural History.[45]

She arrived at the museum wearing a pair of binoculars around her neck. Walking over from her apartment on Fifth Avenue, she had made a quick birding excursion across Central Park, identifying a grackle, starling, junco, and a few others. She wasn't at the meeting long before she learned that Audubon had "dignifiedly stepped aside," as one speaker put it, from responding to Van Name's printed invective. Edge was among strangers, mostly men, many of whom were intolerant of being second-guessed, yet Edge was herself impatient with such intolerance and wasn't going to let the group step aside from the matter on everyone's mind. She raised her hand, and when called on, she questioned Audubon's position on certain birds facing extinction. She suggested that, instead of dodging the pamphlet's criticisms, the organization should take them seriously. She asked so many questions that others objected to her intrusion.[46]

Eventually, the ruckus she stirred forced the board to recommend that Audubon's staff should evaluate the concerns raised in the pamphlet objectively. Afterward, according to Edge, Pearson confronted her, complaining that her meddling had taken up time allotted for showing a new film on birds and had, furthermore, delayed lunch. As it happened, Edge's meddling was only a stone thrown in the pond, and its ripples incidentally coincided with another, more ominous set of ripples stimulated that day, October 29, 1929. By the time the Auduboners sat down for their hot lunch turned cold, there was turmoil down at the tip of Manhattan on Wall Street. The stock market crashed, and many at the board meeting were staring at economic ruin.

Although the collapse diminished Edge's own monetary resources, her fortitude remained imperishable. She followed up the Audubon meeting with a letter to Pearson, wanting to know what action he was taking. She asked specifically about plans, if any, to challenge the eagle bounty. She also met with William Hornaday, who had invited her to lunch to talk about Audubon. Hornaday and Van Name shared a mutual respect, and Hornaday had little for Pearson. From that day until his death in 1937, Hornaday, the sometimes bristly wildlife defender, would be an ally of Edge.

When Pearson's reply to Edge's letter arrived, it contained the usual Pearson evasiveness and doublespeak—what Hornaday had told her to expect. Pearson responded as if nothing had happened at the board meeting. He had decided in the end not to dignify the pamphlet with a response. The public's loathing for the bald eagle was such, he told Edge, that the Alaska bounty should be left alone. What he didn't tell her was that three years earlier, the territorial governor had invited the Audubon Society to conduct an investigation. If it demonstrated the "innocence of the Eagle," the governor believed that "it should be an easy matter to have the law changed." Pearson had declined the invitation.[47]

Edge's counterresponse was to become a full-time activist. Years later she wrote that the "hand of Willard Van Name," not the "hand of God," had ushered her in that direction. Van Name wanted to continue writing critical commentary. He could not put his name to it, he told Edge, but she could. That notion inspired Edge to start a serial publication and an organization, which she suitably named the Emergency Conservation Committee and oversaw until her death in 1962. Holding the Biological Survey responsible for "allowing" the "obnoxious and unjustified" bounty, and censuring Audubon for not stepping up to oppose it, the committee acted as a watchdog that exposed big conservation's deceit. With the assistance of Irving Brant, a respected journalist who had written about Audubon's coziness with hunters and ammo companies and who advised the Roosevelt administration on conservation initiatives, Edge's publication gathered facts and important published material that its counterparts overlooked or outright ignored. Running a dozen or so pages, each edition had its own title. An early one read, "Its Alive—Kill It!" Another read, "The Bald Eagle, Our National Emblem: Danger of Its Extinction by the Alaska Bounty."[48]

Raising the issue of extinction, as Edge, Van Name, and Brant were doing, was having an impact. Print media started giving a lot of space to its threat. "The American Bald Eagle Is near Extinction," read the title of a short but potent 1930 article in *Popular Science*, which drew on Herrick's research and an interview with Miller. Rather than reporting shootings as if each were a score in a nationwide marksmanship tournament, newspapers turned to celebrating the bald eagle and condemning the possibility of its permanent loss. Local women's organizations and the General Federation of Women's Clubs, whose members undoubtedly read some of these news items, adopted resolutions calling for federal legislation to prevent the destruction of the founding bird.[49]

That's exactly what Congress was considering in 1930 after state Audubon societies encouraged their representatives to take up a bill to save America's bird. Some newspapers were reporting that the "majority of Americans would welcome favorable action by Congress" and taking an editorial position in support of the country's living avatar. The Washington, DC, *Evening Star* proclaimed, "God bless every feather on his spreading tail!"[50]

The proposed legislation that came before the US Senate and House of Representatives in January would make killing or capturing a bald eagle and disturbing nests or eggs a federal crime. The House Committee on Agriculture began hearings in the spring, about the time that eaglets were exercising their wings. Neither Edge nor Van Name was called to testify. Both would have argued against an exception in the bill that permitted killing an eagle engaged "in the act of destroying wild or tame lambs or fawns, or foxes or fox farms." They would have also challenged a proposal that Alaska's delegate, Dan Sutherland, introduced. Sutherland wanted Alaska removed from the bill's purview, saying, "[The] rest of it I care nothing about."[51]

Only two other witnesses came before the committee, both from Audubon. One was Pearson. The committee had compiled a list of more than one hundred individuals and organizations that endorsed the proposed legislation. The AOU was not on the list, but Audubon was. Apparently responding to the widespread popular favoritism toward the bald eagle, Pearson had moderated his views on the bird that he had roundly disdained not long before. The first to testify, he presented himself as an advocate of protection. He clarified that stories of eagles kidnapping children were fallacious, and he insisted, as he had not done before, that the bald eagle's toll on farm livestock was minimal and that the raptor was "doing a good economic service in destroying" gophers and other pests. He took no position on the Alaska bounty, however. He instead reiterated what he had been saying all along—that eagles in Alaska took a lot of salmon and herring and their population was healthy.[52]

Theodore S. Palmer followed Pearson at the witness table. Palmer was a current first vice president and a founder of the Audubon Society. His longtime day job was with the US Department of Agriculture as a staff ornithologist. He brought another credential to the table that was crucial to his being there: he had coauthored the Migratory Bird Treaty Act. Palmer touched on the bald eagle's unique stature as a national emblem, noting that the bill had merit in that it would "in effect . . . increase patriotism, and thereby [have] considerable educational value." He also spoke briefly to the decline of the eagle's presence on both sides of the continent, while referring to an ongoing study of the small number of nesting pairs in Ohio.[53]

The Ohio study was surely Herrick's. Testimony from the professor would have availed the proceeding the deep knowledge of a bald eagle expert, which Palmer was not. One can envision Herrick's opening statement beginning with the assertion that bald eagles, as he later wrote in expressing frustration with the committee's hearing, are "one of the outstanding triumphs of organic evolution on this part of the planet." He would have described "bird of prey" as a "sinister phrase" that "has done much, I believe, to incite the indiscriminate anger of people against one of the most interesting groups of birds." He would have spoken with treetop eyewitness expertise in saying that "there is no better parent in the bird world than the Bald Eagle"—a statement that John James Audubon would likely have corroborated. "To destroy" America's bird, he would have asserted in an apt conclusion, "would be a disgrace and unspeakable folly." Bald eagles are "friends rather than enemies of mankind." Finally, if asked about the migratory habits of balds, Herrick would have said with legitimate authority that indeed they are migratory birds.[54]

Whether the subjects of the proposed law were migratory or not was a critical point. In their individual testimonies, Pearson and Palmer claimed that balds were nonmigratory. Why the two clung to this misperception is unclear. Van Name knew they were migratory. Hornaday knew so. Edge knew so. Miller had known too. The president and the first vice president of Audubon should have known. Palmer, after all, was an author of the Migratory Bird Treaty Act. Instead, Palmer maintained that balds were one-state birds. Whether born of ignorance or conspiratorial calculation, the allegation was detrimental. That it was so, he and Pearson both knew. As Palmer put it to the congressmen, protection for a nonmigratory bird under the proposed law would have the "Federal Government encroaching on the prerogatives and rights of the States." Palmer wasn't marching out some obscure legal technicality. Congress's fidelity to states' rights was famously deep. Even committee members sympathetic to bald eagles were likely to question the reach of federal powers. Indeed, the House scuttled the bill.[55]

Still, getting a law before Congress that reflected a degree of empathy for this long persecuted living creature was in itself next to pulling off a miracle. And the Senate passed the bill, giving advocates reason to hope for another shot at federal oversight.

Without hesitation they regrouped and pursued that possibility. Edge published a five-hundred-word plea from Herrick titled, "Can We Save Our National Bird?" In 1934, Herrick also published his long-awaited book *The American Eagle*, based on his research on the Vermilion eagledom, and the press gave it a lot of attention. The *Nashville Banner* called it a "fascinating study." The *Hartford Courant* characterized it as a "striking," "intensely interesting," and "most unusual book." The *New York Times* concurred, saying, "Our 'bird of freedom' has never had such a thoroughgoing, comprehensive biography."[56]

The state Audubon societies kept up the pressure on their congressional delegations, and in 1935 their hard work bore fruit when Congress again took up a bill to outlaw the harming of bald eagles. National Audubon, writes historian Mark Barrow, then "unleashed its publicity machine to promote" the new bill. Edge put out a twenty-eight-page Emergency Conservation Committee issue wholly devoted to the law's merits and taking apart the long history of misinformation. Equally important, the issue emphasized—almost screamed—that balds were migratory birds. Even so, all these efforts failed to generate enough favorable votes. The states' rights question remained the saboteur. Although bald eagles might cross state lines, as the argument went this time around, the state where they nested had "ownership" of them. One could apply that canard to robins that flew up and down the Eastern Seaboard between their nesting season in New Hampshire and declare them outside the bounds of the Migratory Bird Treaty Act, but no one was doing that. Once again, the Senate gave an affirmative nod to protection, while the House shook its head.[57]

ALL BUT SEVEN STATES had some form of statutory armament for the bald eagle, although they could not consistently ensure its safety, and four gave in to farmer and stockmen demands and rescinded protection. Advocates never wavered on their position though. Some newspapers were reporting that the founding bird had a new champion, Maude Phillips of Springfield, Massachusetts, identifying her as the impetus for a third effort with congressional cooperation.[58]

Phillips was president of the American affiliate of an animal welfare organization, the Blue Cross society. A small woman who kept her hair short and in curls, and an 1881 graduate of Wellesley College, Phillips was the author of several studies on English literature. In 1920 she published a groundbreaking book on the humane treatment of animals that expressed no specific compassion for beleaguered bald eagles. If she had written the book twenty years later, she likely would have. That's when she approached two of her congressmen, urging them to sponsor legislation to protect America's founding bird. She also initiated a letter-writing campaign asking constituents to encourage their elected representatives to cast an assenting vote. As a backup in case an individualized law failed, she pushed to have the peripatetic raptor included in the Migratory Bird Treaty Act.[59]

In March 1940, when eaglets were hatching in a shrinking number of nests, the House Committee on Agriculture held hearings on the new bill. This time, there was no dispute over whether bald eagles were a migratory species, no discussion about a particular state's having ownership over a particular bird, no nullifying the reach of federal stewardship. Representative Charles Clason of Massachusetts, who sponsored the House bill, quoted Van Name when he said, "[The] Bald Eagle certainly is a migratory bird." Representative Fred Gilchrist of Iowa concurred, arguing, "It is a matter of common knowledge, isn't it?"[60]

Maude Phillips gave a long statement, offering the representatives a history lesson on the adoption of the bald eagle as a national icon. Unfortunately, her presentation was informed less by truth than by lore. While citing the wrong date, she portrayed the Continental Congress's selection of a symbol as a tussle between Ben Franklin wanting the turkey and others the bald eagle. If nothing, her history lesson was at times entertaining. To the eighteenth-century patriots and congressional delegates, she said, the eagle "appealed to ideals which constituted the fundamental factors in the foundation of their new democracy," while the turkey appealed as the "promise of a good dinner." She wrapped up her testimony with a poignant comparison between two national symbols. Destroy the US flag, she maintained, "and another can be readily supplied. . . . But the eagle is a masterpiece of the Creator which no human hand can fabricate."[61]

Whatever Phillips's contribution, she was building on the momentum established by Edge, Van Name, Miller, Herrick, the Finleys, and others, and nobody was letting up. In one of its greatest triumphs, the Emergency Conservation Committee secured a signature endorsement from President Franklin Roosevelt for a teaching pamphlet on the bald eagle and the danger of its extinction. "I share with you," the president wrote, "the wish to see these birds adequately protected by law."[62]

Big and small dailies gave a lot of ink to the potential for extinction and to debunking egregious falsehoods, which they had once propagated themselves. The list of endorsers of federal status for the bald was again wide and long. It included the US Department of Agriculture, which offered to administer the law's enforcement. That responsibility ultimately, and probably wisely, went to the Department of the Interior. Another endorser was the Fraternal Order of Eagles, a civic organization committed to social justice. It supported protection with a resolution that invoked Old Abe as the "most famous eagle of all time" and a superbly important "living insignia" in time of war.[63]

Symbolism in time of war was the theme that carried the day. Congressional hearings and floor debates in the spring of 1940 made it clear that Americans had decided they could not shoot the bird of their nation and have their symbol too. Hitler and fascism had swarmed Europe and turned into a worldwide threat that worried the country. American patriotism was surging. The US had yet to put boots on the ground in Europe, but it had been providing vital military aid to Great Britain, and once deployed, US troops would again be joined by America's bird. In an editorial titled "Misunderstood Eagle," a Miami newspaper expressed what many people were saying: if Americans allowed their flagship bird to go extinct, "we should be in the embarrassing position of having as national emblem a bird [as] dead as the dodo—not a cheering symbol for any country."[64]

The bird of prey nonetheless had plenty of enemies on the home front. Recognizing that, Congress framed the proposed law around preserving the integrity of a national symbol. Adopting that rationale, the Department of Agriculture wrote in a letter to Congress, "From an esthetic point of view there can be no question as to the desirability of protecting the eagle. Its status as the emblem of the sovereignty of the United States settles that; the bird should be a ward of the National Government." The preamble of the bill read similarly: "Whereas the bald eagle thus became the symbolic representation of a new nation under a new government in a new world . . . the bald eagle is no longer a mere bird of biological interest but a symbol of the American ideals of freedom."[65]

Before the House voted, Alaska's delegate, Anthony Dimond, insisted on the territory's exemption from the law. Oddly enough, Dimond was a member of the Fraternal Order of Eagles. Congress supported the exemption, which forever irked Edge, Van Name, and others. The House then passed the bill, as did the Senate. Killing a bald eagle in the Lower Forty-Eight, harming its nest and eggs, possessing a body part—all of which tens of thousands had done throughout the nation's history—finally got the attention of federal law. Violating that law could bring a penalty of up to six months in prison and a $500 fine.

On the evening of June 15, 1940, President Roosevelt affixed his name to the bill while wearing his signature broad smile. There was never a question that he would. His conservation streak, particularly on preserving woodlands and birds, was as bold as that of his fifth cousin once removed Theodore, and he liked the bird at the center of the new law. Two years earlier, he had personally designed a 6-cent bald eagle airmail stamp, and his aesthetic input was similarly evident in the new presidential Steinway piano. Each of its gilded mahogany legs was a carved rendition of the founding bird.[66]

The Bald Eagle Protection Act was exceptional in American law and American conservation, which was still more focused on resource preservation and wildlife protection. No single animal had ever had its own federal law. Previous legislation had applied to multiple species. Most remarkable, a predator animal had never been the safeguarded ward of the federal government. Despite emphasizing the patriotic value of its ward, the 1940 act delegitimized the parallel universes that had bedeviled bald eagles, and to some extent it erased the ambiguity around the species' place in American society. In their historical relationship with most traditional Native cultures, bald eagles were never anything but revered birds—never predators, never commodities, never objects of sport. With white Americans, eagles had gone from being revered birds in the early days of the republic to reviled birds in the nineteenth century and, in 1940, back to revered birds.

Their salvation wasn't owing to advocates alone. To those who bothered to open their eyes to the real world of birds, bald eagles exhibited their true selves, living their normal lives outside the customary, sometimes demeaning and certainly extreme, references devised within American society. The species was not the anthropomorphized coward, truant, wastrel, and inglorious thief. It would from now on remain "noble" and "majestic" in the American mind, and while those, too, were made-up labels, they were labels bald eagles could live with.

What they could not live with was harmful chemicals that would alter a hopeful future for them within a few years of their salvation by federal mandate.

SEVEN

Birds in a Band

Aloft in the eyrie, safe at home, were two young bald eagles. Five or more pounds each, physically the size of a rugby ball, they were nearly ready to experience the brilliance of flight. Their eyes were dark, their feathers too. The spiky ones on their heads drew back as if a stiff wind were perpetually blowing in their faces. Their legs, feet, and beaks were leather colored, their talons black and long. They were alone together in the nest and safe. At five to six weeks, eaglets are typically left unattended while their parents are off fishing or hunting, giving the young room to romp and exercise. By then, they are too big to be preyed upon by an owl or a hawk, and they have their talons, four on each foot—eight lethal pirate hooks. Eaglets growing into juveniles are to be avoided, not accosted.

Yet there are always violations of what's expected, such as when an odd, unfamiliar creature—no feathers, no fur, no scales—appeared at the edge of the two eaglets' security. The figure rose slowly and deliberately beside the eyrie, not unlike a cautious prairie dog peeping out of its burrow—except this strange creature showed no fear. Its intense blue eyes fixed on the young birds. They backed away to the far side of the nest, beaks open, narrow tongues protruding, hissing. The smaller of the two balanced clumsily at the edge. The odd figure, about to become an unquestionable threat, stretched forward and reached for the teetering eaglet. It jumped. The falling youth instinctively opened its wings and, slack on the air, glided safely into a clump of saw palmettos.

The blue eyes then turned on the other eaglet and reached out again. The eaglet sensed a grip around one of its wings and, fighting back, drew blood but could not free itself. It felt something around one of its feet, small and metallic, clamping but not restraining. The intruder let go, and the frightened bird scrambled back and away, watching as the blue eyes dropped back below the nest.

On the ground, the intruder crashed through stiff, green fronds of saw palmettos, where the second eaglet was hiding. Even if it eluded its antagonist, the young bird could starve or succumb to other dangers, since parents sometimes leave stranded eaglets on the ground. The most immediate danger pushed closer. The grounded bird panted. Fronds opened, and sunlight and the blue eyes fell on the eaglet. It scrambled off, flapping its wings, using them as weapons. The pursuer persisted. The bird was seized, its wings and legs arrested. As with its sibling, something metallic went around one of its feet, but this eaglet was not set free.

A soft cover fell over its head of spiky feathers, and all went dark. For a long period the eaglet was in a blinded, free-floating space, tugged about at times, swaying at others. Finally, light appeared as the cover was lifted off. The eaglet was back in the nest. Its dark eyes found its sibling, and it clambered over and away from the intruder, watching the strange being slip down and out of sight.

Although the two eaglets would not see the blue eyes again, every new brood in the same eyrie south of Sarasota, Florida, would have a similar brief but terrifying encounter with the same visitor. So, too, would eaglets along nearly two hundred miles of Florida's southern Gulf coast, and in various places along the Eastern Seaboard of the US and Canada. Over twenty seasons, 1,240 eaglets would meet those blue eyes.

Those eyes belonged to Charles L. Broley, and they were less intense than bright. Broley, a banker from Manitoba approaching his sixtieth year retired to Florida to do what none of the hundreds of thousands who retired to Florida came to do: climb trees. Up north, Broley had been an active birder and promoter of avian conservation in Canada. He had long participated in Audubon's Christmas Bird Count, he attended Audubon and AOU meetings, and he had met Richard Pough in 1938. A future founder of the Nature Conservancy, Pough worked for Audubon. Upon learning that Broley was retiring and planning to take up residence in Florida, Pough asked him to band eaglets in the southernmost state, the bald's chief breeding ground in the Lower Forty-Eight. Broley agreed to the request, and Pough gave him four bands with the offer to send more if needed. They would be.

Before the exchange between Pough and Broley, scientists had banded fewer than seventy eagles nationwide. Banding one wasn't like banding a sparrow or finch, and getting access to one wasn't anything close to the same as trapping smaller birds with nets or cages. You had to ascend to the nest. Since he was turning sixty, Broley found an agile teenage boy in Florida to climb the nesting trees for him. His recruit had the idea to take an eaglet as a pet, though, and when the first one resisted that idea with its talons, the boy took a stick to it. Horrified, Broley accepted that he'd "have to do my own climbing." That wasn't a ludicrous resolution.[1]

Broley was trim and strong and youthful for his age. He was also bald—a bald banker from Canada who was interested in contributing to the study of bald eagles. To get to their lofty eyries, most of which rode the top of straight-standing longleaf and loblolly pines, he first considered climbing irons that linemen used to scale power poles, but he ruled them out as too clumsy when gamboling along tree limbs. Ultimately, Broley equipped himself with two rope ladders, fishing line, clothesline, hoisting line, climbing pliers, broomstick, catapult, lead sinkers, and bent spoon. The first limb of a nest-worthy longleaf could be as high as seventy feet. Using muscle memory from his days of playing lacrosse, he would sling the weighted fishing line over the limb with the catapult. Sometimes getting the line up and over required ten, twenty, thirty, or more tries. Once he succeeded, he would pull the cord down, followed by the clothesline attached to its end, and then a hoisting line and ladder attached to the clothesline.

Broley's first climbs were brutal on his arm, back, and leg muscles, to say nothing of his sense of security. The rope ladder swayed and twisted unnervingly. Would the branch break? His knots slip? The line unravel? His grip give out? After reaching the crown, he had to rig up a second ladder to get into the nest, where usually a brace of young eagles waited. Some, he said, were "as gentle as kittens and others as wild as panthers." There he reached out with the bent spoon attached to the end of the broomstick to seize one of those kittens or panthers. He never wore gloves, and was never without talon wounds on his arms and hands, or pinesap on his palms.[2]

To keep himself in climbing shape, Broley did push-ups and pull-ups in the yellow morning light in a city park. Eventually, he was climbing as lithely as an aerial acrobat ascending the side of the ladder to his trapeze. Upon reaching a nest and crawling in, he said, he felt "absolutely safe." Not once did he fall, always held up by luck, wit, and some protective angel (Polly Redford said an "ornithological angel"). Scaling trees became second nature. Home was a greater danger. Once, while standing on a chair and putting away bands on a high shelf, he slipped, landed on a table, and knocked himself out.[3]

After Broley moved to Florida with his wife and daughter, they spent the school year and the eagles' nesting season in Tampa, returning to Canada for the summers. An eagle metropolis, Tampa Bay was the second-largest open-water estuary on the US Gulf coast and the largest in Florida—a twenty-two-hundred-square-mile fishbowl for more than two hundred species. The water was clean and startlingly clear. Seagrass and oyster beds were miles long. The bay was all a splendid pantry for nesting bald eagles.

Broley wasted no time in asking locals where to find them. Knowing of the slaughter elsewhere, the locals feigned ignorance of the big birds and their big nests until they got to know Broley. Even cattle ranchers valued the presence of this rare seasonal resident. They anticipated the yearly autumnal return of eagles, as did northerners the fall colors. They welcomed the chittering breaking the silent air, smiled at the toilsome nest-building and rebuilding, admired the constant egg-sitting, listened for the treasured sound of unseen hatchlings and eaglets, marveled at the grab-and-go fishing, and worried about eyries when tropical storms and hurricanes blew through. Looking for intact nests when calm returned was at the top of poststorm checklists. A thistly eyrie in a tree was a source of contentment and pride.

Broley the Canadian thought that the Founding Fathers of the US had made a splendid choice in putting the bald eagle on the nation's seal. They weren't such the bandits as so many claimed. In all his years of climbing into a nest, Broley, like Francis Herrick, found very little farm poultry. What he did see was a lot of fishing line, apparently from caught fish that had eluded the human fisher but not the avian one. Going out scavenging, eagles would haul back epiphytes, conch shells, gunnysacks, corncobs, tennis shoes, fish plugs, skirts, and lace panties. One had brought in a framed photograph of a smiling family.

Although balds had long before impressed people with the energy they channeled into "household cares," people like Broley and Herrick understood that science had much to learn. The more it learned, the more others might come to appreciate the raptors for their "family values." During breeding season, nesting eagles were wholly devoted to their responsibilities to home life. After the eggs were laid, the adults that Broley observed in Florida brought in Spanish moss every day to repad the nesting area of the eyrie. The two shared the responsibility of incubating the eggs, although the female sat more. From the ground, Broley could tell whether she was sitting on eggs or hatchlings by how far her head rose above the nest. Lower meant eggs. Higher and with wings slightly out meant she was giving room for chicks, protecting them from rain or the heat of the sun. When she wanted a spell from sitting, she would whistle for the male. If he were slow to respond, she would kick her head all the way back and let out a sharp, extended *kee-kee-kee*, not unlike the cry of a gull. Once, when a female was working on nest repair and the male was on a stub of a branch sunning himself, Broley watched the female raise a verbal ruckus until the male departed and returned with moss.[4]

After chicks hatched, food came to the nest several times a day, with deliveries concentrated in the morning. A parent would rip a small piece of flesh aggressively with its beak and then feed it tenderly to its young. At five to six weeks old, the eaglets began their gymnastic workouts, and as they developed their muscles and coordination, they would venture out onto branches. At about twelve weeks the eaglets were old enough to fly, but the nest remained their home for another month or so, and the parents continued to stock it with food.[5]

The nests also divulged something quite unexpected and troubling to Broley: By the late 1940s, the number of annual hatchlings had begun plummeting, and it wasn't mites, disease, storms, or anything natural that precipitated their cliff fall. It wasn't derelict hunters either. Broley learned that a new form of human activity was undercutting the ability—some would say right—of the American birds to exist.

POISON RAIN, BLESSED RELIEF

A S HE OBSERVED THE DOMESTIC LIFE OF EAGLES, BRO-ley witnessed their courtship ritual, among the most alluring displays in nature. In the wild, a couple's coming together to foster a next generation is an event we associate with the renewal of life born from the mysterious force of instinct. But the bald eagle's courtship ritual, its grace and elaborateness, the emotion evident in the performance, reveals something else deep within two eagles, which intend to be a pair for life.

The ritual starts from a perch or in the air with one eagle calling loudly to a potential mate. If interest exists, the two will meet in flight. A fleet chase typically ensues, with pursuer and pursued exchanging positions, each barrel-rolling and scissoring their flight paths, or with one flying upside down beneath the other. Sometimes the two will clench talons together and roll horizontally through the rushing air. One might break away and turn into a steep ascent, slow its wing beat, and let gravity pull it into a stall, pitching over and down. With wings folded back, the eagle drops into a hurrying dive. Then, at the last minute before disaster, it soars upward into another ascent, following the first maneuver with another rushing dive, and another ascent, over and over again in undulating flight.

Even more heart-stirring and heart-stopping is yet another performance: the cartwheel display, a veritable aerial waltz. Two taloned feet, sometimes all four, join and become an axis between whirling, somersaulting, pinwheeling eagles—a suspended avian cyclone. Laws of motion pull the courting pair outward. Desire holds them together. They defy separation and, falling together, death.

The release happens moments before the bonded two turn up and away from the water or ground. That death-defying ritual long ago stirred a poem from Walt Whitman, one of his later ones, titled "The Dalliance of the Eagles." Ten lines long and included in *Leaves of Grass*, the poem became one of Whitman's most beloved:

> Skirting the river road, (my forenoon walk, my rest),
> Skyward in air a sudden muffled sound, the dalliance of the
>   eagles,
> The rushing amorous contact high in space together
> The clinching interlocking claws, a living, fierce, gyrating
>   wheel
> Four beating wings, two beaks, a swirling mass tight
>   grappling,
> In tumbling turning clustering loops, straight downward
>   falling,
> Till o'er the river pois'd, the twain yet one, a moment's lull,
> A motionless still balance in the air, the parting, talons
>   loosing,
> Upward again on slow-firm pinions slanting, their separate
>   diverse flight,
> She hers, he his, pursuing.[6]

AFTER A SHORT COURTSHIP of their own, each pursuing the other, Charles Broley and Myrtle McCarthy met at the altar. It was Myrtle's first marriage, Charles's second (his first wife had died of tuberculosis in 1921). The union lasted a good stretch, thirty-five years. Although Myrtle went out in the field with Charles on several occasions, she scaled a nesting tree only once. Her unease scared both of them. It was then, she kidded, that Charles lost the last of his hair. Despite her trepidations, Myrtle was a committed naturalist and birder—had been one for as long as Charles. She was the author of numerous articles in gardening and wildlife publications. She wrote a book on tree swallows and followed that with *Eagle Man*, a biography she published in 1952 relating Charles's adventures in banding.

When Broley started scaling trees, his objective was to pull back the veil between science and the routines of bald eagles by tagging their young. Broley made himself into a one-man research program of a low-tech, devise-and-revise-as-you-go citizen ornithology. The climbing and banding techniques that he perfected became the standard that continued with new generations of scientists. Broley never claimed to be a member of their professional cadre, yet he published in scientific journals, and his work was featured in newspapers and magazines—*Collier's*, *Life*, and *Reader's Digest* among them—and he was in demand as a speaker. He gave forty to fifty public lectures each year. His banded birds revealed how far eagles migrated, where they spent their time between breeding seasons, at what age they matured into their adult feathers, and to what age they might live.

Like so much that is supposedly modern, banding birds dates to the ancients. During the Punic wars between Rome and Carthage, Roman officers tied strings to the feet of birds to send messages to men on the battlefield. In medieval Europe, falconers banded their feathered charges to identify their owners.

In the US, an early recorded banding event occurred in the New York village of Cortland in 1831. As part of the village's Fourth of July celebration, William Bassett, a local engraver and silversmith, released an adult bald from the cupola of the Eagle Hotel. Bassett had fashioned a silver clasp around one of the eagle's legs with an inscription dedicated to Henry Clay, the national Republican presidential candidate. A crowd gathered for the event, the military fired a salute, and the people let out three cheers. Spectators claimed that the eagle took off, heading southwest toward Clay's home in Louisville, Kentucky. The liberated bird of liberty "sailed gracefully," said one spectator, "till out of sight." At some point over the next days, the eagle detoured westward away from Kentucky. It had been traveling a week when someone shot and killed it on the Mississippi River just north of Dubuque, Iowa, at a place later named Eagle Point and where a tavern keeper would buy Old Abe from an Indian.[7]

Bassett may or may not have been the first to band a bird in America. A range of sources maintain that John James Audubon was the first, in 1804, when he was eighteen years old and living at Mill Grove north of Philadelphia. The claim originated with Audubon in a recounting thirty years after his experiment with banding was to have occurred. As he told the story, he tied silver threads around the legs of five eastern phoebe nestlings. When the fledged birds migrated south, they flew off with the rings of thread attached, and when eastern phoebes returned to the area during the next nesting season, Audubon discovered that two were wearing the threads. But a twenty-first-century sleuth-like researcher named Matthew Halley sorted through calendars, memoirs, nesting seasons, and wildlife diaries, and identified numerous inconsistencies in this oft-repeated story, giving reason to call it apocryphal.[8]

It could be, then, that Ernest Thompson Seton, a principal founder of the Boy Scouts of America and resident of Broley's home province of Manitoba, was the progenitor of scientific bird-banding in North America, or of something close to banding. According to *The Canadian Bird Bander's Training Manual*, Seton marked several snow buntings with black printer's ink in 1882 (Broley was three years old) in an attempt to study their migration habits. A more consequential beginning came twenty years later, when Paul Bartsch, a conchologist, launched a systematic bird-banding program at the Smithsonian Institution, as a side project to his mollusk research, using black-crowned night herons. Soon after, tracking migration became part of bird study in the US and Canada. The American Bird Banding Association was founded in 1909, and federal bird-banding offices were established just before the Migratory Bird Treaty Act of 1918.[9]

Broley's banded birds changed the language around bald eagles and migration. The data they provided meant that expert witnesses could no longer testify credibly that eagles typically remained within or nearby their eagledoms. Some did—adults mainly—but others were, in fact, fliers of great distances. Forty-four eaglets submitted to bands during Broley's first season in Florida. The wanderings of one took it up to Columbiaville, New York, on the Hudson River, when just a few weeks out of the nest. In subsequent years, others showed up in the upper Midwest and all along the East Coast, in twenty-four states and provinces outside of Florida. One winged into Broley's former home, Winnipeg. Another got itself to Prince Edward Island, between Nova Scotia and New Brunswick, in the fishy Gulf of Saint Lawrence, sixteen hundred miles from its natal territory.

Broley's science began in the nest and continued through to the end of a banded eagle's life. His eaglets grown to adults indicated not just where they were flying but where they were dying and, in some cases, how they were dying. In her book about Charles, Myrtle Broley used the word "recovered" when writing about banded eagles found in far-off places, which was to say that most had been recovered because they had been shot or poisoned. The poison that killed them often came from a predator wildlife trap. But it wasn't the only poison that was taking their lives.

Broley began his research to track the movement of a species and, unexpectedly, ended up recording its decline. In 1946 he banded 150 eaglets; the next year, 113. The drop in number wasn't worrisome initially, but then in 1948, only 85 young ones met his blue eyes. The next year, 60; the next, 24; the next, the same number; the next, 15. That was 1952, the year Myrtle published *Eagle Man*. Something was clearly wrong. While fewer nests were occupied, Broley also found fewer eaglets per nest than in earlier years. In 1946, eighty-three percent of the nesting pairs hatched eggs. When Broley did the numbers for 1955, he was stunned. A mere sixteen percent produced young that year. The ratios had reversed themselves. Two years later, forty-three couples fledged eight young.[10]

After Broley noticed population numbers trailing off along Florida's Gulf coast, he began corresponding with others who monitored eyries on the Eastern Seaboard and in the Midwest. He learned that eagles elsewhere had fallen into the same odd reproduction chasm as those in Florida. A letter from an eagle-watcher who surveyed eyries on the upper Mississippi River disclosed that fifty-five nests delivered a single chick in 1957. In a three-year span, only one eaglet fledged in a monitored four-county area in New Jersey. Another correspondent reported that nesting activity over the previous dozen years on Mount Johnson Island on Pennsylvania's Susquehanna River had been a "dismal failure." On Florida's east coast, the nesting population in two decade's time had declined by as much as sixty percent. There had been as many as five hundred active eyries in the Lower Forty-Eight's chief eagle state when Broley climbed his first tree in 1939. Nineteen years later, he and others estimated that Florida yielded only eighty hatched young ones.[11]

Skirting the river road, (my forenoon walk, my rest),
Skyward in air a sudden muffled sound, the dalliance of the
    eagles,

SKYWARD OVER HAWK MOUNTAIN on the Atlantic Flyway in eastern Pennsylvania, there were signs of potential trouble. Hawk Mountain had long been a favorite place to watch migrating raptors, even when Theodore Palmer and T. Gilbert Pearson were telling congressmen that balds were nonmigratory. In utter defiance of that impaired assertion, hundreds of eagles flew along the Appalachian chain's Hawk Mountain every fall, coming from New York and New England and disseminating themselves around the mid-Atlantic. They were interstate travelers on a highland thoroughfare, which the Native Lenape called the "Endless Mountain." September was the best month to watch the voyagers, usually later in the day. Their numbers passing overhead were a sight to be seen, as, too, was their soaring flight. Looking up, one could understand why these formidable fliers evoked the reverence of Native peoples, why they were a hallowed species. In Native cultures, the mountain range was a sacred reach into spiritual heights and as wondrously alive as the migrating birds. In the natural realm, the mountain was no inert topographic feature either. It was a maker of air currents and a catalyst of flight. Wind combined with geography gave birds wherewithal.

Soaring species are experts at locating moving air that will carry them on their respective journeys. Some currents rise from the earth's warming during the day, infusing the cool air above with uplifting thermals. Some currents come from two masses of air colliding from opposite directions. Some originate with horizontal wind careening into a hillside or mountainside and turning upward and into soarable lift, as at the Endless Mountain. On prodigious wings, birds ascend on the vertical thrust and then drop into a shallow dive to gain momentum and to advance in a desired direction. Locating another draft, they angle their wings ever so slightly to deflect the air beneath them downward and, as an opposite reaction to this action, bear themselves upward again to start the cycle over.

Essential to this economy of motion are the primary, or outer, wing feathers, collectively known as "pinions." On the bald, primaries measure up to twenty-two inches in length, just shy of two rulers laid end to end. At the wing tip, the primaries narrow to a rounded point that can twist, and when the feathers are opened apart like spread fingers, they reduce drag and produce lift. If the birds catch a tailwind blowing horizontally to the Earth's surface and in the coveted direction, they can simply ride it like a sailing vessel with full sheets on a downwind run. The moving air is propellant, open wings conveyance. The longer a wingspan, the easier and faster the soaring. Extralong wings, however, don't lend themselves well to maneuverability and takeoff. If you've ever seen an albatross with its twelve feet of slow-beating wings launching itself, you know the feat is an unwieldy one that requires a long runway. Even worse, the albatross's landing is something of an avian embarrassment. Eagles—as also hawks, condors, pelicans, vultures, and ospreys—have what is perhaps the ideal wingspan. It enables quick and easy takeoffs and precise touchdowns, high lift, efficient soaring, and tight maneuvering for pursuing prey.

And so they glided over and along Hawk Mountain—eagles, hawks, harriers, falcons, merlins, ospreys, vultures. They flew high, too high sometimes to be identified without a scope or binoculars unless you had a trained eye or could make out an adult bald's exceptional white tail. Eagles and vultures, for example, are similar in size, shading, and wingspan and can confuse the naked and untrained eye. For the Pulitzer Prize–winning novelist Marjorie Kinnan Rawlings, who regularly saw both in the skies doming rural north-central Florida, where she lived, eagles exhibited a discernible poise in flight that separated them from vultures, which she called "buzzards." "Something about the eagle's soaring is more purposeful than that of the buzzard," she wrote in 1942. "The great wings lie in a straighter line on the air, without so much of uplifted curve. He is no more graceful than the buzzard, but the hallmark of the fighting aristocrat is on the flight of one and that of the lazy scavenger on the other."[12]

Along with the objects of their fascination, birders converged on Hawk Mountain in force, testing their skills at identifying high-flying raptors and taking in a thrilling view. The broad-sky panorama at fifteen hundred feet from ridge across valley, heavily forested and generously littered with glacial sandstone rocks, and smelling of minerals and soil, could stir an artist of the Schuylkill or Hudson River School into breaking out brushes and canvas.

Some people, both men and women, broke out guns instead. Hawk Mountain was a popular den for recreational assassins with an if-it-flies-it-dies mindset, blasting away at the caravanners on ancient pilgrimages. There was no purpose to the violence outside of killing for the sake of killing and shooting for the sake of shooting. This was not predator control. The gunners were not hunters of food. The fallen were nothing more than a score to them, a pose for the camera, the sky something to empty.

Determined to put an end to this pointlessness, Rosalie Edge raised money to lease fourteen hundred acres on the mountain and hired a husband-and-wife wildlife warden team to hinder the guns. Four years later, after a heated struggle with National Audubon, which believed it should be the proprietor of Hawk Mountain, Edge bought the fourteen hundred acres with the financial support of Willard Van Name and many others. She then donated the land to the Hawk Mountain

Sanctuary Association, creating the world's first-known protectorate for birds of prey.

Despite thwarted guns, the Appalachian range soon after World War II became a migratory flyway less traveled. While Broley was receiving despairing reports on nest failures around the country, Hawk Mountain birders were seeing the fallout of those failures. Only ten percent of the bald eagle travelers in the fall of 1957 and 1958 were immature birds, representing a drop of twenty-eight percent from the years recorded before and during the war.[13]

Down in Florida, Broley banded only two eaglets in the 1959 season, and not because he was an old man with arthritis and aching bones who could no longer hoist himself up a rope ladder. There were only sixteen occupied eyries that year.

Without its young, the bald eagle was facing a fragile future. Why the species was failing to procreate was a mystery. At first, Broley thought Florida's postwar open season for development was behind the generational loss. "There are just too many people moving to Florida," he observed; the Gulf "coast is soon going to be one long village." Not only were unleashed bulldozers clearing land and trees to advance the madding rush of human transplants from the cold, unsunny North; land-hungry, land-making developers, accommodating the demand for waterfront homes, were dredging up marine habitat, the fishing raptor's kitchen. Out in the water, they were building fingers of land, ironically resembling the eagle's primary feathers—making sodded real estate from nurturing estuaries.[14]

Still, it seemed to Broley that something more was going on here than behavioral responses to human mania. Something within the reproductive system of the eagle had apparently changed, and he thought he might know why. In articles he published in 1957 and 1958, the latter in *Audubon*, Broley pointed to an epidemic in pesticide use after the war. At the front of the decade, a massive fish kill had befallen Tampa Bay. Research had shown that the aquatic victims contained high levels of the chemical pesticide DDT. "By no means promising," was how Broley described the outlook for future generations of bald eagles in the eastern US.[15]

The rushing amorous contact high in space together

IN OCTOBER 1945, RACHEL Carson spent two days at Hawk
Mountain Sanctuary to watch the fall migration. In her field note-
book, she observed of the birds in flight, "They came by like brown
leaves drifting on the wind. Sometimes a lone bird rode the air currents; sometimes several at a time, sweeping upward until they were
only specks against the clouds." Years later, when she was working on
the book that would acquire the title *Silent Spring*, and that would help
deliver DDT's undoing, she wrote the curator at Hawk Mountain,
requesting a copy of its annual migration records. From the sanctuary's inception, one of Edge's objectives was to have the staff log every
species of raptor that passed overhead. By providing a safe flyway, the
sanctuary expected to see an uptick in the birds' numbers. Initially,
they increased, but then they started to fall again. When Edge first
observed this trend, she realized it reflected something beyond hunters
and raptors.[16]

In 1948, twelve years before Carson corresponded with the Hawk
Mountain curator, Edge wrote to the New York Division of Fish and
Game. She had acquired information, she noted, that the golf courses
and country clubs in Westchester County were spraying their grounds
with a pesticide to control mosquitoes and other insects. The recommended mix for the pesticide, she pointed out, called for ten percent
DDT, but the golf courses were using fifty percent. The results were
devastating. Dead orioles and robins were strewn across the precisely
cropped and bug-free fairways and greens and in the yards of nearby
houses. A field agent from the Department of the Interior confirmed
Edge's report of dead birds yet took no additional action.

DDT had become a darling among assorted midcentury
technologies—synthetic rubber, pressurized aircraft cabins, micro-
wave radar, atomic energy, penicillin, plastics, jet propulsion, and
orange-juice concentrate—spawned by military needs of World War
II and destined to redefine modern civilian life, triumphant technolo-

gies that could fix every adversity to humankind. A chlorinated hydro-carbon chemical, DDT had been around since 1874, when a Viennese pharmacology student named Othmar Zeidler first synthesized it. But it went largely unnoticed for sixty years, for lack of a discovered prac-tical application. In the meantime, scientists figured out that vector insects could transmit diseases to humans. Among the most dogged vector-borne health disorders were malaria, typhus, plague, and yellow fever. Insects were also notorious for destroying food and cash crops, as when the boll weevil jumped the Mississippi River from Mexico and the American West in 1909 and began consuming the South's cotton economy. In the 1930s, as Edge was setting up Hawk Moun-tain Sanctuary, scientists across the Western world were furiously trying to develop the ideal bug killer. Insecticides that were inexpen-sive tended to be ineffective. Effective ones—copper, lead, and arse-nic compounds—tended to be expensive and, as one magazine writer noted, as lethal to humans "as an arsenal."[17]

A key scientist on the job was Paul Hermann Müller. Müller was a chemist working for the J. R. Geigy corporation on the river Rhine in Basel, Switzerland, a city of political refugees, pharmacists, and chemists. One of the latter was Albert Hoffman, who was the first to synthesize—and drop—LSD. When Müller was searching for some-thing less hallucinogenic, insects were responsible for a food-crop shortage in Switzerland and a typhus outbreak in Russia. Then, around the time Broley hoisted himself up to his first eagle nest, Müller pulled Othmar Zeidler's DDT down from the shelf. That fateful move would win Müller the 1948 Nobel Prize in Physiology or Medicine. DDT, he discovered, was fast-acting and effective in killing bugs, disease vectors among them. If used sparingly, he believed, it would not harm food crops, livestock, or humans, and it could be offered to consumers at an affordable price.

Geigy immediately patented the chemical and began marketing it in 1940, the year of the Bald Eagle Protection Act. The compound still went by the tongue-twisting name "dichlorodiphenyltrichloroethane." Having nothing of this absurdity, the British conjured up the sprightly acronym DDT and, along with their allies, used the newly pronounceable chemical in the common battle against wartime disease. Typhus had taken an estimated three million lives during the First World War, a horrific death toll that didn't repeat itself in the Second, after Müller's scientific breakthrough. Chemists and military medical personnel lauded the humanitarian benefits of chemical pesticides by arguing that World War II was the first war in which fewer soldiers were killed by disease than by the enemy. The army's chief of preventive medicine called DDT the "War's greatest contribution to the future health of the world."[18]

Industry marketers and scientists reprised that tune in peacetime, crooning that what chemicals did on the battlefield they could do on crop fields—and in homes and schools, at parks and recreation centers, out on sidewalked streets, and in the woods behind your house. Residents of the suburbs, the expanding epicenter of the rapidly expanding American middle class, loved the idea of an environment free of biting, crawling, bloodsucking things. What people were calling the Atomic Age was no less the chemical age—and an over-the-counter age.

Acquiring the wondrous technology was as easy as driving down to the local hardware store in your Chrysler, Chevrolet, or Studebaker sporting rocket-like tail fins, riding on synthetic-rubber tires, the polyvinyl-upholstered back seat jammed with boom-time babies, tummies round with pulpless orange juice from canned concentrate. If you were in a '56 Chevy Bel Air, the chrome-plated hood ornament had the wings and tail of a jet plane and the head of an eagle, spearheading the way to your pick of chemicals for any of an array of purposes—cleaning, sanitizing, scenting, fertilizing. There was bug-killing too, of course. You could buy insecticides in liquid, dust, granular, and, eventually, convenient emulsifiable-spray forms. Most exciting, you could acquire a bomb—not an atomic bomb but an insect bomb—in a compressed aerosol can the size of a water glass.[19]

Although America had emerged from the war as a world superpower, insects, of all evils, were still holding the upper hand. They had always been stowaways aboard the great diaspora of humanity, taking up residence wherever people did. At home and on farms, flies, fleas, spiders, lice, earwigs, wasps, mosquitoes, and cockroaches overwhelmed American life, and people wanted relief. Americans had beaten the Nazis and beaten the Japanese, and they were determined to score a trifecta and beat the bugs. There were countless new pesticidal weapons in the arsenal, many with ridiculously incomprehensible and forgettable names. DDT seemed the most promising. It was cheap, plentiful, and easily dispensed, and it had that simple three-letter name, practically an advertising jingle.

In August 1945, when DDT went on the consumer market, Channing Cope, the farm editor at the *Atlanta Constitution*, tested it on his own place and wrote a prerelease review. He applied it sparingly to his farm animals, as well as the family's itchy cat. It extricated them from the misery of pests: fleas and ticks on the cat and dogs, flies and hornets on the mules, and lice on the pigs. He mixed two ounces of DDT with a quart of kerosene and wiped it with a rag around the edges of doors and windows of his house—"matching wit against wit," humans to bugs—and the house was soon unburdened by the latter. This "amazing" product, wrote the jowly and gentlemanly editor, had the "power to make life much safer for all of us, and more pleasant."[20]

Cope wasn't offering an unconditional endorsement, though. His review was more admonition than approbation. At the outset, he emphasized that DDT was a dangerous product. If used carelessly, he wrote, it had the "power to ruin us." Just as DDT could snuff out undesirable pests, it could do the same to desirable pollinators that were integral to the propagation of wild and commercial plants. "We must remember," he told readers, "that DDT will kill the bees and that means that it will kill the clover (which means, too, that it will kill off our livestock)." Cope was echoing what the War Production Board, the chief beneficiary and most experienced user of the chemical pesticide, had already said: Müller's lifesaving elixir had the potential to upset the "balance of nature." That warning appeared on the front page of the *Washington Post* when the army announced that it planned to sell off its surplus DDT on the open market.[21]

Readers trusted Cope's opinion, but he probably startled a few when he warned that DDT "has greater destroying powers than anything yet known." He wasn't excluding the atomic bomb from his comparison either. DDT lacked "certain safeguards" like those that existed around the nation's nuclear arsenal. Not least of those safeguards was the firsthand knowledge, following Hiroshima and Nagasaki, of the bomb's ominous capacity to turn the world upside down.[22]

The ghastly horrors of DDT would soon reveal themselves after the warnings of Cope and the War Protection Board went largely ignored. The chemical industry preached just the opposite of caution. In the minds of chemists, as well as engineers and scientists generally, nature was flawed as a dwelling place for civilized people, and technology born of superior brains could correct nature's flaws and make life safer. To the latter assertion, there was some truth. DDT ultimately removed malaria from many parts of the world, including the US, by stamping out its vector, the mosquito *Anopheles*. The bug killer also proved an economic worth by putting an end to the South's boll weevil blight.

Unfortunately, the "wonder insecticide," the "chemical that saves millions," was eradicating other life-forms. In modern society, the insistence on managing the environment in and around your home was close to a moral imperative. "Manage" was synonymous with deciding what would live and what would die. Weeds, insects, and rodents were for another world; flowers, grass, and singing birds for this one. While chemical manufacturers remained reticent about the collateral fallout from DDT use, they put the idea of power over life directly before consumers. When buying a pesticide at the local hardware store, you could choose from any number of brands incorporating DDT as the main active ingredient. There were Black Flag and Bolt, for example, and Fly-Go, Fly-Tox, and FlyDED. Their labels, dramatic and emphatic, made the universal promise of effectiveness, each emblazoned with the same accentuated word: "*KILLS*."[23]

By the 1950s, the demand for DDT exceeded all other pesticides worldwide. At the end of the decade, annual use peaked at close to 79 million pounds, nearly a half pound per year, observes the environmental writer Bruce Beans, "for every man, woman, and child living in the United States." It was all safe, insisted the industry, which went to great lengths to convince the public that danger to bugs equaled benefits to humans, that harm equaled relief, that paralysis equaled freedom. To prove the point, makers of an industry promotional film directed kids on bikes to chase the plume of a brutal-looking mosquito truck debugging a suburban neighborhood; they cued chemical handlers with a cannon fogger to blast schoolchildren at an outdoor table, eating bologna and PB&J sandwiches (one kid waits to take a bite at the moment the fog reaches his sandwich; another kid flinches when it does); and they instructed girls and boys in a swimming pool to jump with excitement as a cannon spewed a heavenly white cloud over them.[24]

Never mind that the US Public Health Service concluded and *Time* magazine reported in 1944 that DDT "is 'a definite health hazard'" to humans; that *Science Digest* wrote in the same year that "DDT presumably could send you on a death jag"; that the *Railroad Workers Journal* reported around the same time that "substantial amounts of [DDT] will attack nerve centers and the liver"; that the *American Biology Teacher*, citing the American Association of Economic Entomologists, asserted that "large-scale use might create problems which do not now exist."[25]

That's exactly what sisters Dottie Colson and Mamie Ella Plyler were reporting—that problems were materializing on their farms. Both women were in their forties, born five years apart and not far from where they now lived in southeastern Georgia, where a century earlier John Abbot had collected insects and birds, composed several paintings of bald eagles, and met an aspiring ornithologist named Alexander Wilson. After being steeped in repeated aerial pesticide fogs, the sisters noticed that they and their children were developing sore throats, headaches, and mouth sores that lasted for months. Their cows grew sick, their chickens dropped dead, their honeybees perished. In 1945, Colson and Plyler learned that the army was routinely bombarding nearby Camp Stewart with some kind of chemical. They suspected DDT. Whatever the chemical, it drifted over to their farms as if the army had plotted out their properties as part of the intended strike zone.

Rarely was there any concern for such chemical meanderings, which scientists call "exo-drift." Society regarded pesticides, insecticides, fungicides, and herbicides as socially and economically valuable. So, one man's drift was another's fortune. Besides, containing an aerial application was like trying to lasso the wind. Mass sprayings were so recurrent and widespread, and poured out and down so heavily, that escaping them was virtually impossible. Aside from mosquitoes, there were plenty of kill targets out in the wider environment that brought them on. Since the 1930s, the South had been contending with the interloping red fire ant, which rode to American shores as an unwelcome stowaway aboard a cargo ship from South America before disembarking and crawling across the American South, where it built giant mounds and stung—like fire—hapless animals and people, killing some of them.

The Northeast had accursed, bloodsucking black flies—a.k.a., "buffalo gnats," "turkey gnats," "sons of bitches"—that swarmed every spring, right when people coming off winter cabin fever were getting outside to plant a garden, hike a trail, or paddle a water, forced to suit up like a beekeeper. There was the invasive gypsy moth, which was the regrettable legacy of an amateur entomologist who in 1869 had set it free, leaving it to munch its bedeviling way from the Northeast to the Midwest, denuding in its path millions of acres of hardwood forest. There was the fungus blight, spread by bark beetles, that from coast to coast was relentlessly erasing the iconic American elm, the tree that shaded the avenues of America. And in Charles Broley's Florida, there was nature's cloud of biting midges, locally known as "no-see-ums," and mosquitoes (dubbed the state bird by some wry individual), which afflicted Florida's mainstay industries: tourism and real estate.

Who wouldn't want relief from such malevolence? With the facility of chemical treatments, ordinarily sluggish government agencies turned into rapid responders and sent trucks out and planes up to tuck the country beneath a tidy coverlet of DDT.

In opposition, Colson and Plyler regarded the generally well-intentioned tidiness as perilously smothering. Not long after the army started spraying Camp Stewart, civilian planes began dusting the big peanut farms in the area of Claxton, Georgia, and stillness fell upon the land. A neighbor of the sisters said chemicals weren't saving humanity but wiping out a way of life.

Colson and Plyler wrote scores of letters to the Georgia Department of Health describing the ill effects of the aerial sprayings. The sisters were polite but forthright in their opinion that peoples' health, their livestock, and commercial honeybees—just as Cope had warned—were being sacrificed at the altar of peanut-growing. In response, health department officials maintained that scientific evidence didn't corroborate a connection between DDT and what the sisters "alleged" to be happening on their farms. So Colson and Plyler sent every published report and study on DDT they could find, compelling the department to make inquiries with experts—except the consulted experts tended to be industry scientists. As with the asbestos, lead, tobacco, and petroleum industries at the time, in-house public relations executives and company scientists controlled the public narrative around the safety of their products. Government officials bought into the narratives and accepted self-evaluative claims of innocence, just as Americans had bought into the stories of kidnapping bald eagles.[26]

Georgia health department personnel expressed sympathy for the loss of Colson's bees but otherwise attributed her physical problems to "personal hypersensitiveness to some of the dust." They weren't altogether tactful either, blaming the sisters' vexing campaign against DDT on an emotional-health affliction that medical experts claimed was common in women: hysteria.[27]

Hysteria? Birds were falling out of trees.* Fish were floating belly up in lakes and on rivers across the country, particularly in agricultural regions and downstream from them. Dying small animals were lying on the ground trembling as violently as an Arctic explorer who had lost his coat. House cats that took advantage of dying animals lost that advantage by losing their nine lives. The industry did its unmitigated best to ignore the deaths.

Nothing seemed to irritate the makers of DDT more than claims of trace elements of their product in the food supply. In 1949, the Food and Drug Administration announced the discovery of DDT in cow's

---

* Today, pesticides kill an estimated sixty-seven million birds a year, according to the US Fish and Wildlife Service.

milk, which was consumed mostly by children. The news naturally made headlines. DDT was later found in fruits, vegetables, fish, and meats. To defend itself, the industry paraded out its most eloquent, most emphatic lab-coated experts to write papers and testify before boards. The FDA had established sufficient parts-per-million tolerance guidelines for DDT absorption in food, the experts argued, so there was no need for banning the pesticide's use on farms. It was a matter of weighing costs against benefits, and the little bit of DDT in the body fat of Americans (an average of three to five parts per million, according to one study) wasn't hurting anyone.[28]

In 1945, the debut year of both DDT's commercial use and a new generation, the baby boomers bumping along in fin-tailed sedans and station wagons without seat belts, the National Pest Management Association recommended that every DDT product bear the label "POISON." The recommendation didn't translate into policy. For the next two decades and beyond, the unlabeled poison infiltrated the American diet, including baby food. Seventy-three million American baby boomers entered the world during that period. As they were reaching their senior years in the twenty-first century, medical research scientists learned that the level of a DDT by-product was nearly four times greater in the blood of patients suffering from Alzheimer's disease than in healthy individuals. Similarly, a six-decade study concluded that women who had grown up with exposure to high levels of DDT were at significant risk for breast cancer. Another study, released in 2020, indicated that health threats had passed down from exposed women to their children and grandchildren.[29]

The poison was also getting into the wild food of wild birds. For the feathered lot, the DDT decades were years of great mortality.

The clinching interlocking claws, a living, fierce, gyrating wheel

THREE-QUARTERS OF A CENTURY after Walt Whitman paid tribute to the eagle dalliance, Broley published the *Audubon* article that disclosed his suspicions about DDT. Chemicals of the DDT sort were central to the Green Revolution of the 1950s and '60s. "Green" referred not to the green of the Earth but to the green of farm crops. Green involved not the retention of natural systems but their conversion into artificial ones, using, in part, irrigation systems, engineered crops, synthetic fertilizers, and chemical pesticides. So, DDT was set loose, and it didn't drift only through the air and fall only on the neighbor's place next door. It wandered thousands of miles down rivers and streams and across lakes from one part of the country to another. It traveled on the full course of the continent's second longest waterway, the Mississippi River, from the upper Midwest and America's breadbasket—the plowed and planted front line of the Green Revolution—down to the Gulf of Mexico and over to the part of Florida where Broley's eagle counts were falling off.

Coinciding with *Silent Spring*'s publication in 1962, the brown pelican disappeared from a favored habitat, the upper Gulf coast. One of the seashore's most commonly seen creatures—gliding millimeters above the water on seven-foot wingspans, skydiving at damn-the-torpedo speeds for a pouch full of fish, mooching amiably at fishing piers for a free one—had slipped out of sight, nowhere to be found. It remained on the state seal and state flag of Louisiana, but not on the state's shores, and not on those of Mississippi, Alabama, or eastern Texas. No one could figure out what had happened to it. Southern California was losing squadrons of pelicans too—white ones—and researchers were determined to learn why. What they discovered was that since 1947, the Montrose Chemical Corporation, the country's largest manufacturer of DDT, had been illegally discharging millions of pounds of the chemical into the Los Angeles County sewer system. From there, DDT streamed into the Pacific Ocean and through the food chain to the industrious fisher and genial moocher, dooming its reproduction.

The tribulations of one plucky fish eater turned out to be those of another. DDT—or more precisely, DDE, a metabolite of DDT—was responsible for the bald eagle's demise. The problem, though, wasn't sterility in one or both of the adults, which had been Broley's theory. The problem was ill-formed eggshells and embryos. In 1961, while

Carson was in the midst of writing *Silent Spring* and, coincidentally, fighting breast cancer, the US Fish and Wildlife Service launched a two-year study that involved capturing twenty-five eagles along the Chilkat River, once adulterated with Alaskan bounty hunters, and feeding them fish laced with varying doses of DDT. Those that consumed higher doses died within a few weeks, and those taking in lower doses, which resembled levels found in wild fish, survived.

The problem in the wild, though, wasn't so much death associated with high levels of DDT; it was nest failures—eggs not being laid and laid ones failing to hatch. Among the members of that surviving test group of eagles, the metabolite DDE formed in their bodies while DDT gradually diminished. Studies dating from the late 1940s to the mid-1960s showed that egg weight among raptors and large waterbirds experiencing significant population declines—murres, merlins, egrets, cormorants, prairie falcons, ashy petrels, red-tailed hawks, great blue herons, white pelicans, and brown pelicans—had decreased over time. In Florida, an intrepid graduate student convinced egg collectors to let him measure the bald eagle eggs in their possession. Some of them were contraband. Eggs taken in the decades before 1947 were virtually all the same in structural integrity: robust and originally hatchable. Those from a succession of years after 1947, when DDT began pouring down on the avian landscape, had thinner shells than eggs laid the years before. In a separate investigation conducted in the late 1960s, DDE and DDT residue turned up in *every* egg that researchers took from active nests in Wisconsin, Maine, and Florida.[30]

Here's what was happening. DDT infiltrating bodies of water, by means of runoff, discharge, or intentional spraying to control mosquito larvae, found its way into bacteria and plankton. Fish that ate the fish that ate the plankton carried the DDT up the food chain, to the unsuspecting raptor that ate the contaminated fish. From the DDT, DDE was metabolized in either the fish or the eagle, or both. Where there was DDE, there was aberrant nesting behavior. Eggs weren't laid, or those that were laid were thin-walled and misshapen. Calcium is an essential building block in eggs, and somehow, researchers learned, DDE was impeding the transfer of the building block from the mother's bloodstream to her shell gland, undermining the integrity of her eggs. And even when an embryo developed and the egg held together and the chick hatched, the baby bird's chances for a debilitating condition were great and for survival not so. In investigations of other raptors, researchers learned that the parents sometimes ate or discarded a weak chick. It is reasonable to assume that the same behavior probably went on in the eyries of eagles. The renowned family life of the distinguished bird was in peril—another among the kaleidoscopic risks of being a bald eagle.[31]

⌢

Four beating wings, two beaks, a swirling mass tight grappling,

THERE WAS ONE PLACE in the US during the height of the DDT era in the 1950s where the bald was able to live a normal life and raise healthy families: Alaska. As Americans were spreading DDT like cake frosting across the contiguous states, Alaskans went mostly without. There wasn't much of a market for it. Big farms didn't exist in the cold, big land. Mosquitoes existed, and they were big, but their season was short, Alaskans were tough, and malaria and yellow fever were diseases of other places. There was the pernicious eagle bounty, yes, but at the very moment in the 1940s when DDT was getting its start elsewhere, the eagle bounty was approaching its end.

It didn't go down easy. Bounties were an extension of the Alaskan way of life with hunting, fishing, and herding. Conditions that anticipated bald eagle control date to the 1700s, when Russians introduced fox-fur farming to the Aleutian Islands. The trade took off for the Americans in the 1880s, and in 1913 President William Howard Taft designated the Aleutian chain a federal reserve for reindeer herding, fisheries, fox-fur farming, and native birds, yet not his country's emblematic bird. Four years later, the territory implemented a bounty for both of the country's eagle species. The wolf was the only other recognized predator in Alaska with a price on its head, effected in 1915. Hair seals, an accused salmon filcher, followed wolves and eagles on the bounty list in 1927; coyotes in 1929; Dolly Varden char (once confused as trout and regarded as a threat to young salmon) in 1933; and wolverines in 1953.

At the end of 1940, six months after Congress passed the unprecedented Bald Eagle Protection Act with its Alaska exemption, a hunter whose name has escaped history turned in a set of talons to the territorial treasury, bringing the number of paid Alaskan bounties to eighty thousand. To its supporters, retaining the bounty was a matter of Alaskan independence and pride, and of conserving individual rights and rugged self-reliance. Some years before the number eighty thousand, the Canadian and Alaskan ornithologist George Willet reflected on the statutory killing: "Unfortunately, the question has stopped being one of conservation and has become economic, in that many Indians and some whites . . . have come to consider the eagle bounty as part of their income."[32]

One such person was Walter J. Larson, a fish counter for the US Bureau of Fisheries and a professional bounty hunter. Larson wrote an article in 1929 about his eagle exploits for *Harding's Magazine*. That was the year Rosalie Edge executed her challenge against Audubon for ignoring the Alaskan way with eagles. Larson offered a few ornithological observations that experts like Herrick and soon Broley were starting to record: Breeding eagles were loyal to each other. The male did most of the hunting when eggs and eaglets were in the nest, doing so in the early morning and evening and taking two salmon during each shift. If the male dallied for too long, the female shrieked at him and even flew out from the nest to move him off his fat perch. Adult eagles knew the sight of a gun and, upon seeing one, cackled to new-flying young a loud warning to flee. After describing this lesson on survival, Larson noted that an "eagle lives to an old age unless something happens to shorten its life." Larson was that something.[33]

Ornithology wasn't the central subject of Larson's piece for the sporting magazine. Shooting was. Larson packed a .300 Savage rifle and a sharpshooter's proficiency. He said he once hit an eagle from two hundred yards out. With one bullet from 150 yards, he knocked down two sitting side by side on a branch. The previous summer, he had killed seventeen sequentially, "without a miss." The article includes a photo of Larson posing with a long stringer of talons that hangs in continuous loops along the gunwale of a boat and over and around his neck like an exaggerated—and flowerless—Hawaiian lei. Clean-shaven and smiling, he resembles a proud father of maybe a half dozen young ones more than he does a grizzly last-frontier hunter showing off a swag of thirteen dozen severed pairs of claws. That's roughly $156 to be paid from the territory's bounty budget, or $2,200 in 2020 dollars. That was decent pocket money for a forty-six-year-old who had no children, lived alone on an island in southern Alaska (where almost certainly he fished and hunted for his food), and had another job. In that other job that summer, for the US Bureau of Fisheries, Larson counted 1,936,135 salmon running up Anan Creek—perhaps a few more than he would have, if not for his .300 Savage and his eagle eye.[34]

If Larson was still gunning for eagles in 1940, he and other vocational hunters were soon hit in their wallets when the bald eagle secured its federal entitlement to live in the Lower Forty-Eight. Although Congress exempted Alaska from the Bald Eagle Protection Act, the governor thought the territory should comply with the law anyway. So he rescinded the bounty's appropriation. The governor was Ernest Gruening, who was born in New York City and educated at Harvard, where he earned bachelor's and medical degrees. He left doctoring early on to seek a more exciting life in journalism. From 1912 until 1934, he worked as a reporter, editorial writer, and managing editor at some of the country's biggest publications. Then, after President Franklin Roosevelt ramped up the New Deal, he accepted a position with the federal government. He had moved up to director of the Interior Department's Division of Territories and Island Possessions when Roosevelt appointed him as Alaska's seventh territorial governor. He was known for his professional integrity, and his ears. They were big. His nose was big too, his eyebrows robust and smile perpetual. He would serve as governor until 1953—fourteen years—and when Alaska gained statehood, voters elected him as one of their original US senators.

Before Gruening's appropriations veto, the prospect of losing the bounty hung as a menacing cloud over fishers, trappers, herders, and bounty hunters. There was the possibility that the legislature would abolish it or the governor would defund it, as Gruening did. Under the same cloud was the patriotic argument that Congress and eagle defenders had been making: Alaska was the only place in America that was paying to kill the living form of the national symbol. Appealing to patriotism on behalf of "marauding" birds, however, was a hard sell to those who believed eagles were plundering the warehouse.

The word "marauding"—actually, "marauding things"—was one of many pointed descriptions rendered by an editorial in the *Seward Gateway*, the newspaper that served the Kenai Fjords region just south of Anchorage. Sympathetic to fox-fur farmers, the anonymous editorial writer blamed the bounty controversy on "dreamers and sentimentalists in the East who see in the killing of the eagle a menace to democracy." The editor clearly wasn't going for the patriotism argument. "There is no group of people anywhere," emphasized the editor, "who are more loyal to the institutions of liberty and good government . . . than can be found in Alaska." Hardworking members of the "fox fur business can see no connection between patriotism and a savage predator bird that makes a practice of wrecking legitimate undertakings."[35]

Gruening was an easterner, and an Ivy Leaguer at that. But there was no questioning his patriotism. As governor, he would serve Alaska as a reassuring leader during the forthcoming war after the Japanese twice bombed Dutch Harbor, and in the previous war he had proudly worn the bald eagle on the brass buttons of his army uniform. He was consummately loyal to the territory and its people—a loyalty that was tested by his belief that wild places and animals were essential to Alaska's identity. On long drives across the long territory, he habitually counted the bald eagles out his window. They belonged in Alaska; they were part of what Alaska was. The claim that the white-headed raptor was a "curse to the rest of the animal kingdom," as one newspaper put it, never sat right with Gruening.[36]

Before vetoing the bounty appropriation, Gruening read up on the bald's feeding habits in the Aleutian Islands, in a research report written by biologists with the US Fish and Wildlife Service. The territory's bald eagles were the biggest in the country, yet they weren't big enough to pull off some of the heists for which they were allegedly responsible. The researchers believed that eagles were wrongfully accused in most cases. For example, even though fox cubs fell within the raptor's lifting range, researchers witnessed eagles out on that string of summer-green islands roosting around a number of fox families without instigating any disturbances. The report noted that "evidence shows that eagles are not a serious menace to the blue foxes." Balds fed primarily on spawned-out salmon, which in their dying state were happy grabs for the raptor commons, yet lousy product for the fish market.[37]

When a new territorial budget came before Gruening in 1943, he again struck out funding for the bounty. Two years later, he lobbied for and succeeded in having it repealed altogether. In his reasoning for abolition, he steered clear of the national-symbol argument. He didn't go with sentimentality either. There was nothing to gain in reminding opponents of abolition that he was an easterner. He preferred that his action be associated with his fervor for statehood, which Alaskans widely supported. Gruening knew how the federal government operated and understood Washington bureaucracy and politics. One way, then, for Alaska to demonstrate its preparedness for statehood was to show that it could fall in line with the national agenda. Drawing on that reasoning, he argued that if the territory didn't abolish the bounty on its own, the federal government would very likely step in to do it. Washington's meddling was the last thing that individualistic Alaskans wanted.

Despite their official, though momentary, reprieve, eagles remained on hit lists that stimulated a lot of uncompensated killings. Poachers shot them, saved the talons, and hoped for the bounty's restoration. They had elected some formidable eagle antagonists perched in the territory's capitol. A most active one was William Egan. Likable and pleasant, Egan was an Alaska-born son of a gold miner who came from Valdez and a constituency that attached its livelihood to the land and water. As far as he was concerned, by sustaining the bounty he would be sustaining a way of life. From the moment he was elected to the territorial house, at age twenty-six, he championed both the bounty and statehood. For some, Gruening's argument that eliminating the bounty would better Alaska's chances for statehood made little sense. The argument should be flipped on its head: statehood would better the chances of preserving the bounty by giving Alaska voting representation in Washington.

Gruening nevertheless remained ever a patron of the bald eagle. He wrote an old New Deal friend in Washington, saying he agreed with what Willard Van Name had recently said in a letter to the *New York Times*, in which he had derided the "eagle-killing fraternity" and the "wanton and inexcusable persecution of the bald eagle in Alaska." Many Alaskans not only wanted the bounty restored after it was canceled in 1945; they wanted more money for a set of talons. Egan shepherded a bill through the legislature in 1949, awarding a $2 bounty. Only four members voted against it. One was Frank Barr from Fairbanks, who was quoted as saying, "For the privilege of seeing [eagles] impressively in flight, I'm willing to throw them a few fish."[38]

Judging the one-sided vote in the legislature, Gruening believed that, rather than vetoing the bill outright, he could make a more powerful statement by not signing it, exercising a pocket veto. As expected, the legislature overrode it, and the bounty became active again. The press started calling Egan "Eagle Bill," and a caustic Associated Press reporter wrote, "The Alaska Legislature, hoping to become the 49th state under the wings of the eagle, nevertheless voted Wednesday to place a bounty on eagles."[39]

Statehood wouldn't arrive for another ten years. Gruening would go to the US Senate, where he would remain an influential friend of the eagle, while Egan would go to the capitol building in Juno as Alaska's first governor, where he opposed federal initiatives to create national parks in the new state. In the meantime, Congress took the very action that Gruening had warned about. In 1952 it voted to amend the Bald Eagle Protection Act for the specific purpose of nullifying the Alaska exemption. (The National Audubon Society portrayed itself in the press as a principal force against the bounty.) Pressured to comply with federal law, the territorial legislature revoked the bounty.

Beginning on the first day of the new year, no one could legally kill a wild bald eagle anywhere in the US. The Alaskan bounty was dead.

To Van Name and Edge, and for the law's beneficiary, safe harborage came 128,273 Alaskan eagles too late.[40]

⌣

> In tumbling turning clustering loops, straight downward falling,

THE BALD EAGLE PROTECTION Act brought blazing guns under control yet did not eliminate them altogether. You make laws and then you make lawbreakers too, and there were plenty of the latter in Alaska and throughout the contiguous states. Yet Americans were tolerating the killings less and less. Expressing a growing sentiment among fellow citizens, a man in Aberdeen, Texas, said, "A person who would kill a bald eagle would trample the flag in the mud." After the original passage of the Bald Eagle Protection Act and the 1952 revision, the increasingly winsome raptor continued to inspire Americans to make the sky and nesting and roosting places safer for it. New levels of relief materialized every decade.[41]

In 1962, Congress amended the act again, this time by adding the golden eagle to its provisions. Over the centuries, predator control had hit the golden with a full volley of lead and poison. Like the bald, the golden was an accused livestock thief, economic competitor, and seasoned baby snatcher. In truth, the golden was a potential economic asset. If its population had been allowed to thrive and go about its business keeping snake, rabbit, and prairie dog populations in check, western farmers and ranchers might have adopted the golden as a best friend rather than a worst enemy. As one syndicated columnist put it, "If the golden eagle goes . . . the rodents will eat the crops." Yet during the Senate hearing to consider the new bill, Texas was noted as a strict no-fly zone. A representative of the state's Sheep and Goat Raisers Association insisted that "we cannot have anything but open season on the golden eagles." According to an Audubon Society investigation, one bounty hunter who worked for ranchers, and who sold feathers and body parts to Indians, proclaimed that he had killed twelve thousand eagles in his twenty years on the job. An untold number of those eagles were balds. Some were deliberately targeted in violation of federal protection, and others were shot when immature balds were confused with goldens.[42]

Mistaken identity was ultimately what brought the golden to Congress's attention. According to many who testified at the hearing, including representatives from the National Audubon Society and one from Hawk Mountain Sanctuary, young balds living in close proximity to goldens tragically flew in the crosshairs of hired guns. Supporters of the bill, which included the influential voice of the new senator from Alaska, Ernest Gruening, argued that the way to avoid the fallout of mistaken identity was to make the golden a ward of the state. Language in the bill codified that notion, stating that the golden eagle deserved sanctuary because it was "threatened with extinction," it benefited "agriculture in the control of rodents," and its security would "afford greater protection for the bald eagle."[43]

Congress passed the Bald and Golden Eagle Protection Act in 1962 with strong support. The bald of the wild had helped rescue a fellow eagle. Being associated with the founding species had its advantages. It had become a bird of leverage and influence. It had clout. It held sway.

Haliaeetus leucocephalus and other raptors had held their sway with Rosalie Edge, who died on the last day of November that year. For decades, she had maintained a file labeled "Poison." In it were the earliest documents on DDT. One was a Fish and Wildlife Service news release written in 1946 by Rachel Carson, who worked for the agency in public affairs. The first words in the news release were these: "A warning."[44]

During the 1962 hearing, pesticides were not cited as a major intrusion in the avian world, although prepublication excerpts of *Silent Spring* were appearing in the *New Yorker*. The full impact of Carson's opus was felt when it hit bookstores three months and a day after the hearings concluded. The harmful chemical pesticides instantly became a topic of national discussion. Carson referred to the reports and articles of the 1940s that had sounded early warnings—warnings that tragically had failed to change attitudes about the chemical age. If they had, they might have saved the country from the national landscape of death that followed. "If we are going to live so intimately with these chemicals—eating and drinking them, taking them into the very marrow of our bones," Carson wrote in a chapter titled "Elixirs of Death," "we had better know something about their nature and power." We knew a good bit, she argued, but we disregarded our knowledge and upon our vaunted perch held ourselves above nature and partook in our deleterious tinkering with it.[45]

Carson exercised a voice of reason, cautioning against reckless use of and addiction to chemicals but never calling for their outright ban. The industry launched a smear campaign against her anyway—which continues today—once again parading out a suite of tedious scientists in lab coats and horn-rimmed glasses, each with an equity stake in producing marketable chemicals, each saluted as "Dr." They called Carson an emotionally overwrought spinster who lacked the scientific credentials to speak authoritatively about her subject. Science, they were saying not so obliquely, was the intellectual realm of men.

Carson drew such frothing animosity because she was an unmarried woman without children who supposedly had no stake in the future, because she was a detached observer and brave writer, because she had assiduously documented her evidence, and because she was indeed a trained scientist—one who bore the credentials of a best-selling author who could translate technical information for general consumption, and one who aroused widespread public concern about chemical pesticides.

Carson wrote about Hawk Mountain Sanctuary, citing the records she had requested. She wrote about Broley and his work. She spoke of the "dying out of the race" of bald eagles. She identified numerous other imperiled birds, but if ever there were a feathered harbinger, it was the bald eagle. The year after *Silent Spring* appeared, spotters counted only twenty of the bellwether birds flying over the sanctuary, the Endless Mountain, in the fall of 1963. A North American inventory of nesting bald eagle pairs led by the US and Canadian Audubon groups totaled a despairing all-time low of 487 in the forty-eight states. More than likely, only a single-digit percentage of those nests were free of life-curtailing DDT and DDE.

While chemical companies were casting aspersions against Carson, the *New Yorker* writer E. B. White was pecking out verse castigating the industry's self-appointed post above nature and the wild beings for which it showed spectacular indifference. Beloved by generations of readers, White had written the immensely popular young readers' novels *Stuart Little* and *Charlotte's Web*. In them, he had dared to give animals voiced opinions about their plight in the world. The grandson and grandnephew of three Hudson River School artists, White preferred his low-tech farm in Maine to his high-rise office in Manhattan. He prefaced his poem "The Deserted Nation" with a few facts, including the state of eagle nests in Maine in 1965. Fifty-three active ones produced four eaglets. Two lines from the poem say it all:

> Chemists and farmers flourish at their peril:
> The bird of freedom, thanks to them, is sterile.[46]

In those fifty-three nests in Maine, there had been many more eggs than the rarefied four from which eaglets hatched. Seeped into the forty-nine addled others were DDT, DDE, and additional poisons.

～

Till o'er the river pois'd, the twain yet one, a moment's lull,

IN *SILENT SPRING*'S CHAPTER 8, titled, "And No Birds Sing," Carson wrote that in climbing trees and banding eagles, Charles Broley had achieved "ornithological fame." Neither Charles nor Myrtle lived long enough to see those words or to know the full savagery of DDT.[47]

Myrtle died in 1958 at age fifty-six. The heading in her newspaper death notice read, "Wife of 'Eagle Man' Dies," a diminishing acknowledgment for a woman who had been an accomplished naturalist and writer. Six years after publishing the biography of Charles, she had a heart attack and passed away in the middle of the Florida nesting season.[48]

Later that spring, Charles returned to Canada for the first time alone since he and Myrtle had been married. As they had always done, he spent the summer on Lower Beverley Lake at their white clapboard cottage on Whiskey Island in Delta, Ontario. The lake was a good place for fishing, whether you were equipped with hook and line or wings and talons. There were pike, bullhead, stickleback, bowfin, carp, bass, perch, trout, and salmon. For as long as he had been banding birds in Florida, Broley had been doing the same around the lake, and he wasn't finished with his life's work. That fall, he made his ritual snowbird's commute to Tampa for the twenty-first time. He would turn seventy-nine in the middle of the season, and despite the accretion of years, he hardly slowed down. Push-ups and pull-ups in the park were still part of his morning routine. In April, he mounted what he estimated to be his eleven hundredth tree and then went north again to Canada.

If there was anything that he could always depend on back on Whiskey Island, it was the leaves from the previous fall awaiting his attention. He raked them from around the cottage, as he usually did, and burned them in a pile. The chore done, he pushed off in his boat and began rowing across the lake to run errands in town. Hidden flames beneath the ashes of the leaf pile erupted into a grass fire onshore and caught his attention. He came about in the boat and hurried back. Beating out the blaze, he fell to the ground unconscious. He never woke. The fire marched across his body and on toward the cottage, consuming both.

"Tragic" is how most would describe these last moments of Broley's life, but perhaps his was a fitting departure from this world. He had ritualistically tidied up a place of personal meaning and significance, as one might prepare an altar. After resurrected flames called him back off the lake, the natural funerary pyre gave tribute to his worthy life. Smoke and ashes rose to where the spirit bird flew.

A motionless still balance in the air, then parting, talons
loosing,

TWENTY YEARS AFTER THE eagle bounty came to an end, thirteen after Broley's death, ten after Rachel Carson published *Silent Spring*, the newly established Environmental Protection Agency pulled DDT off the domestic market, permitting its use only in cases of extreme public health threats. Gruening was the only one of the bald's more prominent early best friends around to witness the ban. He was eighty-five and would make it to eighty-seven, dying twelve years to the day after the 1962 congressional hearing to secure a safer future for the bald by protecting the golden. But this was the 1970s, and now there was no shortage of friends of nature.

Americans had become more aware of bleak environmental conditions, largely because media attention to polluted water and air and dying wildlife was more pronounced. The television, technology in nearly every home, broadcast upsetting images into living rooms and dens. There on the screen and on newspaper front pages was the Cuyahoga River running through Cleveland, a foul-smelling, foul-looking, dead river on fire, flames reaching five stories high, a greasy-black eclipse of smoke as tall and wide and dense as a mountainside. *How the hell does a body of water catch fire?* people wondered. The answer: floating oil and chemicals from local industries using the river as a sewer line. That happened in 1969, and it was the twelfth time in a century that the river had lit up into a rising furnace smearing the sky with black.

Everywhere around the country, industry was doing the same, using the public commons as a personal dumping ground for its unwanted waste, foisting it on the American people, looting the public trust, and profiting from the abuse. And it wasn't just businesses offending the natural heritage. Virtually every municipality in the country was pouring raw or poorly treated human waste into rivers, lakes, bays, bayous, sounds, and coves. In 1970, scientists announced that Lake Erie was dying. The lower reach of the Hudson River running beside Gotham City was already a dead zone, auguring a financial collapse that would threaten the survival of the city by mid-decade. Estuarine environments around the nation's coasts lost sixty, seventy, eighty percent of their seagrass beds and oyster bars, and the marine and avian life that was of those environments vanished like a coin in a magician's palm. So, too, did the carbon-sequestering that estuaries perform.

Nothing hit the TV watcher and magazine reader in the gut like the visual of an oiled bird struggling and drowning in a syrupy black slick. That's what was happening to thousands along the Santa Barbara coast in the winter of 1969, four months before the water caught on fire in Ohio. The equivalent of 100,000 barrels of oil spewed from the drill hole of a Union Oil offshore platform, opening out into the largest US oil spill to that time. By day eighteen, according to congressional testimony, contaminated water reached across "many hundreds of square miles." It wasn't just birds that the sludgy onslaught grabbed. Dolphins, sea lions, elephant seals, and an unimaginable number of fish died.[49]

The country had lost touch with its natural heritage. Ruin and exploitation had for too long been blemishing the complexion of national identity. A flaming river and greasy sea, the fetid and decaying environments where people lived, played, and worked—they were all too much. The plundered, rotting human habitat, writes historian J. Brooks Flippen, turned a "concerned public into an activist one."[50]

The next year, Americans on April 22 launched Earth Day, a celebration of the planet and a demonstration against what humans had done and were doing to it. These mobilized citizens were mostly white, but they weren't mostly social elites, as those representing the old conservation movement had been. They were twenty million Americans, one-tenth of the US population, and they were saying they wanted the desecrated Earth cleaned up and detoxified. Those twenty million expressed themselves in parades, festivals, marches, teach-ins, and protests. One hundred thousand people in New York City alone came out. Down in car-clogged Miami, mothers pushing baby strollers led a procession of "pollution floats" in a Dead Orange Parade, which followed the traditional route of college football's Orange Bowl Parade. Everywhere, people held up signs declaring the pollution of adults as the debt of children. Schoolkids cleaned up litter and planted trees. Congress went into recess so that its members could join constituents in Earth Day events.[51]

Unsettled by the ongoing protests against the Vietnam War, President Richard Nixon had the FBI surveil Earth Day activities on college campuses. Still, he couldn't appear out of step with the largest civil display of public sentiment in US history, so he and First Lady Pat Nixon planted an evergreen on the South Lawn of the White House.

Three months later, the president signed an executive order creating the Environmental Protection Agency. It was the beginning of the end of DDT. The chemical had a number of vocal and theatrical defenders. In demonstrations before the public, industry entomologists ate DDT to prove it was safe for humans, and a California couple took DDT capsules, or what they claimed were DDT capsules, with their lunch for ninety-three days—and survived, he for ten more years, into his eighties, and she into the twenty-first century.

Yet even before Earth Day, Americans had been letting the leadership in Washington know that they were tired of the wrack and ruin. Disturbed by the die-off of raptors, a small group of scientists and lawyers founded a nonprofit organization, the Environmental Defense Fund (EDF), for the specific purpose of bringing about the legal prohibition of DDT. The group used New York as a test case, got the state to outlaw the use of DDT, and then focused on national prohibition. Responding to the public's growing disfavor for harmful chemicals, the USDA had already suspended DDT use on fifty food crops, a number of livestock—including beef cattle—the elm and other shade trees, ornamental turfgrass, and flowers, as well as in commercial buildings, residential homes, restaurants, and food-processing plants.

That wasn't enough for the EDF and a coalition it had organized, which included the Izaak Walton League, the National Audubon Society, and the National Wildlife Federation. Not long after Nixon created the EPA, the federal regulation of pesticides shifted from the USDA to the new agency. The EPA also inherited a lawsuit brought by the EDF for a complete ban. Complying with an order from the District of Columbia Court of Appeals, the EPA announced its intent to revoke registration for most other DDT uses.

Thirty-one formulators of DDT petitioned for an administrative hearing on the phaseout. The hearing began in August 1971 and concluded in March of the next year. With the marshalling of 300 evidentiary documents, a 9,312-page transcript, and 125 expert witnesses, the case against the chemical pesticide was damning, although not damning enough for the hearing examiner, Edmund M. Sweeney. Sweeney concluded that DDT was not a "hazard to man." Nor did it "have a deleterious effect on freshwater fish, estuarine organisms, wild birds, or other wildlife," yet he had no explanation for the massive fish kills, the disappearing pelicans, and the dwindling eagles. He recommended that the EPA vacate its latest and all proposed restrictions on DDT.[52]

The matter ultimately went to William D. Ruckelshaus, appointed by Nixon as the EPA's first administrator. A forty-year-old Republican from Indiana, Ruckelshaus had come over from the Justice Department, where he had been an assistant attorney general. If central casting had to send up a Hollywood version of a government bureaucrat, Ruckelshaus would fit the bill. Everything on him was straight: the dark, horn-rimmed glasses on his face, the part in and comb of his hair, the thin-lined mouth and facial expression (or lack thereof), the drape of his arms, the seam lines on his suits. His demeanor as a public servant was identical: straight-as-an-arrow forthright and honorable (subsequently, when he was a US deputy attorney general, he resigned rather than following President Nixon's instructions to fire the independent special prosecutor investigating the Watergate affair). Aside from building a new agency from the ground up, one of his first tasks as EPA administrator was overseeing execution of the significantly bolstered Clean Air Act of 1970.

Ruckelhaus's best-known action as administrator, however, was that which he took against DDT in 1972. Nixon, sensing the political winds, backed him, even though the industry had every expectation that the business-friendly president would put an end to this nonsense about neutering its most effective pesticide. Ruckelshaus instead put an end to DDT's use in the US, excluding limited public health applications. He recognized the profound link between the chemical's introduction into the environment and the threats that followed. The costs to society and nature, he concluded, were significantly greater than the benefits. Given the "knowledge at the time," the potential of a wrong guess and the accompanying risks to human health were, he said with remarkable prescience, "quite high." He laid out a bullet list of seventeen crucial facts related to DDT. Number eight in category four, "Toxicological Effects," read, "DDT can cause thinning of bird eggshells and thus impair reproductive success."[53]

The pesticide's defenders would forever criticize Ruckelshaus for his betrayal. Wildlife would forever benefit from it. Throughout its centuries-long battle royale with unrestrained malice, cruelty, and indifference, America's bird had shown its mettle. It would go on to navigate the post-DDT years by rising from the past and reclaiming its place in nature.

> Upward again on slow-firm pinions slanting, their separate
>   diverse flight,
> She hers, he his, pursuing.

Fledgling leaving the hack tower at Quabbin Reservoir, Massachusetts, ca. 1983. (MassWildlife / Bill Byrne)

Part Four

RESTORATION

EIGHT

✦ ———— *Eagle Lady* ———— ✦

BULGING OUT FROM BETWEEN THE UPPER BRANCHES OF A loblolly pine, a large fingerlapped arrangement of sticks formed the familiar aesthetic of an industrious eagle couple. For some unknown reason, the pair had not returned for the 1979 nesting season. Staring up, Doris Mager was aware of the centrality of nests in the lives of bald eagles. Those compositions of meticulous labor, enigmas of intricacy and strength that marry art with utility, are essential to the renewal of life. The identity of few birds is as closely attached to their nest as the bald eagle's is to its eyrie. None in North America build larger or stouter ones. The balds' are emblematic of their species' resilience. Nests had been a key variable in determining the population's decline, and they would be imperative to its revival. Without them, Mager knew, there would be no birds.

Mager was also aware of the violent spontaneous weather that frequented Central Florida, and at that moment dark clouds filled the sky to the west. Standing at the foot of the loblolly, one hand hesitantly on a climbing ladder hanging down from the height of a fire lookout tower, she was intent on spending time in the nativity of the former occupants.

Mager had never scaled a tree before, much less in a storm. She reached over and touched an ominous-looking lightning scar running down the tree's trunk to the ground. Pushing ahead of the storm, the wind pulsed, and the green needles trembled in the branches high above. One eyewitness described the tree as "spindly." Another called it "wind-whipped." Jeff Klinkenberg, the outdoor editor for the *St.*

Petersburg Times, is the one who used the word "spindly." "Here she was," he reflected decades later, "fifty-three years old and climbing a ladder I would not have dared to climb at my age then, thirty."[1]

Before putting herself at the mercy of the swelling wind, Mager tied a red bandanna around her head of silver hair, which she had had cut and styled in a new hairdo for the occasion. Owl earrings dangled beside her cheeks, and, retaining the raptor theme, a spread-eagle necklace wreathed her neck. She wore black jeans, a denim shirt, and gray running shoes. Yet her jogging routine had been inconsistent of late. In relating that detail, she confessed to Klinkenberg, "I've got fat little legs, and I probably shouldn't be that far off the ground at my age." She slipped into a safety harness secured to an upper branch. Alongside the harness line, the grounding cable of a lightning rod chased down the side of the tree. A number of precautions were taken that day, and Mager added one of her own by swallowing a motion sickness pill. "I get airsick and I get seasick," she again confessed to Klinkenberg, "and I'm probably going to get nest sick."[2]

Mager put one foot up on a lower rung and followed that with the other on the next rung. Grabbing a third at eye level with both hands, she stared nervously into the tree's rust-colored scaly bark and coaxed herself toward a fifty-foot summit. Whenever the wind kicked up, the tree creaked like an old door. When it swung like one, she would pause, grip the ladder tighter, and take a deep breath. She shouted to a friend below, "Get down on your knees and pray, Viola!"[3]

When she got to the first branch, she dismounted the ladder and stepped into the loblolly's outspreading crown. Still ten feet below her objective, she used the hoisting rope to pull herself up and over into the nest. She caught her breath and looked out across the distance, over the sprawl of the russet and green pine barren, out toward the faraway silence of wetlands and the shimmering silver light reflecting off Lake Harney, the fishing hole of the previous tenants. The storm in the west was moving to the north of her, and she hollered to a crowd that had gathered around the base of the tree, "You don't know what you're missing! It's gorgeous up here!"[4]

Mager was planning to "live" in the tree for six days and six nights.

She explained to a reporter, "I always wanted to sleep overnight in a nest." She was interested in knowing the eagles' residence firsthand, seeing what they saw, experiencing the motion of the tree they experienced, feeling the touch of the sun they felt, listening to the pitch of the wind through the woodland canopy and the stick nest.[5]

Mager wasn't there merely to take a joyride in a swaying tree. She was a vice president at the Florida Audubon Society, overseeing raptor research and rehabilitation. A donor had recently provided the start-up capital for the construction of an aviary at the group's headquarters. It was a good sum of money but not enough. Mager then announced plans to stage her "nest-in," as some were calling it, to bring attention to the plight of the bald eagle and to raise $20,000 to add to the aviary fund.

Klinkenberg said that Mager was "funny and quotable" yet also "heroic." Divorced, and the mother of a grown son, Billy, she had devoted the previous seventeen years to rescuing and rehabilitating injured and orphaned raptors. Like Charles Broley, whom she knew of but never met, she had no formal science background and came by her education and work on her own. When she was in her twenties and living in Haddam, Connecticut, she was a victim of the "old wives' tale," as she put it, that eagles were known to abduct babies. A nesting couple that fished on the nearby Salmon River frequently flew over her yard and Billy's sunning bassinet. That's when Mager learned to recognize the call of an eagle and to run outside at the sound of it. "Don't you come down here," she would say. "I like you very much. You just keep going." She soon learned that her fears were unwarranted, that eagles were not the birds of storybook tales and ranchers' recriminations, or even certain ornithological lore.[6]

After moving to Florida in the early 1950s, Mager began working at the gift shop of Florida Audubon in an old farmhouse outside Orlando. In 1963, when fewer eagles were nesting around the country than ever before, someone walked into the gift shop holding a red-tailed hawk with an infected foot. Mager didn't know how to help the bird and was reluctant to try, but she took it home. She fed and watered it and soaked its foot, recalling what her father used to say: "Use common sense and Epsom salt for everything." Once the hawk was healthy again, she released it.[7]

That's how raptor rehabilitation was in those days: improvisational or nonexistent. Once word got around that the woman at the Audubon gift shop knew how to restore the health of birds of prey, more people brought in injured hawks and owls. Mager called on an ornithologist who oversaw a raptor rehabilitation program on the Gulf coast for guidance on caring for birds, and she took to her new endeavor like talons on a wholesome speckled trout. Injured birds she kept at her house, and those that recovered yet remained flightless for a bad wing or lost eye she took to the store to help promote the cause of raptors.[8]

At one time she had as many as seven balds living in her backyard. Mager and Florida Audubon decided that a facility equipped and designed to revitalize the rescues would be more appropriate, and they would need to raise money to build it.

The nest Mager was now sitting in was located near Oviedo, an old citrus town known for its curious demographic of street-roaming chickens—of which, on occasion, one or two likely were lost to local bald eagles. Mager knew all the nests in Central Florida, and she chose this one because, although abandoned, it remained sturdy. The US Fish and Wildlife Service assented to her nest-in on the condition that if eagles began hovering about, she and her entourage would evacuate immediately. Mager wanted to give the event a patriotic theme appropriate to the American bird, so she made her ascent on Flag Day, June 15, and flew the Stars and Stripes from the side of her perch.

Mager's publicized plans brought some one hundred people out for her climb, turning off State Road 419 and driving a mile down a sun-scorched sandy road through open pine and scrubland. The usual faithful Audubon members were there, along with journalists, television film crews, and general spectators, a good many of whom had never seen a bald eagle in the wild.

Mager had a support crew with her too, though housed in the slightly snugger, stale-air quarters of a trailer. Helpers had rigged up a pulley system to send food and water up to her. In return, she would periodically send down a Maxwell House coffee can repurposed as a chamber pot. Someone brought her a steak with mushrooms one day, and the ice cream man came with a treat every day. On some evenings, there was live music and dancing down below, as well as pizza parties, when dinner came to her by the slice.

The dinner leftovers of the previous residents—fish heads and animal carcasses rotting in the plait of sticks and moss—didn't bother her. She was protected, she said, by a "bad smeller." She was also equipped with a walkie-talkie and a makeup kit. Clean clothes were sent up daily. Blankets and a pillow were hoisted at night, to go atop the patio-lounge cushion used as a mattress, all easily held by the nest's six-foot expanse. "I can stretch out from north to south," she cheerfully told a reporter, "and still have a good foot or foot and a half left over." She wasn't lounging around, though. She wrote thank-you cards to donors and tried to finish up that year's nest survey for the state.[9]

News and photos of Mager and the nest-in ran on television and in newspapers across the country. *Life* magazine printed a full-page color photograph of her waving topside. She was in bare feet and a pink shirt, wearing Jackie O sunglasses and her owl earrings and eagle necklace. The iconic "Hello, Americans" radio personality Paul Harvey did a segment on his nationally syndicated *Paul Harvey News*, which reached upwards of twenty million listeners. Supporters and curious people came out to see the "Eagle Lady," as the press dubbed her. One day, a girl barked through a bullhorn, "Do you like it up there?" Mager shouted back, "Honey, I *love* it." How could she not? She would wake each morning to watch the ground fog dissipate as the sun rose, and she thought she was in heaven.[10]

As planned, Mager descended the tree on June 20, the 197th anniversary of the Great Seal of the United States. Flapping her arms like an eagle, she admitted that she was a bit of a "nut." The nest-in raised over $6,000, and donations kept arriving in the mail. Florida Audubon met its fundraising goals, built the new aviary and education center, and opened it in October of that year. It was the first of its kind in Central Florida, adding another to the handful around the country, which would quickly double in number, and soon after that double again.[11]

Mager was among many who ushered in a new era in the country's rapport with its favored raptor. Here was "the rest of the story," as Paul Harvey always said at the close of his radio spots, in the bald eagle's improbable journey with America. The descendants of those who tried to eradicate the species began protecting it with unprecedented zealotry. Significantly, they created a welcoming and healthy environment to enable the bald eagle to flourish once again across the continent.

PERSISTENCE AND RESTORATION

THEY CAME TO BOSTON ON A DC-9—TWO EAGLETS between six and seven weeks old, and five and six pounds healthy. A male and a female, they were in special wooden transport crates borrowed from the wildlife folks in New York. They flew first class on Republic Airways from the Midwest at an altitude that only the rarest of birds can fly.* The plane landed at Logan International Airport on June 11, 1982, nine days before the bicentennial of the Great Seal of the United States.

Only the Stars and Stripes exceeded the national symbolic importance of the irrepressible bird on the front of the Great Seal. Yet by comparison, the image of the white-headed eagle was everywhere and on everything. And it had been since George Washington visited Boston in the fall of 1789 and led his presidential procession beneath a ceremonial arch crowned with a large carved bird of America. A block away thirteen years earlier, Bostonians had taken possession of the Massachusetts Town House, seat of the British colonial government, and claimed it as their statehouse. In a celebratory moment, they gathered below the building's east gable and, using ropes, pulled down statues of the royal lion and unicorn, symbols of the detested empire. When the statehouse was restored in 1882, the centennial year of the Great Seal, the architect installed a gilt figure of a bald eagle alighting on a globe high on the west-side façade. A century

---

* The highest recorded altitude of a bird on the wing is thirty-seven thousand feet, credited to a Rüppell's griffon vulture over Ivory Coast in 1973.

later, the city had a plaza, a neighborhood, and two streets named "Eagle." The whole of Massachusetts had 205 streets named "Eagle," 13 of which were specifically "Bald Eagle." The state also had two recently adopted eaglets.

After jetting into Logan, the pair joined Boston's panoply of revolutionary patriotism when someone named them Betsy and Ross, creating a witty if unintended allusion to the flag and eagle as one. The living species was gaining the same level of acceptance as the emblem. Indeed, the popularity of the species was, in part, what brought Betsy and Ross halfway across the country. There was also a revolutionary purpose behind their presence in Massachusetts, which, if all went as planned, would become their new home. They were initiates in an innovative bald eagle restoration project.

Three years earlier, an article in *Life* magazine, from the same issue that featured the photograph of Doris Mager in her Jackie O sunglasses, announced in bold headline font, "The Bald Bird Lives! America's Endangered Eagle Is on the Increase." The magazine wasn't printing hokum. The number of nesting pairs in the Lower Forty-Eight had risen modestly during the 1960s and '70s. To a large degree, the bald eagle's proclivity to survive was responsible for the upward trend. Yet no one was certain to what extent the population would recover and whether eagles would or could return on their own to places where they had stopped nesting.[12]

Along with state partners, the US Fish and Wildlife Service developed restoration plans for five regions of the country. When the plans were conceived in the 1970s, significant impediments lay in their path. In 1972, Congress updated the Bald and Golden Eagle Protection Act with heavier penalties, yet days after these revisions went into effect, someone shot and killed a bald eagle near Lake Moultrie in South Carolina.* When Mager was occupying the nest in Central Florida,

---

* Under the new law, if found and convicted the shooter would have faced a maximum $5,000 fine and up to a year in prison. For a second offense, the fine could be doubled. (Under Canadian law, those convicted could be fined up to $300 and have their hunting license revoked.)

she had five eagles at home in rehab, gunshot victims all. Eagles were similarly doomed to be hit by cars, electrocuted by power lines, caught in traps, roughed up by dogs, and separated from nests that were destroyed illegally. The ravages of DDT remained evident in skies without wings and trees without nests, and those skies weren't likely to come alive again as long as the general ecological health of the environment continued to decline.

None of these hazards were fated to go away completely, but they could be mitigated. In the late 1960s and early 1970s, Congress passed a raft of farsighted laws for the benefit of wildlife renewal and a cleaner environment. But a viable foundation for these initiatives first required science—which informed the laws that improved conditions for wildlife, which made the resurgence of *Haliaeetus leucocephalus* possible—to catch up with the times.

WHEN FRANCIS HERRICK WAS still pitching tents atop steel platforms in the 1930s, American science was decades away from fully understanding the ripple effect of removing an apex species from its environment. Killing off top predators can throw an entire ecosystem out of whack. Remove the wolf and mountain lion, and you are likely to detonate a population explosion among their prey—deer or elk, for example. The precipitous expansion of either of the two can instigate the spread of disease and overgrazing of plant life. Thinned-out flora can then abet rain and wind in carrying off topsoil and organic matter to a stream or river, clouding the water and shutting out the sun to aquatic growths. If the plant life dies, the aquatic animal life leaves or dies, and fishing prospects for raptors significantly diminish.

Take out raptors, and rodents have the run of the place, eating more plant seeds than when their demographic is kept in check—seeds that would normally sustain song- and game birds. More rodents mean more food for snakes, the number of which raptors would ordinarily control. But the raptors are gone because the fish are gone. And what do snakes on the loose like to eat? Bird eggs. Some species, such as herons, find a security in living near eagles, even though young herons might occasionally end up as a meal in an eyrie. But if the aquatic life is healthy, the eagle is more likely to eat fish instead of herons.

American science was slow to understand this complicated web, in large part because the American academy was recalcitrant to accept ecology as a valid field of study. The old-time naturalists studied individual species, typically apart from other species and the host environment. Their bounded focus was evident in their paintings of birds adrift on a blank space absent any sort of matrix beyond a perch. Even the naturalists, like Catesby and Audubon, who added background to their paintings were creating an aesthetic rather than a symbiotic habitat. Insects crawling about and berries dangling weren't necessarily those to which the avian subject had a dietary connection.

Connections within nature formed the guiding matrix of ecology. Around the time when birds were sitting on DDE-softened eggs and Broley was banding fewer and fewer eaglets, American ecological science was finally piecing together the interconnected world that existed in nature and that had long been familiar to Native peoples as the circle of life. In Native thought, the circle must remain unbroken. Nothing is separate from the circle—not the spirit bird, not the king of birds, not the American bird. In the simplest of science speak, no species is independent of other members or parts of an ecosystem. Indeed, if the environment is suffering, top predators are among its first constituents to indicate decline, to reveal a rupture in the circle.

That circle and the web-of-life principle in ecological science informed the activist and legal environmental agendas that emerged around Earth Day. Beginning in 1969 with passage of the National Environmental Policy Act, federal agencies were required to prepare an ecological impact statement to go along with proposals for "major actions significantly affecting the quality of the human environment." An important phrase here is "human environment." Placing people in an affected area, making it more than about plants and animals, provided stimulus for resuscitating natural systems.[13]

What people and eagles both needed were wholesome habitats, starting with safe water. Cleaning up murky bays, bayous, lakes, rivers, and coastal waters was like bypass surgery for the country. All across

America, waterways had been clogged with unwanted substances of factory, farm, and city. When introducing the Clean Water Act of 1972, legislation drafted to remedy a bad situation, Senator Edmund Muskie of Maine painted a bleak but not inaccurate picture for Congress, stating, "Today, the rivers of this country serve as little more than sewers to the seas." He pointed across the way and reminded his colleagues that pouring into the Potomac River, the "Nation's River," the river beside which George Washington's Mount Vernon rose, the river that supplied his colleagues' drinking-water, were fifteen million gallons of untreated human waste—every day.[14]

The foulness was local, and it was everywhere. One would have to embark on a long and hard search to find a municipal public works that wasn't excreting virulent stormwater, unrefined sewage, and insufficiently treated wastewater into the liquid heart of America. Exchanging clean water for dirty was a Faustian bargain of the highest order—environmentally, socially, and economically. If a cosmic villain were looking for a wicked strategy to reduce a viable civilization to a failing one, wrecking clean-water sources was as good as any. In a disquieting reality of postwar America, civilization itself was the villain. A species of nature, *Homo sapiens*—corporeally seventy percent water—was sabotaging the very nature of its own self.

Then, Americans made some smart, constructive moves. One of the wisest was to provide some $60 billion in grants to local and state governments during the 1970s and '80s to upgrade old and build new water treatment systems. Prompting those grants was the 1972 Clean Water Act, which set quality standards for surface water (although not groundwater) and regulations for the discharge of pollutants. Even though EPA administrator William Ruckelshaus urged Nixon to sign the act, the president vetoed it, arguing that the nation could not afford the cost of compliance.

But many believed that the nation could not afford to live without it. No one benefited from dirty water—not, ultimately, even polluters, who lived in the same human nest they fouled. Bad water, for example, destroyed nearly one-fourth of the country's 1969 shellfish crop, which translated into a $60 million loss to the economy and compelled taxpaying workers to apply for tax-supported unemployment benefits.

Two hours after President Nixon vetoed the bill, the Senate overrode him with a firm bipartisan vote. The House matched it with a 247–23 spread, with nearly forty percent of the yes votes coming from Nixon's own party members.

Congress adopted three broad goals related to its clean-water initiative: to eliminate toxic releases; to restore all US waters to a condition safe enough for drinking, swimming, and fishing by 1983; and to end every form of polluted discharge by 1985. Barely one-third of the nation's waters at the time were swimmable and fishable.

To date, none of these goals have been met, and quality-impaired water continues to stream across America. Still, the Clean Water Act was one of the most important and life-changing measures of post-war America. Its enforcement, primarily by the Environmental Protection Agency, was responsible for significantly decreasing pollution discharges. Dead or dying waterways and water bodies—emptied of life—rebounded. Vegetation came back. Marine life came back. Commercial and recreation fishing came back. Birds came back. They could not have done so without Congress's foresight. Americans could ban DDT from the land, attach stern penalties to the Bald Eagle Protection Act, and fly scores of eaglets into Logan, but if they didn't mop up the mess in the country's waters, the prospects for species continuation were little better than those of a lone bird on a lonely island whistling for a mate that didn't exist.

It's a sad truth, but those who came to dominate the land after the Lenape, Cherokee, Aleut, Salish, Pueblo, and Luther Standing Bear's people broke the circle. It's a happier truth that they subsequently sought to mend it. The Clean Water Act, with its drinkable, fishable, and swimmable goals, was built around human priorities, but that doesn't matter. All things are connected. The benefits that clean water brought to people it also brought to plants and animals.

The benefits flow in the other direction too, as in the case of the Endangered Species Act, another farsighted piece of legislation. While making life more livable for plants and animals, it upgraded the quality of the human environment.

ON APRIL 12, 1983, ten months and a day after Betsy and Ross landed at Logan, the pop artist Andy Warhol opened a new exhibit at the American Museum of Natural History. The bewigged, peacocky Warhol, who preferred tall buildings and crowded sidewalks to tall trees and open spaces, had never before shown at the museum, and on virtually any other occasion his style would seem out of place in the Romanesque Revival spaces filled with skeletal mammoths and mastodons and mounted birds and bears. His latest exhibit, "Warhol's Animals: Species at Risk," meshed with no other venue better. The New York *Daily News* explained that "after years with the jet set, the gossip salons and the literary lions," the "pop artist and celebrity register resident" had "turned his interest to endangered animal species." Since Charles Willson Peale and *The Exhumation of the First American Mastodon*, more than five hundred continental species had joined the mastodon in oblivion. Well over a hundred more were at risk of the same. At the time of Warhol's show, a wildlife scientist asserted that the biological diversity of the nation, the very robustness of North American nature, is "part of our natural heritage." Warhol's exhibit was in some measure a statement about the teetering existence of that heritage. Its dismal state in 1983 made the US no longer so unique among other nations.[15]

Worldwide, species foundered—thousands of them. Wanting to give them memorable faces, Warhol executed ten thirty-eight-inch-by-thirty-eight-inch silk screens of emblematic animals. His trademark unmodulated colors and shades, with emphatic accent lines and outlines, represented each animal's singular energy. There were the Siberian tiger and giant panda of China; the elephant, black rhinoceros, and Grévy's zebra of Africa; the orangutan of Indonesia and Malaysia; the San Francisco silverspot butterfly of the western US; the pine barrens tree frog of the eastern US; the Sierra Nevada bighorn sheep of North America. Rounding them out was the essential bald eagle.

Although hundreds of birds were at risk across the planet, the bald eagle was the only one in Warhol's group of ten. It was an obvious choice and one of the most popular members of the series, done as a bust profile from the shoulders up, with the eagle's impressive head turned to its right. Seen from below, the subject is in command of its own image. It's an unwavering one that holds the viewer's gaze. Beneath the eagle's orbital ridge, the yellow eye is decisive and stern. The white head feathers are thrust defiantly backward. The hooked upper beak, an evolutionary design common to raptors that enables them to rip apart flesh, conveys self-determination and survival. Magnifying these effects, Warhol drew thin red lines that sweep through and around the contours of the eagle's features and profile, and he set all against an electric-blue background.

Eleven years after DDT went off the American market in 1972, the founding bird was thriving in British Columbia, Canada's westernmost province, while only thirty US states reported couples domestically engaged—approximately 1,250 total—with miserable sums in most of those states. Yet Warhol's bald eagle reveals, as perhaps the artist intended, no hint of being a vulnerable species. It is determinedly alive and staring down on the future with a damned-if-you-will look.[16]

Warhol printed 150 sets of the series and donated them to conservation groups to auction at fundraising events. After the New York premier, the exhibit traveled around the country, showing at art and natural history museums, coinciding with the tenth anniversary of the Endangered Species Act (ESA), one of the most important environmental measures of the 1970s. The ESA was the evolved and more celebrated version of the 1966 Endangered Species Preservation Act, itself groundbreaking. The earlier legislation had introduced "endangered" as a new federal classification for animals. The Department of the Interior's initial tally of endangered animals—those at a high risk of extinction—smacked of a preboarding manifest for a modern-day ark, totaling seventy-eight mammals, reptiles, fish, and birds. Winged creatures, including *Haliaeetus leucocephalus*, accounted for nearly half.

"Endangered" in this legal context was an admission, a confession, even an act of self-absolution. Since the late nineteenth century,

with varying degrees of commitment and success, federal stewardship of wildlife had stood on the two pillars of *protection* and *conservation*. The new law added a third: *propagation*. The humpback whale
and American alligator regained numbers without the assistance of
planned restoration programs. But with many species of plants and
animals and many ecosystems and wild places on the edge, protection
and conservation could not alone rectify desperate conditions. For the
whooping crane (down to forty-four birds), California condor (down to
thirty-eight), and bald eagle, as well as the grizzly bear, black-footed
ferret, and gray wolf of the Lower Forty-Eight, the path to salvation
required the added thrust of population restoration.

Three years after Interior compiled its first list of endangered animals, Congress in 1969 upgraded the species preservation law by man-
dating a minimum of $15 million for buying conservation land and
funding restoration. Such initiatives were imperative to the bald eagle's
future, and their effectiveness depended largely on the support of the
Interior Department, where Walter Hickel served as secretary. A Nixon
appointee, Hickel had not been the choice of environmentalists. He was
an Alaskan real-estate developer who defeated the bounty supporter
William "Eagle Bill" Egan in his reelection bid for the governorship
before the president tapped him for the cabinet post. To environmentalists, apocalypse seemed a certainty with his Senate confirmation. But
Hickel, much like Nixon, surprised environmentally concerned citizens.
A year into the job, he ordered the department to change its practice
of issuing blanket permits to states and counties to kill golden eagles
and instead limit permits to ranchers and farmers on a case-by-case
basis only under the "most critical circumstances." The "responsibility
for protecting golden eagles," Hickel said, with novel discernment, in a
press release, "outweighs the need for controlling eagles."[17]

The Nixon administration included many who were sympathetic
to improving environmental quality. The president's chief domestic
advisor, John D. Ehrlichman, a well-regarded land-use attorney who
many considered a "covert Green," created within the administration
a workable admixture of earnestness and politics that tended to go
along with, if not forge, environmental policy. Much of it originated

on Capitol Hill with a conservation-minded triumvirate consisting of Senators Henry Jackson of Washington, Edmund Muskie of Maine, and Gaylord Nelson of Wisconsin, who represented states that had retained fairly wholesome eagle populations. Nelson was the progenitor of Earth Day, and he and his two colleagues constantly challenged the White House to do better. In a future time, Nixon, with his place in history tarnished by the Cambodia lie, civil protests, and Watergate, would embrace a redeeming environmental legacy.[18]

Together, the White House and Congress fattened federal statute books with a baker's dozen of major environmental laws. Among those thirteen was the 1973 Endangered Species Act, which, advancing the objectives of the antecedent legislation, added a "threatened" category to go along with the "endangered" category. With some exceptions, ornithologists were no longer officially recognizing the separate southern and northern bald eagle subspecies. Without that technical distinction, the seven thousand breeding pairs in Alaska might have bumped the bald eagle off the list. Besides, America's bird was an obvious candidate to be kept on with the new law. It was a flagship species, a poster bird that resonated with Americans and carried considerable clout in Washington.

In return, bald eagles got an added layer of protection. The enforcement guidelines were more clearly defined under the ESA than under the Bald and Golden Eagle Protection Act. The ESA, for example, established a 330-foot buffer between a bald eagle's nest and human activity. Someone now harming America's bird could be charged for violating three federal laws: the Bald and Golden Eagle Protection Act, the ESA, and the Migratory Bird Treaty Act, which Congress amended in 1972 to expand its purview to hawks, owls, and eagles. The private, nonprofit National Wildlife Federation, its logo emblazoned with a bald eagle in flight, contributed to the cause by offering a $500 reward for information that led to the conviction of eagle killers, creating a bounty that turned history on its head.

It was more than a coincidence that these advances occurred as the country's bicentennial was approaching. While Americans were looking forward to honoring the memory of the Founders and Patriots, no one wanted to be reminded of the still uncertain predicament of the founding bird. Although its population was inching upward, a reversal remained a possibility. Two years after the bicentennial, Fish and Wildlife, charged with compiling the ESA's list, identified the bald eagle as threatened in five of the lower forty-eight states—Michigan, Minnesota, Wisconsin, Washington, and Oregon—and endangered in the remaining forty-three.[19]

IN MASSACHUSETTS, NO ONE had seen a nesting pair since 1905. So, in the year of the Great Seal's bicentennial, two chosen eaglets, soon to be christened Betsy and Ross, were flown in to Logan International Airport.

Once their crates were taken off the plane and opened, aides misted them with cool water, and a veterinarian examined them as a doctor would returning space shuttle astronauts. The eaglets came from Michigan's Upper Peninsula, where they had hatched in separate nests along the Menominee River. The year before, Michigan had had 102 known nesting pairs, the third-highest number among northern states, behind Wisconsin and Minnesota. The protocol for removing the chicks from the wild was to prioritize the continuing production of the donor nest. Wildlife officials took only one eaglet from each harvested nest, and only from nests that had at least one other healthy chick. They also acquired the smallest. The theory was that the larger sibling had a better chance of enduring in the survival-of-the-fittest milieu of the eagle eyrie, while the runt's odds for reaching maturity improved in supervised captivity.

After the two Boston arrivals revealed themselves to be wide-eyed and just fine, wildlife personnel drove them to Hanscom Field in Bedford. From there they flew on a Cessna 172 floatplane to Quabbin Reservoir in central Massachusetts. Built during the Depression to provide drinking-water to the urban and suburban masses in and around Boston, Quabbin was the state's largest inland body of water. A lush green valley bisected by the Swift River had existed in its place before engineers flooded twenty-five thousand acres and four towns in

1945 to create the reservoir—an unfortunate matter for thousands who lost their homes. Quabbin was a clean environment. Its primary inflow came from New Hampshire, where farming had tapered off significantly over the years, and it was flanked by a fifty-five-thousand-acre wildlife area, parts of which were off limits to fishing and hunting. At its center and giving the reservoir a V shape was Prescott Peninsula. Halfway to the end of the peninsula, down an eight-mile hard-packed dirt road through two locked gates, and waiting when the Cessna skidded onto the water and taxied up to a gravel shore, were journalists, wildlife personnel, and something known as a "hack tower."

Betsy and Ross were entering a program that utilized a technique in the sport of falconry called "hacking." The term dates to the Elizabethan era, when handlers took young, unfledged birds of prey and kept them in "hack wagons," which were enclosed or semi-enclosed. The wagons were parked on high ground in the birds' prescribed hunting territory, and the birds were fed by a handler while imprinting on the territory—which is to say, developing a visual relationship or attachment to it. Once old enough, they were permitted to fly and hunt for their food. The assumption was that, given this freedom, falcons would grow swift and strong on the wing and develop superior hunting skills. The handlers would eventually recapture the raptors and train them for falconry.

In the early 1970s, ornithologists adopted a modified version of falconry hacking to reintroduce raptors into healthy environments that they had abandoned sometime before. With DDT's presence in the environment subsiding, the Fish and Wildlife Service introduced an experimental program using ospreys. Wildlife personnel removed eggs from nests at the tip of the Florida peninsula, where the native population had mostly escaped pesticides, packed them in heated and cushioned containers, and shipped them to Georgia and the Carolinas. There, the migrant eggs replaced cracked and collapsed eggs in active osprey nests. The adults instinctively carried on in their roles as parents, hatching and raising babies from the foreign eggs. This was egg translocation, not hacking, and its success was encouraging.

The translocation approach to species preservation inspired New

York congressman Brownie Reid and restaurateur Pete Kriendler, part owner of the famous 21 Club on West 52nd Street, to meet with Interior and propose a similar program for bald eagles in their state. With its vast apple orchards and summer crops, New York over the years had been soaked in DDT. Only one known nesting pair survived in the state, and year after year its eggs failed to hatch.

Interior gave Reid and Kriendler's proposal the go-ahead and authorized Fish and Wildlife's Patuxent Wildlife Research Center in Maryland to set up an experimental program. Patuxent possessed the wherewithal in physical resources and institutional history for the job. Established in the middle of the Depression by the Franklin Roosevelt administration to study wildlife in their natural habitats, Patuxent was the first of its kind in the US, and vaunted for its living laboratories of woodlands, wetlands, and prairie land, as well as a blue-ribbon staff of scientists. Early members of that staff were among the first to identify the harmful effects of DDT to bird and aquatic life, and few places had as much cumulative experience in the study of raptors.

Initially, team members at Patuxent intended to tap the Alaskan eagle population for healthy eggs to be hatched in the East. Eventually, they gravitated to the idea of hacking eaglets instead. Under this new plan, they would take young from nests in Alaska, relocate them to New York, and raise them from behind a blind in artificial nests, or hack boxes, located in the wild. Once the adoptees were ready to fledge, they would be released.

In the year of the nation's two hundredth birthday, bald eagle hacking was born in an initial trial using two eaglets raised at Patuxent and moved to pesticide-free Montezuma Wildlife Refuge in central New York, once prime eagle habitat. After their flight feathers grew in, the juveniles fledged, signaling to their human ground crew that it was time to shift from trial to program stage.

Officials at the Alaska Department of Fish and Game were happy to share eaglets with their counterparts in New York. When he was governor back in the 1950s, Ernest Gruening had repeatedly advocated, in *Audubon* magazine and in Congress, the use of Alaskan bald eagles to replenish numbers in needy states to the south. One way "Alaskans could share the bald eagle," he wrote, "would be by transplanting some of the birds to the lower 48 states." Gruening died in 1974 and would not witness the Alaskan bailout. But it almost didn't happen. Still subscribing to the existence of northern and southern subspecies, the American Ornithologists' Union filed a lawsuit claiming that relocating Alaskan eagles to New York would wash out distinctive gene pools. Challenging the suit, Fish and Wildlife would eventually prevail in court, but in the meantime it took birds from Michigan, Minnesota, and Wisconsin.[20]

In the first five years, New York released twenty-three eagles, and the original bicentennial pair returned upon maturity to nest. These promising results encouraged Fish and Wildlife to expand reintroduction to Massachusetts.

AMONG THOSE ONSHORE AT Quabbin waiting for Betsy and Ross was Jack Swedberg, the person arguably most responsible for making the day happen. If you were to call the Massachusetts hacking program any one person's, you'd call it Jack's. Swedberg wasn't a biologist, or any kind of scientist. He was the senior wildlife photographer for the Massachusetts Division of Fisheries and Wildlife. For decades he'd been traveling the state, tromping through wild areas, up and over moss- and lichen-covered granite rocks, across frigid snowmelt streams, through slush and mud, chancing thin ice on a pond—under the worst of conditions, yet also at times the best, to get the unexpected picture, that perfect shot. Swedberg was the rare individual who had turned his lens on the few bald eagles spending time in Massachusetts. The year before the launch of the Quabbin repopulation effort, avian census takers counted nineteen balds in the state, none of them nesters.

Bald eagles could not have asked for a more effective "pitchman," to borrow a description from one of his colleagues, than Swedberg. Brandishing a tanned and chiseled face, graying temples, and squinting eyes, and vested in a regulation khaki shirt with pocket flaps and shoulder straps and a MassWildlife patch on the upper left sleeve, Swedberg looked like a field officer in the military who, with sleeves perpetually rolled up to the elbows, got things done. He lobbied more aggressively than anyone to convince the higher-ups at division head-quarters to cooperate with the feds when they were looking to expand New York's successful eagle-hacking initiative to other states. Hacking in Massachusetts would be good for the species and good for the image of the state, demonstrating to its citizens that it cared about wildlife and the environment, that Quabbin was more than a drink of water tapped from a lost valley for Boston.[21]

Virtually all of New England at the time was a sink for eagle num-bers. No chicks hatched in five of the six states the year before, 1981. Maine was the exception, with thirty-four nests yielding hatch-lings. The two states in the US with hacking initiatives fully up and running—New York and Georgia—were generating eagle life, and new programs were coming online in Missouri and Arkansas. A Mas-sachusetts program would give New England its first. Swedberg and others knew that since eagles typically return to breed in the general area where they fledge, some of the birds hacked in Massachusetts would help repopulate New England by nesting in neighboring states.

After being brought onshore at Quabbin, the transport crates with Betsy and Ross were taken to the hack tower. The tower was essen-tially two large cages positioned thirty-five feet up on top of six power poles, which the Massachusetts Electric Company had donated and set in the ground at water's edge. The cages were five feet tall and sixty-four square feet, with an observation corridor running between them. Resembling elevated lion cages, the boxes were barred on two sides and the top with galvanized electrical conduit. This open design exposed the eaglets to the elements as they would be in a natural set-ting. They would feel the wind, rain, and sun. To stave off boredom and stimulate hunting instincts, they had a bird's-eye view of the res-ervoir and potential future prey on the other side of the bars—fish in the water, ducks on it, loons nesting beside it, and songbirds whirling around it. In each of the cages was a replica eyrie fabricated with sticks that MassWildlife personnel had cannibalized from an abandoned beaver lodge.[22]

Researchers decided to put Betsy and Ross in the same cage to replicate an eyrie with a typical brood of two. Unless the pair became quarrelsome, they would stay together throughout their residency. To avoid their imprinting on humans, they were not to come in contact with handlers. The corridor between the cages was fitted with one-way glass that enabled researchers and caretakers to observe activity in the nest. Still too young to walk on their taloned feet, the eaglets would stay in the nest unless one flopped out. After settling into their new home, recalled Bill Davis, a biologist with MassWildlife, they "panted, preened, and pecked at a white sucker" fish left for them. On the other side of the glass, media cameras whirred. The Michigan transplants appeared "entirely unimpressed with the whole event." A genuinely pleased Swedberg announced, "Everything is fantastic." Within an hour, the new arrivals were eating, exercising, and "having a ball."[23]

Betsy and Ross would remain in their hack box until they were released into the wild. To feed and protect them, to ensure they stayed out of trouble with each other and anything else, they required a full-time babysitter. That became the job of Dave Nelson, a biology graduate student from the University of Massachusetts, who would observe the birds' behavior and track their growth. In the real wild at this stage, the chicks would have been spending most of the time alone. Their parents would roost away from the nest while dropping off food, door-delivery style. They were providers more so than protectors. At most they would circle overhead, screeching. And parents did not prevent one sibling from killing another, or ensure that each in the nest was fed equally. The law of the eyrie was that the most aggressive was the most served.

Betsy and Ross, in other words, weren't missing out on quality family time by having Nelson as their invisible parent. They were on their own. Aside from occasionally spying a fleshy hand dropping fish down the feeding chute, the eaglets did not see humans either. Then Ross, at about eleven weeks old, pulled an unexpected stunt.

The smaller of the two birds, Ross pushed through the jail bars like a mouse through a locked cupboard. Maybe he saw enticing fish in the reservoir or baby loons on it. Maybe he was older than estimated and was stir-crazy, having reached the age when instinct prompted him to fledge. Whatever the case, he took his maiden flight. Fortunately, he wasn't interested in long-distance travel and stayed in the area. Swedberg and Nelson tried baiting leg snares to recapture Ross, but he didn't go for it. What did draw him in was food at the tower, his nest as far as he had ever known. Nelson got Ross back by baiting the vacant second cage and closing the door behind him when he stepped in for a bite. Even after eagles start flying, they return to where their mom and dad have always left food, and they continue to do so until they leave the area for the season.

Ross would have saved everyone, including himself, the aggravation of a fugitive bird hunt had he waited to test his wings eight days later. That's when he and Betsy had been scheduled for release. In preparation for their official bon voyage, Swedberg and a veterinarian banded the two and fitted each with a near weightless radio transmitter to a tail feather. Scientists would track the travelers until they flew out of range or molted the tail feather. On July 29, the eight-mile road to the hack tower turned into a thoroughfare conveying MassWildlife officials and press people coming to witness the launch.

After the doors of the hack boxes were raised, Ross was like a sprinter coming off the starting blocks. He lifted from the platform with wings open, dipped down a bit, turned head-on into the wind, grabbed a swell of rising air, flapped his wings, and disappeared behind a line of trees. Betsy stepped up to the platform edge, opened her brown-and-white wings as if she knew what she was doing, and leaped. She, too, turned into the wind, but neither it nor her beating wings kept her heavier weight airborne. She circled over the water, spiraling downward, and went in for a gentle splash landing. Nelson had been standing by with the project's thirteen-foot Boston Whaler and darted out to rescue her.

Back in the cage, Betsy dried off, regained her composure, and responded when the door opened again. This time she took flight on her six-and-a-half-foot wingspan. She circled north, winging toward a leafless roost tree that Ross had discovered, and joined him. No longer eaglets, the full-fledged eagles, Davis wrote soon after, "were free."[24]

∿

FOR THE NEXT SEASON, the hack team added a second floor to the tower, giving it four cages, with the capacity for eight birds. In the two summers after Betsy and Ross were freed, nine more transplants, less one that died before the release day—from Michigan, Nova Scotia, and Manitoba, Broley's home province—took flight at Quabbin. These weren't quite the numbers seen in New York, which released twenty-three bald eagles in four seasons, but the Empire State was five times larger than the Bay State, and Massachusetts was on pace to achieve natural reproduction within a few years.

Eight chicks arrived at Quabbin in April 1985 from Nova Scotia. Dianne Lefrancois took over as the live-in babysitter. Swedberg had recruited her in anticipation of Dave Nelson's graduation and departure. Growing up in Worcester, Massachusetts, Lefrancois wore a dimpled smile and bangs to her eyebrows. Both aspects stayed with her into adulthood, as did her love of animals. In the fifth grade she developed a lifelong relationship with an Audubon teacher who taught the kids about raptors and migrating monarch butterflies, and at home she kept company with an array of small wild critters. In high school, she was the only girl in a special program in plant and animal sciences, and on weekends and after school she volunteered at the nature and science center in Worcester.*

After graduating from high school, Lefrancois delayed college to take a full-time job as a zookeeper at the center. When Swedberg asked her to join the hacking project, she was like Doris Mager: she had no college diploma in an age when such a credential had become increasingly important. Yet she had worked at the science center for six years, she was a licensed falconer—the only licensed female in the state—and she and her husband at the time founded and operated It's an Animal's World, an educational program with live animals that they presented to scouting and civic groups. Lefrancois's work and her "sixth sense with raptors," as Bill Davis put it, earned the respect of credentialed scientists.[25]

---

* It was called the Worcester Science Center, which was renamed the New England Science Center in 1986, and the EcoTarium in 1998.

Lefrancois had one other quality that stood out: she was a licensed wildlife rehabilitator. One of the Quabbin eaglets had died the previous season, and Swedberg believed that Lefrancois's experience observing and treating animals with a variety of health issues would enable her to recognize a troubled bird and restore its health. Still, lacking a degree, she suffered from impostor syndrome, and the social expectation of her sex didn't buoy her confidence either. Although women in science, like eagles in the wild, were expanding their presence, the residuals of male dominance were profound. At every turn, patriarchal society signaled that women were supposed to see themselves, whether degreed or not, or experienced or not, as little more than the proverbial girl Friday who made the coffee, washed the test tubes, cleaned the cages, and deferred complicated assignments and intellectual authority to men. Lefrancois could not always free herself from this sociopsychological burden. In a letter home when she was deep into her tenure at Quabbin, she quoted Erma Fisk, an amateur ornithologist who had experienced a crisis of confidence while banding birds in Arizona for the Nature Conservancy a few years earlier. "Damn, damn," Fisk had written, "they have sent a girl to do a man's job."[26]

Lefrancois's job would be the same as Nelson's, although she would care for eight eaglets, a full house at Quabbin for the first time. Except weekends, when a substitute covered for her, she lived full-time near the tower beneath white pines in an aluminum-and-canvas pop-up camper shared with nosy mice and piercing mosquitoes, where at night she watched the stars and fireflies and listened to owls, coyotes, and loons. Her primary responsibility was feeding the eaglets twice a day, and they ate a lot. At one sitting, a nine-pound eaglet could fill its crop with a pound or more of meat. She net-fished for their food—bass, perch, bluegill, bullhead, pumpkinseed, landlocked salmon—providing something on the order of twenty pounds a day. From the start, she noticed that the established practice of slicing open the fish didn't work well in preparing a meal for the chicks. They had a difficult time ripping the meat into sizes small enough to swallow. In the wild, the parents did that for their brood. As the surrogate, Lefrancois began cutting the fish into bite-size chunks. The task added hours to her already full workday and knife cuts to her hands, but the eaglets ate ravenously. Whenever she fed them fish that had been frozen and thawed, she would lace it with vitamin B—another first at Quabbin.

For the initial couple of weeks, the chicks used their wings like crutches and hobbled around on their heels (the joint that looks like the backward knee). Their toes and talons, proportionally large for their still-growing bodies, flopped out in front of them like clown shoes. They were comic and pathetic and adorable. As their bodies filled out and they rose up on their toes, they began what Francis Herrick had described as gymnastic exercises: spreading and flapping their wings and bouncing into the air, like a cartoon character having a tantrum, going higher and higher each day. They would hop in and out of their nests, onto the perch pole, and over to the electronic perch scale, enabling Lefrancois to record their weight. She kept detailed notes on their workouts, food consumption, and feces content; the brightness of their eyes; and what those eyes watched outside their barred cages—gathering all this information without being seen by the brood.

She also recorded their general disposition and interactions with their cagemates. The eaglets had been paired by the closeness of their size to lessen the possibility of one dominating another. Battles for position still happened. There were peaceful moments of rubbing beaks, exercising together, perching and snoozing side by side, and one preening the other. But some of the birds were predisposed to aggression and dominance. Lefrancois ultimately named one caged pair Bert and Ernie. Bert, if you know your *Sesame Street* characters, was the ornery one. Bert the eaglet would push and shove Ernie around, peck at him, bicker with him, claw him, and steal his food. Over in another cage, Attila earned her name by routinely doing the same to her nest partner, Marjorie. When one cagemate deprived the other of food, Lefrancois would wait until the bully gorged itself and then drop in more food for the bullied one. The domestic violence continued until the birds' liberation on August 9.[27]

Release day for Lefrancois brought the kind of emotion that parents experience at their child's graduation. Opening the cage doors was a commencement, a new beginning for the class of '85. Lefrancois

was excited for their rite of passage into eagledom, for their graduation from nestling to brancher, from eaglet to eagle. Yet she also worried about the future of her "babies," as she would continue to call the
Quabbin eight until long after.[28]

The event was preceded by a line of cars and trucks kicking up dust
down the eight-mile road, gates unlocked and opened for the occasion.
The cage doors were raised simultaneously. Attila dashed out and vanished around a copse of trees to the north. Snoopy, who had endured
the wrath of Woodstock, bolted. Woodstock followed and then, after
a momentarily graceful flight, crash-landed on a beach. Marjorie lifted
off and flew across the way for a perfectly executed beach landing. Dick
and Jane hung back for a bit before taking flight. Bert and Ernie seemed
befuddled, jumping around in their cage, Bert leaping from perch to
perch. He then left. Ernie, the passive one, stayed behind, alone in the
tranquility. Sometime after the ceremonial excitement died down, he
made his exit.

Having graduated from fledglings to branchers, Lefrancois's babies
slept in roost trees at night and stayed around Quabbin into the fall.
She, Swedberg, Bill Davis, and others left food along the shore of
the peninsula and Loon Island, straight out from the hack tower.
While keeping track of the birds with telemetry, Lefrancois continued to pull nets for fish. If someone happened to come across roadkill
while commuting to and from Quabbin, they would bring it out for
the eagles—and the ravens, kingfishers, coyotes, and any hungry other
that might happen along. The eagles were turning into able fliers and
landers, and they were magnificent to watch. Some of the squabbling
that Lefrancois had witnessed in the hack cages continued in the free
spaces. It was no surprise to her when Attila chased another eagle from
her roost tree. That eagle was her former hack-tower mate, Woodstock.
Mostly, the Quabbin eight fought over food. Whenever one carried
off a fish left on the shore by hooking and gripping it with talons,
Lefrancois knew the young eagles were learning how to take care of
themselves.

In late September, the branchers started transitioning into migrators. Snoopy was the first of the eight to leave the reservoir. Ernie was

the last; his telemetry signal fell out of range on October 25. Lefrancois left a few days later.

⌢

THE MASSACHUSETTS HACKING PROGRAM continued for three more summers, with Bill Davis taking on the babysitting. In seven seasons, ending in 1988, the program released forty-one eagles. Most seemed content to stay in the upper northeastern states when migrating in the fall. On occasion, a Quabbin bird would make an appearance as far away as the Chesapeake Bay.

One, endowed with nothing more than its project name, S3, flew three hundred miles to the Conowingo Dam. Completed in 1928, Conowingo was a hydroelectric generating station on the Susquehanna River in northern Maryland, just below the Mason-Dixon Line. S3 wasn't the only Yankee bird frequenting this southern dam, or the only hacked eagle or descendant of a hacked eagle. For balds, Conowingo was a Gershwinesque haven where the livin' was easy. When the gates to the hydrogenator tunnels were open, which was most of the time, fish above the dam were sucked through and deposited below, left stunned or dead, turned into floating offerings for waiting raptors. Before the long tyranny of DDT, Conowingo in the late fall and early winter gave the appearance of a daily food festival for thousands of cormorants, gulls, and eagles. Wedges of ducks and Canada geese flew in for vegetarian offerings growing around the dam's riprap and along the river's edges. Then, in a matter of a few years, the number of birds dwindled to almost nothing. The dam and its transmission lines and towers stood alone, buzzing and humming in the unwilded silence.

Once hacking programs began sending their graduates out into the world, migrating eagles rediscovered Conowingo, eventually turning out each year in the hundreds. All the waterbirds of before and an occasional peregrine falcon contributed to the avian cacophony. Black vultures showed up as well, in greater visible sums than eagles, hunching over as vultures do in their roost trees, conserving energy and patiently waiting for leftovers. Historically denigrated like eagles for their scavenging ways, vultures suffered the added insult of being characterized as repulsively ugly. A kindhearted dam visitor once offered a rare expression of compassion for them, saying, "They're really kind of pretty," then followed with a qualifying afterthought: "when they fly."[29]

People came too, and in the same swelling numbers. They came with thousand-dollar cameras front-loaded with multi-thousand-dollar super-telephoto zoom lenses, some of the latter the size and shape of a cheerleader's megaphone. The chittering cries of *Haliaeetus leucocephalus* echoing in the dam's concrete canyon signaled a promising day to veteran picture takers. A good many wore camouflage jackets, pants, and hats. Their lenses, tripods, and gear bags were camouflaged. The only thing not camouflaged was their excitement over a flyby. More exciting was an eagle pitching down to the water for a fish. More exciting than that was another eagle flying in for a steal, often resulting in a twirling aerial tussle, a nondalliance engagement. There's no telling how many times, over the years, that S3 and other Quabbin alumni passed in front of the wide-eyed lenses, helping to make Conowingo one of the most popular eagle-watching venues east of the Rocky Mountains.

Florida eagles traveled to Conowingo too. The regional migration of the Quabbin birds was very different from that of the Florida juveniles who winged up the length of the East Coast. MassWildlife personnel saw a lot of them at the Massachusetts shore, where the agency put up nesting platforms for ospreys and banded their young. No one could conclusively say why the Florida juveniles made such long journeys, which were often suicidal. Many of them arrived on the Massachusetts coast exhausted, starving, and at the threshold of death. For every one that MassWildlife personnel were able to save, they lost close to the same number. Perfecting fishing skills can take a few years. In the meantime, juveniles have to depend on scavenging and theft. As wild waters warm, fish move to cooler temperatures below, reducing the catching and stealing prospects of fishing birds. Southern eagles will then head north, where water temperatures are still cool and fish are still comfortable near the surface. Eventually, those waters warm up too, and the eagles keep going north, burning more fat that must be replenished for the bird to avoid a tragic end.[30]

By 2004, all of New England except Maine, which was never at grave risk of losing breeders, had instituted hacking programs. Vermont, where the bald eagle didn't make the endangered list until 1987, was the last to gear up, bringing in eaglets from Maryland, Maine, and a raptor rehabilitation center in Massachusetts. Following strategies laid out by Fish and Wildlife, every region of the country was involved in eagle hacking in some manner, either dispensing chicks to other parts or receiving them to rebuild the local population.

SHIFT TO THE SOUTHEAST and, for a moment, back to the early days of DDT. The town of Ocean Springs on the Mississippi Gulf of Mexico coast got a pair of eagles in 1951. They were adults and came not from a restoration initiative but from the imagination of an artist. Walter Anderson, who lived in Ocean Springs and had studied at the Pennsylvania Academy of the Fine Arts, the fountainhead of the Schuylkill River School, was the artist. He painted not landscapes so much as individual living things of nature, almost exclusively those of the Gulf coast. If one had to attach him to a movement, he came closest to expressionism, influenced by what he realized emotionally, aesthetically, and intellectually from his subjects.

When town officials decided to spruce up their twenty-five-hundred-square-foot community center, Anderson volunteered to paint a mural for the nominal sum of $1. The mural ultimately covered all 250 feet of the community center's four walls, from floor to ceiling, and its doors and door trim. On one wall he depicted the aboriginal peoples of the coast, on another the French arriving to settle it in the seventeenth century. He devoted the rest to the cosmos, what he called the Seven Climates, and to Gulf-coast nature. Among the latter was the dalliance of two eagles. A reader of Whitman, Anderson shared the poet's sensibilities of nature. "Mr. Whitman," he once wrote to himself, "be my aid—friend of the wind I am."[31]

Anderson spent weeks at a time alone on the wind-clipped barrier islands off Mississippi and Louisiana. Each was unpeopled, yet none was a deserted island. They were exuberant sovereignties of crawling, stalking, burrowing, and flying beings. They were fascination and inspiration. Anderson called himself a "privileged spectator" of all the living that happened on the islands. On Horn Island, an eight-mile row or sail out from Ocean Springs in a skiff of some uncertain seaworthiness, he had witnessed the mating ceremony of the American bird, and referred to the sight of it as a "regal gift." Horn was a ten-by-three-quarter-mile dune-humped sandspit giving residency to assorted vegetation, including a grove of pine trees. It had nesting pelicans, ospreys, and eagles, to name just a few species.[32]

In the 1950s, Anderson noticed that the presence of those birds was falling off. The last pair of nesting eagles in the area had been seen in 1940 on Ship Island, eighteen miles west of Horn. By the 1960s, pelicans, ospreys, and eagles had disappeared altogether. Anderson suspected chemical pollution traveling down Gulf-bound rivers. He was right. He was witnessing the shock of DDT, as was Broley on the Florida Gulf coast.

Anderson died of complications related to cancer surgery in 1965, just months after surviving Hurricane Betsy, a category 4 crusher, as it trained its destructive eye on Horn Island. (He twice refused Coast Guard rescue to stay on the island to observe how nature would come back from destruction.) Twenty years and six months later, Horn was once again a dwelling place for *Haliaeetus leucocephalus*. Its new feathered residents came courtesy of the George Miksch Sutton Avian Research Center in Oklahoma, a number of unsuspecting donor eagles in Florida, and squirts of superglue.

❦

THE LAST ANYBODY HAD seen bald eagles keeping house anywhere in Mississippi had been sometime in the 1950s. In Alabama it was 1962. In Georgia 1979. In North Carolina between 1963 and 1971. They had left Arkansas in the doldrums of 1929, the year of the stock market crash. Nesters endowed Tennessee with nothing throughout the 1960s and '70s. Enduring slightly better in Texas than in other southern states, the birds managed in a rare year after 1945 to get a few eaglets out of nests. Discouraged, the Texas Parks and Wildlife

Department noted in its magazine in 1970, "Saving the Bald Eagle may be beyond our powers." Balds similarly maintained hardly anything more than intermittent family life in South Carolina during the DDT years. In Louisiana, with Lake Pontchartrain, the Mississippi River, and nearly as many square miles of coastal marshes as feathers on an eagle, not to mention endless tangles of swamps and bayous, breeding couples marshaled but four nests during the late 1950s. They retreated altogether from reproduction between 1960 and 1974.[33]

Compared with the Northeast, repopulating the Southeast presented a significant challenge. Candidates for reintroduction from the North would not do. When wildlife managers hacked Alaskan eaglets in North Carolina in 1987, six of seven died of avian malaria, a mosquito-borne disease to which the southern relatives apparently had an immunity. Even if the big birds could handle the molesting mosquitoes, they weren't going to tolerate the sweltering temperatures down south. The region also didn't have the equivalent of an Alaska, Michigan, or Canada that could donate birds. Florida had the healthiest native population, with more than three hundred couples partaking in domestic rituals by the 1980s. Still, the state ran on growth as much as on a tourist economy. The economic recession of the early 1970s had brought construction to a cadaverous halt. Unfinished buildings stood along roadsides, skeletal wall studs turning gray in the sun. But Florida had seen these lulls before. By the end of the decade, the development onslaught was steamrolling again, with heavy equipment turning under thousands of acres of habitat every day. The future of the eagle population wasn't certain enough for Florida to sacrifice eaglets to other states.

The Southeast required a more creative approach than conventional hacking. The executive director of the Sutton Center in Oklahoma, Steve Sherrod, came up with an idea. A Cornell University–trained ornithologist with extensive experience in peregrine falcon reintroduction, and whose bald pate allied him with the name of the bird he sought to aid, Sherrod proposed combining egg translocation with eaglet-hacking. Eggs would go from the center of sylvan Florida to the edge of the tallgrass prairie in Oklahoma, where at the Sutton Center the eggs would be hatched. Sutton personnel would then transfer the eaglets to hack sites across the Southeast where eagles once had flourished.[34]

Nobody had ever pursued this approach with eagles, and understandably, it had skeptics—US Fish and Wildlife for one, the Florida Game and Freshwater Fish Commission for another. The central question was whether the Florida breeders would produce a fertile second clutch of eggs after researchers took the first from their nests. Despite apprehensions, everyone agreed to give Sherrod's plan a try. Mike Collopy, who was head of wildlife and range science at the University of Florida, and a self-confessed doubter, collected eighteen eggs from Florida eagle nests in December 1984. The eggs successfully hatched at Sutton. At eight weeks old, the unfledged juveniles were relocated to hack sites in Oklahoma and two other states, and after they fledged, they took off to carry on life in those states. The Florida eagles that had unwittingly donated eggs indeed double-clutched, and their new eggs hatched. This exciting development convinced everyone that such "egg-recycling" was possible.

At the start of the next season, Petra Wood, a graduate student who worked with Collopy, conducted a weekly aerial reconnaissance of fifty to sixty nests in north-central Florida. Counting nests in eagle states from the air every January had become a common practice in the previous decade. In the early 1970s, at the urging of Doris Mager, Florida wildlife officials started nest surveys, which had for some years been a part of waterfowl research. Conducting a census from the air was more efficient and revealed nests that were hidden from the ground, especially in areas hard to reach by foot. You couldn't find "every nest out there," said a Florida Game and Fish biologist who flew aerial surveys for decades, but you could establish trends in the population: whether it was growing or shrinking and in what areas.[35]

Petra Wood flew with Florida Game pilots, surveilling the piney backwoods of a six-county area in north-central Florida. Unlike other parts of Florida, not much below the plane had changed since William Bartram had traveled through the area two centuries earlier and complained about the thieving ways of bald eagles. In the 1920s, Marjorie

Kinnan Rawlings made a home in the region at Cross Creek, and set many of her stories and books in the rural hinterland, one of which, *The Yearling*, won the 1939 Pulitzer Prize in fiction. Rawlings described the bald's eyrie ungraciously as a "ragged cluster of sticks in a tall tree." Yet of the hamlet of Cross Creek, she said, "There is no magic here except the eagles."[36]

Ace fliers, the pilots would circle one of those ragged clusters of sticks while keeping the nose of the Cessna pointed on it. Wood, trying to keep her last meal down, would look for eggs and the practicality of access for a ground crew, which would include a climber and a small contingent from the Sutton Center. The tree jock would strap on climbing spurs and ascend each host tree. The vibration referring up the trunk from his spurs daggering into the bark and sapwood was usually enough of a fright to flush the sitting parent. Upon reaching the nest, the climber would put on a surgical mask and gloves to prevent contamination and carefully place the eggs in a padded cylinder cinched to a long rope. "Eagles pick the most majestic views in the world," one intrepid tree jock told a writer for *Life* magazine. "Up there, surrounded by the wind, a sense of awe comes over me. I feel like I should recite some sacred incantation."[37]

Someone might have been reciting one for getting the eggs to Oklahoma. Transporting them nearly thirteen hundred miles was a risky undertaking. The Sutton team had come down in a motor home and driven the twenty-plus-hour return trip straight through. Rotating duties, one person would drive, another would sleep, and the third would sit with a pillow on their lap with the incubator on top. The eggs were kept at 99.5 degrees Fahrenheit, with no more than a half-degree deviation. Every three hours an alarm would ding to remind the team to turn the eggs, as do the parents, to keep the embryos from sticking to the inside of the shell.[38]

When eggs arrived at Sutton, those with thin shells stayed in an artificial incubator. Others were placed underneath Cochin hens, large chickens with springy feathers that fall down over their toes like downy boot gaiters. They are accomplished sitters. Still, a shell would occasionally crack, and the staff would rush in like all the king's horses and all the king's men with superglue and mend the egg. Famous for sticking fingers together, if nothing else, the product worked in this rare application, and the repaired egg was moved to an artificial incubator. Whether occurring under a chicken or in a hatching machine, incubation ran approximately thirty-six days, and hatching took approximately thirty-six hours. As with other birds and many reptiles, eagles have an egg tooth they eventually lose after using it to chip, scrape, lever, and squirm their way out of an egg. Once free, each hatchling was moved to its own black rubber tub qua nest lined with straw and warmed by a heating pad.

Starting out at three to four ounces, chicks reached about five pounds in four weeks. As at Quabbin, humans screened themselves from the eaglets and fed them using a hand puppet with a latex eagle head and black sleeve. The eaglets ate well, sumptuously treated to a mixed diet of Japanese quail, rabbits, lab rats, and roadkill deer. In their eight-week stay at Sutton, each consumed the equivalent of eight hundred quail. On most days, staff would put on dark cotton work gloves and a "ghillie suit" (essentially a bedsheet-size camouflaged mesh thrown over the head that gave them the look of a Halloween ghost haunting a combat zone) and carry each tub and occupant out to what was known as the hardening yard. There the eaglets became accustomed to the sights, sounds, and smells of the outdoors. At about eight weeks of age, when their dark-brown juvenile feathers were grown in, the eaglets were moved to hack sites in one or more of five states: Oklahoma, Alabama, Georgia, North Carolina, and Mississippi.

Walter Anderson's Horn Island was one such site. Ospreys and pelicans had made a comeback around the Mississippi Gulf coast by the 1980s, but bald eagles needed succor. The local biologists heading up operations on Horn were with the Gulf Islands National Seashore, a division of the National Park Service that managed Horn and other islands off the coast of Mississippi, Louisiana, Texas, and Florida. The Gulf islands were a busy research zone. Every spring, the park service banded specimens of the millions of song- and shorebirds that cross the Gulf from South and Central America. Wildlife researchers also for a time studied the impact of red wolf predation on Horn Island,

and managed an osprey program that involved banding, census-taking, and controlling predators, such as raccoons, which were pugnacious egg thieves.[39]

For the elevated base of the hack boxes, Ted Simon, who was overseeing the hacking project with another biologist, found a surplus radar tower at nearby Keesler Air Force Base. The seal of the air force headlines a spread bald eagle, and by association the military branch had a stake in the Horn Island operations. The command at Keesler dispatched a Sikorsky Skycrane to airlift the tower and boxes out to the island before the eaglets arrived in early February. It also sent a C-130 Hercules transport aircraft out to Oklahoma to pick them up. Having been given excellent care at Sutton, the eaglets arrived in tip-top shape. Simon wanted to keep them that way during the month they would reside in the hack cages.[40]

After the first season, when four eaglets were hacked, the numbers were increased to twenty or more in the seasons thereafter. The feathered charges were released from mid- to late March, although not all at the same time. If a particular three were especially active and seemingly eager to go, Simon would make the call to open the door to their box. Like the Massachusetts birds, some took off immediately, and some lingered in the cage or on the perch limb outside the door. Some flew like they knew what they were doing, and others like they didn't know what wings were for. One, eagle #60, rose from the perch, wobbled on its wings like a regular tottering out of a pub, continued for about a hundred yards, careened headfirst into a pine tree, and fell to the ground. After a few minutes, it got on its feet and turned and walked the hundred yards back to the hack tower with its head down.[41]

Eventually, #60 flew with some aptitude, but it was an ill-fated bird. Days later, an alligator ate it. The Horn eagles had discovered a freshwater pond. Still trying to figure out how to fish from the air, some of the eagles would wade into the pool up to their neck to catch fish in their beaks as a duck browses duckweed. One day, a ranger happened to see an alligator slide up silently to the side of a wading eagle and use its muscular tail to slap the bird into its mouth, like a kid tossing a Milk Dud into her own.

The program lost three eagles to the alligator that year, 1989, before a trapper came out and removed the eagle's primeval counterpart to another island. The project was down only two birds from the original total, however. Partway into the hacking season, Simon acquired a rescue from Louisiana after a storm toppled its nest. The new bird, #95, was weeks younger than the rest, and the last to leave the tower. But it was a precocious one. It had not been flying long when one day it swooped down and pulled a fish out of the Gulf. The target was likely a dead one floating on the surface, but for a fledgling the maneuver was impressive. On another occasion, #95 forced an osprey to drop its catch—the kind of sight that the artist Walter Anderson had witnessed out on Horn many times decades before.[42]

The fledglings learned to be scrappy early on. They had to compete with each other, as well as second- and third-year juveniles that returned in the spring to the memory of free fish in the hack boxes. The older juveniles subjected the freshman eagles to a form of avian hazing, which amounted mainly to stealing their rations and dominating the territory. There were also fifty active osprey nests on the island that were like a silver-laden Spanish galleon was to freebooters. The ospreys kept an aerial river of fish flowing between water and nests, exposing themselves to the cavalcade of sky pirates.

When the birds started panting more than usual, the weather was turning uncomfortably hot and they would soon leave. In late April and through the month of May, their numbers steadily decreased. Most took off for the Great Lakes region, some winging on to Canada, flying and scavenging during the day and sleeping in a quiet branch at night. One ended up on Long Island, where it learned to take handouts from sportfishers. Between 1985 and 1992, Horn and another coastal hack site on the property of a nearby DuPont plant graduated ninety-eight eaglets, all of them hatched at Sutton, all of them from Florida eggs. The survival rate before release was one hundred percent.[43]

IT WAS ON AN island, one-thirty-sixth the size of Horn, at Quabbin Reservoir where the Massachusetts hacking program first revealed that it might bear fruit. In 1987, a few building sticks showed up in the highest crotch of a red oak. With an expansive view of the water, the oak was prime real estate for piscatorial raptors. But were they ospreys or eagles? And would they come back the next season to finish the nest? They did, and they were eagles. Although they finished the nest, no eggs were laid. Eggs were the crucible for the success of the hacking program, which ended that season. When the Quabbin staff and volunteers dismantled the hack tower, the most pressing question on everyone's anxious mind was whether there would be chicks the next season. There would be—three total in two nests.

One of the parties responsible for the new chicks was Ross—of Betsy and Ross. This bird that had been so eager to squeeze through the bars and escape from his cage back in 1982 turned out to be, as MassWildlife's Bill Davis put it, "a homebody." Every year after his release, Ross apparently strayed no farther than southern Massachusetts in one direction and the Vermont border in the other. His stepsister, Betsy, the more hesitant of the two, and initially the less skilled flyer, had not been seen by the Quabbin staff since 1982, and last near Ottawa, Canada, where Jack Swedberg had tracked her by plane. That's where he lost radio contact with her transmitter. Some wondered whether she had been older than expected when brought to Massachusetts and had imprinted on her natal environment in Michigan's Upper Peninsula. Perhaps she ended up back there and eventually found a mate and built a nest.[44]

The mate Ross paired with was Marjorie, Attila's tormented punching bag from Dianne Lefrancois's cohort. At the southern end of the reservoir, Woodstock and a 1984 Quabbin bird built a second nest. They had the honor of hatching the first eaglet in Massachusetts in nearly a century. When a tree climber lowered the chick to the ground for a medical checkup and banding, Jack Swedberg was there wearing a sweeping smile. He took the chick in his arms and called it his "grandchild." Ross and his mate brooded two eaglets, and they returned season after season. In 1995, they had triplets. In the next years, more Quabbin eagles came back to nest. The reservoir was turning into a wintering mecca by the 1990s. On one particular day, Bill Davis counted forty-eight balds perching, hunting, scavenging, and quarreling.[45]

The Quabbin birds ended up scattered across the state and region. They showed themselves to be true waterbirds. They nested on lakes, reservoirs, and rivers—the Connecticut, Housatonic, Farmington, and Merrimack, to name a few. In 2000, MassWildlife personnel counted thirty-two pairs of nesting bald eagles in the state. Beginning with the first three Quabbin chicks of 1989, they had banded 376 eaglets. Down south, the Sutton Center ultimately sent 275 birds to its five partner states. Eagles were nesting where they had not been seen since Old Abe's day and before. Restoration, said an assistant at Horn Island, was scientific "problem-solving at its best." It was "organic" too, he added, which is to say that the American people were rediscovering cultural roots in the land.[46]

Which is also to say that the bald eagle, that opportunistic bird, saw in revivified environments the opportunity to regenerate and took it.

NINE

◆ ——— *Bird on Top* ——— ◆

ON A GRAY FALL MORNING EARLY IN THE TWENTY-FIRST
century, on Skagit Bay in northwestern Washington State, a bald eagle
sat atop a tall, weathered piling—watching. Its white head pivoted,
and its eyes blinked when, one hundred yards away, a man approached
the bay's edge carrying a Winchester pump-action 12-gauge shotgun.
The man hunkered down in the eelgrass along the soggy tidal flats. He
was wearing a jacket, pants, and hat the color of the gray-green and
amber surroundings, and almost disappeared into them. But the eagle
could see him as easily as it could see a loon on the open water.

The predator bird's ancestors had learned to be wary of someone carrying a gun and to fly quickly from the scene. But the eagle at Skagit
wasn't fleeing; it was anticipating. A person like the one on that fall
morning was reason to stay put. The man's name was Paul Armbrust,
which made no difference to the eagle. What made a difference was
that Armbrust was ensconced in the eelgrass, waiting to shoot a duck.
The eagle sensed what might follow his presence and remained atten-
tively positioned on the piling.

Armbrust was alone and without a retriever, which meant he would
have to slosh out into the water to fetch the duck if he scored a hit.
He was good with that. He'd been hunting and fishing since he was
a kid. Getting and staying wet, shivering in the raw cold, was part of
the adventure. He would score a hit too. Rarely did he return home
without dinner in his bag.

When his first duck fell, it landed in the water as expected. Then,
before Armbrust could put aside the 12-gauge and stand up, the eagle

swept down, lowered its legs with toes and talons open, clamped them around the floating limp muscle of duck, and rose skyward on pumping wings—all in a single, breathless motion.

Armbrust stared in disbelief as he watched his evening's roast duck with cherry-rosemary sauce being flown away. He swore half under his breath, the kind of "son of a bitch" swearing that reflected both annoyance and awe. There was the theft, yes, but there was also his witness to what had occurred, seeing close-up the gleaming white head, implacable yellow eyes, and horizon-broad wings; feeling the squall of air that came off the feathered flight during the downward glide and upward thrust; and being present for the feat and the aesthetic beauty of motion.

Before all this happened, Armbrust had noticed the eagle over on its piling when he first settled in the eelgrass. He momentarily admired the bird, its beautiful white tail feathers draping down below the piling's rim. He even smiled, but then thought nothing more of the eagle. Letting it slip from his consciousness, he later realized, had been a mistake, for Armbrust had never left the eagle's.

If balds routinely spend time around a body of water—and Skagit Bay had phalanxes of them—there is surely a perch somewhere along it, most often a tree with bare branches and an open view of the territory. An eagle sitting in such a tree in the morning or evening is likely hunting, as Armbrust was hunting from the grass. The man had his gun; the bird had its wits, magnified vision, and bullet speed.

There beside the bay, the piling had become the prime perch of the day. Sometime after the duck abduction, Armbrust saw the feathered hunter return and take up its post again. But maybe this wasn't the same eagle as before. Maybe another opportunist had come in the absence of the first. Would eagles start lining up for a fast and easy meal, like Armbrust might later have to do at a restaurant drive-through if he ended the day empty-handed?

Eyeing the bird, Armbrust swore to himself that he would not be outsmarted by an eagle a second time; he would reach the next duck before his savvy opponent did. The eagle was watching him too, seeing the man's mindful glances toward the piling, the muscle twitching in his jaw, the determination in his squinting eyes and puckered lips. Armbrust welcomed the contest. The eyes of man and bird were on the same prize.

Before the next duck hit the water, Armbrust was up and moving. So was the eagle. Splashing through the eelgrass and muck, Armbrust was prepared to dive for the downed prey if he had to, like a base runner sliding on his belly and reaching for home plate. The bird was fast, but the man bested it. Armbrust got the prize and crowed his victory. He had dinner in his hand, and he had a story to tell next time he was tippling an IPA at a backyard barbecue. Maybe he would host one himself, and soon, just so he could tell the tale. Who knows what the eagle was thinking as it banked and returned to its perch. All part of a routine workday in nature.

The eagle's endeavors reminded Armbrust that the world did not exist for his benefit alone. Out on the tidal flats he was in a community of others; he was a lowly outsider among so many insider species. People from his grandfather's generation and before would not have tolerated the brazenness of a duck-snatching eagle; they would have shot it before it swooped down for a free lunch. Armbrust could not imagine doing that. For one, he worked in law enforcement and knew the penalty for killing an eagle. For another, harming America's bird would be like burning the American flag.

Armbrust's restraint was compelled by more than his steadfast patriotism, though. He had been passionate about wildlife for as long as he could remember, and he understood the importance of top predators to the food chain and to the functioning of an ecosystem. When he visited his in-laws at their farm on the other side of the continent, in rural New Hampshire, he trapped weasels, skunks, and raccoons that preyed on the poultry they raised, and did the same with squirrels and woodchucks that raided the garden. He knew he wouldn't be fighting off so many barnyard raiders if raptors were persistently circling the sky or perching in trees.

New Hampshire wasn't like the Northwest Coast, where the population of seaside eagles survived DDT and poaching. To everyone's knowledge, no bald eagle eggs had hatched in New Hampshire between 1949 and 1988. Eventually, as milestone conservation policies and laws from the previous century began producing positive outcomes and as attitudes about birds of prey softened into appreciation, the populations of raptors recovered. Armbrust was witness to their expanding presence in the second decade of the twenty-first century. Their ubiquity started to resemble that which people must have known before the virulent times of the nineteenth century. The difference in a few years was particularly noticeable in the fresh sighting of bald eagles around his in-laws' farm. On early summer mornings when he went out fishing on a nearby lake, he would see one in a perch tree waiting for him to catch fish and throw back undersized ones. Before a released fish could scurry below the surface, it was sky-bound in a set of talons and ferried over a tree line to a nest on a neighboring lake.

A pair of eagles had started the nest around 2007. The female died of unknown causes in 2011 when she was nineteen years old, and the male took up with a new mate. The active nest was located on Nubanusit Lake, and the people in the three towns around it bubbled with excitement as their days increasingly filled with eagle sightings. They talked about them at the post office and general store, and emailed family and friends after spotting one of the adults or a fledged juvenile. A bald in a tree outside a window sent people racing for a camera, and chipmunks for safety in the chinks and cubbies of stone walls. If an eagle was scouting the water while they were kayaking or canoeing, lucky observers fumbled for their smartphones and prayed they wouldn't drop them over the side. A white-headed, white-tailed raptor crossing above outdoor dinners excited crescendos of *Look! Look!! Look!!!*

What Armbrust and the townsfolks didn't know is that the male from Nubanusit Lake had a name assigned to it by the US Fish and Wildlife Service—Gold W-84 ("Gold" was not for "golden")—and he was the oldest known nesting bald eagle in New Hampshire. He had hatched in 1997 in a nest at Quabbin Reservoir in Massachusetts, an offspring of one of the hacking program's alumni.

W-84, his two female mates, and their broods were the embodiment of a twenty-first-century appreciation for the bald eagle and the legacy of restoration. Not least of all, they epitomized the resilience of their species. The new century also, however, signaled new challenges for a burgeoning bird population and a swelling human population trying to live agreeably with one another.

NEW CENTURY, NEW AGE

T HE DAY BEFORE INDEPENDENCE DAY IN 1999 WAS A milestone in the history of the bald eagle's relationship with America. It was also a special day for a twenty-one-year-old named Lavar Simms, who was introducing President Bill Clinton at a ceremony for America's bird on the South Lawn of the White House. A trim five feet, eleven inches, wearing a buzz cut and wire-rimmed glasses, Simms got to shake the president's hand and say a few words to him before the speeches started. Now, to an audience of a couple hundred, he was saying more than a few words, 613 to be exact, delivering them from behind the cobalt-blue presidential lectern, two microphones in front of him, a crop of American flags behind him. On the face of the lectern was the presidential seal with its bald eagle. Simms told those gathered in the bright sunlight that to be there that day was "thrilling," "wonderful," and "exciting." Of the 613 words he spoke, he devoted most of them to talking about his life, a river, and bald eagles in Washington, DC.

Simms had grown up in the projects a few blocks away, where many of his friends got lost in an unfortunate world of drugs and crime. He was wearing an oversize T-shirt plastered on the front with the logo of the Earth Conservation Corps (ECC), a public nonprofit founded in 1989 that engaged at-risk urban youth in environment-related voluntary service, education, and job training. Thanks to the ECC, Simms had eluded the lure of the streets and ended up instead on the Anacostia River, restoring its vitality for people and wildlife.

Reaching eight miles from Maryland to the Potomac River in the heart of DC, the Anacostia was the capital's forgotten river, and a metaphor for its similarly forgotten youth. People knew even less about bald eagles that once lived on the river. The last producing nest, on the Anacostia's Kingman Island, fell silent in 1946 when Truman was president. Raw sewage, urban stormwater, and the Washington Navy Yard, designated a Superfund site in 1998, turned the river into a viscous gray-brown gumbo of floating garbage and toxic substances. Neither eagles nor kids could fish or swim in it. Green crusader Robert Kennedy Jr. called the Anacostia the "deadest" river in the East. "Those birds became our canaries" in a coal mine, Simms proclaimed in his speech, "and we didn't listen."[1]

The ECC and Anacostia Riverkeeper, cofounded by Kennedy, spent years clearing out debris dumped in and around the river and the connecting creeks, among which were five thousand auto and truck tires. As the river began showing life again, Fish and Wildlife in 1994 issued permits to raise and release eagles at the National Arboretum. Calling themselves the "Eagle Corps," ECC kids installed a hack box in an oak beside the river in preparation for four eaglets brought down from northern Wisconsin. The Eagle Corps cared for its charges as would any professional wildlife manager, releasing the first in spring 1995, and four birds each year thereafter through 1998. The kids named several after twenty-six ECC members who had been murdered. One eagle was named "LB" for Leroy Brown, who had been like a brother to Simms. "For eight weeks, I nurtured, fed, and got to know this eaglet," Simms told the audience. "Releasing LB made me think about life and what it means to be endangered. This Corps believes that everyone should have a clean, wonderful, beautiful river for the eagles and people to fish. . . . What I've learned from working with the birds is that we should give life a chance."[2]

Life was given a chance just south of where Simms was speaking. On the other side of the Anacostia and near the Potomac, on the grounds of the old St. Elizabeths Hospital in Ward 8, one of the most dangerous neighborhoods in the city, a pair of balds—Monique and Tink—was building a nest. The nest was in an eighty-foot oak affording a lavish view of the Washington Monument and National

Cathedral. In March of the next year, Monique and Tink hatched the first eagle in DC in fifty-four years. Two other couples eventually built eyries at the Metropolitan Police Academy and the National Arboretum.[3]

"By saving the bald eagle and bringing it back home to the nation's capital," President Clinton maintained, after following Simms at the lectern, "these young people have honored our past." Simms and the others stood on the front lines of the American reconciliation with the founding bird and, by association, the country's natural heritage. A representative of the new era was sitting alertly on a perch next to Clinton. Named Challenger, he was a ten-year-old male rescue that the Tennessee-based American Eagle Foundation had brought to Washington for the ceremony.[4]

The main purpose of the South Lawn event, aside from celebrating the ECC youth, was to announce that the Department of the Interior would soon initiate removal of the bald eagle from the Endangered Species List (ESL). Four years earlier, Fish and Wildlife had upgraded the status of the bald eagle in the forty-eight states outside the region around Nevada and New Mexico from "endangered" to "threatened." "The return of the bald eagle," Clinton declared, "is a fitting cap to a century of environmental stewardship." The previous couple of decades had accomplished much: clean water, revitalized habitat, rejuvenated bird song and flight, and new priorities.[5]

After the ceremony's formalities ended and guests came up to meet the speakers, Clinton rested his left hand on Challenger's perch, and the raptor eyed the wedding band clicking against the wood. Watching film footage of the event, you can see what's coming. The bird of America thrusts its head and beak down and takes a harmless swipe at the hand of the president of the United States. It was as if Challenger were making a point: it's not quite time for delisting.

Biologists at Interior were questioning the timing too, as were some environmentalists. The latter wondered whether the eagerness of the Clinton administration to move a high-profile conservation effort to the win column overshadowed a full consideration of threats that persisted. Public sightings of white-feathered heads and muscular wing beats were a reassuring reflection of a trajectory that had been arcing steadily upward since the 1980s. Interior had recently issued a press release titled "The Bald Eagle Is Back," and the department's secretary, Bruce Babbitt, repeated that phrase during his moment at the lectern on the South Lawn, adding, "The bald eagle joins a growing list of other once imperiled species that are on the road to recovery." Twenty-eight plants and animals, including the peregrine falcon, had come off the list, yet nearly three hundred more species were awaiting inclusion on it. And not all of those listed were turned toward recovery. A number of them had been forced down dead-end roads to extinction.[6]

Yes, the eagle population was healthy, but would it stay so without ESL defense? Security was still afforded by the Migratory Bird Treaty Act and the Bald and Golden Eagle Protection Act. Both said you could not kill, possess, or "take" the safeguarded birds. But conservationists and wildlife researchers wanted to know exactly what, in legal terms, "take" meant. Did that language protect the bald eagle from potentially disruptive activities within a certain range of a nesting tree, as the ESL did? Or would a developer be permitted to cut the trees in a habitat right up to the one bearing an eyrie and then surround it with rooftops, power lines, paved streets, traffic, and general human din? Many who were assisting with restoration wanted more time to understand which shields would remain between pressures from human population expansion and the bald's own impulse to procreate.

They ultimately got that time, thanks in part to sluggish government bureaucracy and the new presidential administration's preoccupation with war in Iraq. But the grace period didn't sit well with everyone. One vocal opponent of delay was Edmund Contoski of Minnesota, who owned seven acres on Sullivan Lake north of Minneapolis that he wanted to subdivide into home lots. The problem was that his property had become an eagledom. A couple had established a nest in a dead, barkless oak handsomely bleached by the sun. Under the rules of endangered species legislation, construction could not encroach within

330 feet of an eagle's nest. Complying with the law severely restricted Contoski's development plans.

Contoski was an inveterate libertarian—indeed, the founder and president of American Liberty Publishers and an active member of a free-market think tank for limited government and personal liberty. He was fond of saying that he, not eagles, paid the taxes on his lakefront parcel, yet the government was impinging on his constitutional freedom to do what he wanted with his property. He complained that he could not "even trim a tree or cut a tree that's not the eagle's tree." His family had owned the property for seventy years, he said, while the bald eagles had been around for a dozen. "We're not Johnny-come-latelies."[7]

Contoski's protests got the attention of the Pacific Legal Foundation, a libertarian public-interest law firm that specialized in defending against government overreach. In 2006, Contoski's lawyers filed suit requesting that the federal government comply with statutory guidelines and follow through with its 1999 determination that the bald eagle was ready for delisting. The court agreed with Contoski and ordered Fish and Wildlife to either remove the bald eagle from ESL protection by June 29, 2007, or show just cause for continued delay.[8]

On the day before the deadline, Challenger, the rescue from the American Eagle Foundation, was back in DC for a second delisting ceremony. The time had finally come to set free the bird of liberty, Interior Secretary Dirk Kempthorne announced from the steps of the Jefferson Memorial. Someone at Interior apparently credited Thomas Jefferson with leading the selection of the bald eagle for the Great Seal. Kempthorne said as much, standing behind the department's red-Formica-and-wood-veneer lectern, Interior's seal attached to the front with, at its center, a bison, the future national mammal. The speech was, nevertheless, a thorough summation of history and progress. Kempthorne even pointed out that the delisting was occurring in the centennial year of the birth of Rachel Carson. Challenger then took a whirling celebratory flight around the audience before returning to his handler's glove.

Environmentalists were generally cool to Kempthorne and the

George W. Bush administration for which he served. During his three-year term as secretary, Kempthorne added no new plants or animals to the ESL. Critics complained that he pursued an agenda to weaken rather than strengthen or sustain wildlife protection and water quality. But environmentalists were with him on the bald eagle's graduation from the ESL. The second coming of the American bird might serve to quiet disciples of deregulation who attacked the ESL as being extreme, too expensive, a detriment to economic expansion, and overly influenced by bleeding-heart animal lovers.

Three years earlier, the Environmental Defense Fund (EDF) began advocating the bald's removal from the ESL. Having come into being in the 1960s specifically to remove DDT from the market, EDF posted a "Back from the Brink" page on the internet that encouraged readers to sign an online petition requesting that President Bush expedite delisting. Although they were recording a depressing decline in the number of game and songbird species across America, the Audubon Society and the American Bird Conservancy were equally enthusiastic about the bald eagle's improved status. Fish and Wildlife estimated the 2007 population of nesting pairs in the Lower Forty-Eight to have reached 9,789—twenty times the number that had existed at the 1963 nadir. "The recovery of the bald eagle," said the EDF president, "is proof positive of the potential of America's conservation and restoration efforts."[9]

The Center for Biological Diversity, founded in 1989 to safeguard plants and animals at risk of their wholesale disappearance, also ran the eagle numbers for 2007. It came up with an even more encouraging result than Fish and Wildlife had: 11,062 pairs. Minnesota posted the highest count, with 1,312 nests, followed by Florida with 1,166 and Wisconsin with 1,065. Fish and Wildlife and its state counterparts continued to monitor the status of the bald eagle population for five years after delisting, and most states then stopped budgeting for annual census surveys.

However promising the numbers were, the bald eagle had not reached a full accord with its sole predator. Many old threats to America's bird were diminishing, and others, like pernicious myths, had disappeared.

Yet some old ones endured, and new ones came along. Awareness and education helped mitigate dangers, although not always.

ILLEGAL PREDATOR CONTROL AND habitat loss were the leading persistent challenges. The penalty for violating the Eagle Protection Act had remained the same since Congress updated the law in 1972. Conviction carried a maximum $100,000 fine and a year in prison. Violate the law a second time, and you could be looking at five years and $250,000 in fines. The monetary penalty for organizations was double that for individuals. Nearly four decades later and even without the backup of the ESL, the Eagle Protection Act remained a formidable deterrent. Yet some people in parts of America, out on farms and ranches, lived by the law of the land, as they put it, and willingly violated federal mandates. Private bounty hunting survived, for example. The preferred attack machine had become the helicopter, and in the absence of aircraft and hired guns, carrying a firearm at the ready in one's truck or car was a way of life in farm country.

The most dangerous skies were over sheep ranches in the West. Eagles didn't carry off sheep, as they had once been accused of doing, but goldens and balds were sometimes guilty of killing lambs on the ground. After a Texas rancher was indicted in 2017 for allegedly killing two bald eagles, he said he had been losing $200,000 a year to predator birds. His attorney observed, "There is the juxtaposition of this national symbol . . . and this man's livelihood. . . . That's in addition to him being branded as a criminal."[10]

Eagles had long been branded criminals and condemned for pursuing the only livelihood they knew—eating, procreating, and surviving—even when they partook of their preferred fare around a newer kind of farm. In Massachusetts, the owner of the Mohawk Trout Hatchery, located twenty miles from Quabbin Reservoir, never installed protective netting over his fish as did other hatcheries, and instead dealt with predator birds the old-fashioned way. Cruising his property in a golf cart with a rifle, he treated the hatchery as a carnival shooting gallery. He was a decent shot, but he didn't walk off with a teddy bear prize. Someone reported him. Officials discovered the carcasses of over two hundred great blue herons, several ospreys, and a bald eagle. The hatchery owner pled guilty in exchange for $65,000 in fines, five years probation, and six months in a halfway house. He also relinquished his right to bear arms.[11]

Some ranchers indirectly killed eagles while trying to control other predators, or used that argument when arrested. Where ranches existed, so did poisons. (In 1970 alone, the US government provided 850,000 poison bait traps for western states.) Eagles weren't typically the target of the poisons, but they became collateral damage of the chemicals' use. In 2020, a federal magistrate ordered a North Dakota bison stockman to pay the maximum penalty under the Eagle Protection Act after workers spread thirty-nine thousand pounds of the pesticide Rozol across his property to kill prairie dogs. Controlling prairie dogs wasn't illegal (their burrowing holes can cripple a bison), but workers had been careless in following the pesticide's application guidelines, and six unlucky eagles ultimately got caught up in the prairie dog roundup and died.[12]

The greatest collateral fallout, and a troubling one, came from lead poisoning associated with game-hunting. Hunters commonly gutted their kill in the woods, believing they were contributing to the food chain by leaving something for scavengers. What they also left, unwittingly, was fragments of lead from their bullets or shot, and virtually all scavengers were poisoned by it. Studies released in the late 2010s revealed that one in five California condors had to be treated for lead toxicosis, and the same malady accounted for twenty-five percent of all eagle deaths, which spiked during hunting season. Lead is a neurotoxin, and a fragment the size of a grain of rice can kill a full-grown eagle. Symptoms are obvious even before clinicians run tests: The adult more or less regresses to a young nestling, although without the nestling's energy. Lethargy is significant. The head hangs to one side. The eyes have turned dull. The talons are clenched shut, and the bird hobbles on its haunches. Juveniles are disproportionately affected because, as unskilled hunters, they rely more on carrion.[13]

The hunter's family and friends who share in the meat from the kill absorb lead too. After lead toxicity became a growing concern in the twenty-first century, the National Rifle Association, with its eagle logo, fought proposals to abolish lead bullets in game-hunting. Abolition smacked of gun regulation to the organization—a slippery slope that would, opponents believed, result in more restrictions on gun ownership and use.

Yet there was precedent for eliminating lead in the outdoor sport without affecting gun rights when, in 1991, Fish and Wildlife had prohibited its use in waterfowl-hunting. Ducks and geese and the like were dying after consuming lead shot that sank to the bottom of lakes and ponds. Losing their game birds, sportsmen petitioned for the ban, and after it was imposed, they converted to steel shot. Following that precedent, and compelled to reduce the domino effect on eagles, President Barack Obama's administration instituted a phaseout of the use of all lead ammunition and fishing tackle by 2022 in national wildlife refuges. But then, on his first day on the job, President Donald Trump's secretary of the interior, Ryan Zinke, repealed the ban, declaring that it was unfair to hunters who could not afford to convert to copper bullets, which was the best alternative to lead and thirty percent more expensive. Taking care of the land had long been a tradition among sportsmen, and, despite the cost of copper and Zinke's ruling, many thought that removing lead from the environment represented an act of good stewardship, for wildlife and people alike. California legislators thought so when in 2019 they instituted the country's first statewide prohibition of lead in hunting ammunition.[14]

Another lingering offense originating in the past—DDT—was not altogether a surprise, but the magnitude of its impact sometimes was. This was most true in California, a former DDT hot spot that remained a hot spot. Early in the twenty-first century, eagle counters could identify only two hundred breeding pairs in the country's third-largest state. That was a significantly low number, compared with the 1,305 that nested in California's Pacific coast neighbors, Oregon and Washington, where their shorelines together equaled only sixty percent of California's. And neither of those two states had been aided by restoration programs. Nest numbers in California had fallen to zero in the

1960s, and it had been the beneficiary of over twenty years of hacking efforts, longer than any other state.

The grave conditions had a lot to do with the Montrose Chemical Corporation, the manufacturer that had dumped some eighteen hundred tons of DDT-laced chemical refuse into the Los Angeles wastewater system. In 2000, after filing a suit under the federal Superfund law and spending ten years in court, the US government won a $145 million natural-resource-damage assessment from Montrose, the Los Angeles County Sanitation Districts, and other polluting corporations. DDT, as well as PCBs, lingering on the ocean floor in southern California limited the success of the hacking initiatives. Both chemicals can take as long as a century to break down, and when eagles started nesting again, many of them laid thin-shelled eggs as if they were living in the 1950s or '60s.[15]

Montrose, it turned out, had employed more than the LA stormwater and sewer system for disposing of its private-property waste. From 1947 to 1961, the chemical manufacturer contracted with the California Salvage Company to barge fifty-five-gallon steel drums of DDT-impaired refuse ten nautical miles out to a designated dump site near Santa Catalina Island, which was surrounded by kelp forests and visited by pods of gray whales. The crew axed holes in the drums and rolled them over the side. "Deep-sixing," the practice is called. Montrose and California Salvage were deep-sixing more than two thousand drums a month. Remarkably, they were breaking no laws, engaged in a practice that was common worldwide, following notions of dilution being the solution to pollution—and out of sight being out of mind.

The latter was pretty much the case. Hardly anyone outside the companies knew about the drums. A few scientists in the 1980s found corporate records alluding to them, but their disposal never came up in the court proceedings associated with the sanitation system dumping. Evidence remained tucked in file cabinet drawers. Not until about 2010 did researchers confirm the underwater refuse. Using sonar scans and deepwater robots, they discovered a wasteland of fifty-five-gallon drums, perforated, rusting, and leaking in the fashion of slow-release fertilizer seeping persistently into the environment. The estimated number of drums was a half million.[16]

Bald eagles nevertheless insisted on nesting on Catalina Island. Who could blame them? The fishing was good, even if the fish weren't. To help the eagles out, wildlife managers began pulling eggs from nests before they broke, hatching them in incubators, and returning the chicks to their parents. They continued to lend assistance for nearly twenty years, until 2009. By then, DDT levels were down enough for eagles to carry on unaided. California's breeding population showed a steady rise. A 2016 census put the number of occupied nests at 375. Two years later, Catalina Island reported seven active eagle homesteads.

Just as the eagles had reasserted themselves across the land, the human population had nearly doubled since 1950.* When restoration efforts began in the 1970s, no one was quite certain whether the no-nonsense bird would tolerate an increasingly tight squeeze with humans.

Fifty years earlier, Francis Herrick had himself been uncertain until he mounted his observation platform in the Vermilion, Ohio, eagledom. He had imagined his subjects "essentially as they must have existed in the days of the Indian or before," which is to say, overseeing a dominion of nature. Then, when he stood atop his platform for the first time and looked out from the level of the monarchs' high perch, he was "amazed at the remarkable contrasts" in the avian reality of the modern age. The eagles' view took in steamboats and low-riding freighters on the lake, the train chuffing along its fine edge, power lines that paralleled the burnished rails, roads where automobiles churned and bellowed, chimneys of the harbor city of Lorain's "great steel-works"—the "smoke and gases of which often reached us"—mail planes that groaned across the sky, and a whalelike dirigible that once navigated sightseers over the big nest. Many of Herrick's photographs of the Great Eyrie reveal the neighboring farmer's furrowed cornfield carpeting the panorama to which the eagles awoke in the mornings. Still, the birds persisted, wary of men who erected observation towers

---

* Nevada and New Mexico remained the exception. Neither experienced much growth in their nesting populations—an anomaly that scientists could not explain.

but tolerant of familiar sights, sounds, and smells. "Where could one find," Herrick wrote, "evidence of greater adaptability than this among the diurnal birds of prey?"[17]

Charles Broley made the same observation in Florida a few decades later. If their trees were left undisturbed, nesting couples went about their usual business. Broley found a nest on a golf course, one behind a high school, one in a "fashionable suburb," and another on the campus of a veterans' hospital. The most astonishing attachment to place was a nest on the perimeter of an army airfield with a practice bombing range. As sand-filled sacks (the bombs) rained down from roaring airplanes overhead, the occupying couple continued with its domestic routine.[18]

To test the question in the restoration years of whether eagles would live around people, researchers went to coastal Washington's San Juan Island. Over the course of a decade starting in the late 1960s, they observed eagles building new nests as people built new homes. The total number of nests might have been greater if not for people taking up space; still, bulldozers and hammer blows did not prevent a net increase in eyries. With eagles essentially living on top of people—in other words, doing something scientists once thought they would not do, like coyotes and raccoons did—wildlife officials had to become more vigilant in preventing conflict.[19]

The more crucial question was whether Americans were willing to accept living near eagles. Confrontations between people and nests were most frequent in states where the natural habitat had reached its capacity to accommodate breeding pairs, forcing the birds into areas where people worked, lived, and played. Long a developer's paradise, and longer an eagle's, Florida became center stage for conflict. As early as 1980, the year after her nest-in in Central Florida, Doris Mager thought about the inevitable changes ahead—the wetlands drained, roads cut, and buildings built*—and became depressed. "We know we're going to lose nests, no matter how hard we fight," she lamented. "I just don't know what the devil is going to happen to those birds."

---

* Florida filled and drained nearly half of its inland wetlands in the second half of the twentieth century.

What happened to one particular bird species seven years later—golf courses, tourist venues, shopping meccas, and waterfront condominiums destroyed the pitiful remains of the endangered dusky seaside sparrow and turned it into a ghost of nature—had the potential of happening to the hallmark species.[20]

During the forty years after Mager voiced her concerns, the state's residential population doubled, and by 2020, eighty million people were visiting each year. The number of eagle couples by then had peaked at fifteen hundred, and Florida had little wild space left for them. An eagledom requires a range of approximately thirty-six hundred acres, the equivalent of ten to fifteen average-size housing developments. Cramped quarters were most common in the built-up coastal regions. In 2005, developers in separate cases in Collier and Sarasota Counties, both on Gulf shores, pleaded guilty to illegally removing eagle trees to make way for their artificial environments. The defendant in the Collier case, Stock Development Company, paid a $356,000 penalty, the largest the bald eagle had ever elicited.[21]

Two years later, after Edmund Contoski's own court victory, the returning bald eagle came off the ESL. When it did, it lost the required 330-foot "buffer distance" between human activity and nesting activity—a provision that the Bald and Golden Eagle Protection Act lacked. Recognizing the buffer's effectiveness in inhibiting developer encroachment, Fish and Wildlife transferred the rule to the Eagle Protection Act—an action that frustrated Contoski and a considerable number of developers. Despite delisting, nothing had changed for him or others who wanted to remove a nesting tree or build near one. Contoski died in 2020, never able to develop his property in the manner he had wished.[22]

Prosecuting cases required evidence of wrongdoing, though, and for lack of evidence many eagle nests went missing. In the 2010s, those on a planned 522-acre housing development on Sarasota Bay in Bradenton, Florida, kept disappearing—one mysteriously, one not so mysteriously, and one by way of what many observers believed was political favoritism. In 2013, a horrified boater reported seeing a bright-red helicopter drop down over the property and use its rotor blades to blow apart an eyrie stick by stick. Two years later, a county commissioner discovered a second nesting tree that had been cut down under the cover of night. A year later, a third nest in a tree vanished. Someone found an air rifle left behind at the base of the tree. The eagles that had been using the nest returned in January and rebuilt their home.

In the meantime, the principals of the development applied for a federal permit to take the nest tree. Fish and Wildlife at the time was under pressure from the Trump administration to ease up on restrictions imposed against businesses and in March 2018 approved the application. Language in the permit read with uncompelling circular logic. The "pine tree containing the eagle nest," it instructed, "must be removed as an avoidance measure to prevent future bald eagle nesting attempts at this location." Local activists trying to save the nest were dumbstruck. Forcing eagles to abandon their habitat was not a legitimate way of protecting them. A last resort for raptor advocates was a caveat in the permit that "removal activity" should stop if the nest proved to be active. Pictures of eagles in the area were produced, but the tree came down in May.[23]

Local governments weren't entirely innocent either. When a pair of bald eagles flew over Ed Smith Stadium in Sarasota, they saw an opportunity to build a nest on the skeletal lighting structure above right field. It happened that Ed Smith was the spring-training home of the Baltimore Orioles, and the visiting eagle team wasn't welcome. Preparing to launch a $31 million renovation project at the publicly owned sports facility, the county applied for a take permit in 2010, arguing that the electrical lights endangered the eagles. If someone had looked around, they would have learned that ospreys, their eyries near duplicates of eagles', had been nesting atop stadium lights around the country for years. Other ballparks welcomed ospreys by building nest platforms on light towers. Fans loved the birds and their broods, but Sarasota officials were no ballpark raptor fans. The county got its permit and hired a crew to evict the eagles and their nest. Their season snatched from them, the birds were forced to forfeit two eggs. Both went to the American Eagle Foundation, Challenger's home in Tennessee. One hatched.

The next season the eagles staged a comeback and built a nest on the upper deck of a cell phone tower across the street. Everyone seemed fine with the eagles' new home turf. Then, a few years later, in 2014, the nest vanished. Its disappearance looked like the foul play of maintenance workers, but investigators wrote the incident off to natural causes. When the eagles returned to an empty tower two months later, they moved back to Ed Smith Stadium and started building on the left-field lights. The county again acquired a permit to eject the eagles and sent out workers to dismantle the nest. The eagles circled the field from above, watching everything they had built being destroyed and thrown in a dumpster, and then they left for good.

⌒

FIVE YEARS LATER, SOMEONE took down an eagle's nest from a cell tower across the bay in Tampa. The willingness of the once maligned raptors to nest in the built environment opened them both to new dangers and to being labeled a nuisance—a worrisome echo from the nineteenth century.

Few displays of behavior prompted the "nuisance" aspersion more than a congregation of these avian scavengers at the local dump. For many people, the sight of bald eagles atop heaps of refuse, ferreting out rotting food from beneath other rotting waste was difficult to stomach. Crows, gulls, and vultures—that was fine. They were doing what they had always been known to do. But for America's bird, the proud stalwart species, scrounging around in America's garbage was beneath its dignity. At Dutch Harbor, locals who weren't bothered by eagles on the fishing docks weren't keen on having them on the other side of the bay convening in dozens at the town dump. There was a particular enticement out there for the feathered piscators: fish guts and heads from the local processing plants. And there was an enticement for others too. Visiting the dump to see eagles became a quirk in the tourist trade, which had traditionally centered around hiking, sportfishing, scenic vistas, and wildlife in natural habitats, balds at the docks notwithstanding.

Down in King County in the state of Washington, solid-waste personnel were dealing with a couple hundred bald eagles fishing around the county's landfill for, as one local put it, a "free lunch." Washington had one of the largest nesting populations in the lower forty-eight states, and some people were coming to resent the comeback birds' presence. Among the complaining were residents of the Seattle suburbs who sent their rubbish to the landfill, many of whom were finding the same returned to them by passing birds dropping it in their neighborhoods. At a public meeting, one man held up a ziplock bag with, inside it, a biohazard container filled with blood, presumably human blood. Bald eagles were blamed (although not the person responsible for the improper disposal of a biohazard), and the suburbanites wanted the "aerial bombardments" to stop.[24]

Pyrotechnics and noise cannons, also known as "bird bangers," to scare but not harm the eagles were a remediation option, but their deployment subjected residents to booms and blasts. Some landfills elsewhere had found success in using drones to chase away other kinds of birds. But drones didn't work with birds of prey. When small, non-military commercial and recreational drones became popular after 2010, operators were chagrined to learn that eagles attacked their whizzing, multipropeller, camera-wielding flying widgets as if they were predators. Among drone aficionados, eagles became known as "Angry Birds," borrowed from the popular video game of the same name. Goldens were so good at such aerial assaults that the French military trained them to take down terrorist drones. In Manitoba, the Royal Canadian Mounted Police employed a drone fighter named Eddie, a dashing—in more ways than one—bald eagle, to keep airspaces secure from bird-size aircraft. In the summer of 2020, a bald eagle strafed a drone over Lake Michigan. The thousand-dollar air robot belonged to the Michigan Department of Environment, Great Lakes, and Energy, a conservation agency that, ironically, went by the acronym EGLE.[25]

Just as landfills and eagles together were a new realization with the amassing human population and the rebounding eagle population, so, too, were the appreciably visible deaths by power line electrocution and automobile strikes. Near Jacksonville, Florida, in 2016, an adult bald flew in front of a moving Saturn and found itself stuck in the car's grille. The bird was still alive, and after dislodging it, rescuers took it to a rehabilitation center, where it recovered. In Henrietta, New York, the year before, another of its kind was not so lucky when it collided with a car. It had been banded in a nest near Bemidji, Minnesota, thirty-eight years earlier, making it the oldest-known bald eagle living in the wild.

Collisions of this sort put drivers at risk too. An eagle in flight is capable of crashing through a windshield, not unlike a deer or moose, and such meetings between human and wildlife are not infrequently fatal. Airplane pilots know of such encounters firsthand. A bird the size of a bald eagle traveling thirty miles an hour that collides with a plane moving at two to three hundred miles per hour or more becomes, as one safety expert said, a flying elephant. It will bend a plane's propeller, pierce the aircraft's nose cowling, and smash through windshields. Planes went down. In 2010, a Seattle-bound Alaska Airlines Boeing 737 had to abort on takeoff as its wheels were lifting off the ground after a bald eagle flew into the violence of its left engine, which exploded into flames. None of the 134 passengers and 5 crew members were injured, and later that afternoon they boarded a new plane to resume their travels. Jetting down the runway, the second plane also hit an eagle. The impact was a glancing blow, and the plane continued to Seattle. The eagle lay behind, dead, beside the runway.[26]

General bird strikes were becoming more common as flight travel increased and new technologies made jet engines quieter. Between 1990 and 2020, ill-fated balds accounted for less than 1.5 percent of the 250,590 wildlife strikes. That number seems minuscule, but bald eagle strikes in that same period increased by twenty-two hundred percent. There were no reported direct human fatalities, but the average collision repair cost ran approximately $400,000.[27]

Even outside airports, flight paths, and gunfire, the flagship bird's airspace wasn't entirely safe, particularly around wind farms. Here was the epitome of the environmentalist's dilemma: green energy rubbing against wildlife preservation. The 110-foot-long, six-ton turbine blades took down as many as a half million birds a year. The instances of striking and killing a bald eagle were comparably small, about a hundred cases annually. That number didn't stop President Trump, who was an outspoken apologist for polluting energies such as oil, coal, and gas, from saying in a speech that wind generators chopped up untold numbers of America's bird: "You go under a windmill; you see them all over the place. . . . Not a good situation."[28]

One could argue that the bad situation started under the Obama administration. In December 2016, Fish and Wildlife announced that it would issue thirty-year permits, for a fee of $36,000 each, that allowed wind farms to kill up to forty-two hundred bald eagles. And as long as power companies implemented measures to reduce chances of electrocution, they could kill an unlimited number of goldens with turbine blades. So, if you stopped electrocuting them over here, you could slash them over there. The Trump administration, which took office within days of the new policy going into effect, did not eliminate the $36,000 kill permits and the death one might see at the base of a wind generator.

The industry was, nevertheless, taking mitigation measures to prevent fatalities. Experts recognized that, in lacking natural predators, eagles were on the lookout for terrestrial fare, not aerial dangers. "They're like teenagers on their cell phone," said Tim Hayes, an industry specialist and veteran of Indiana's hacking initiative of the 1980s. "They're distracted." Some of the mitigation measures included building facilities outside of flyways and away from large concentrations of birds, slowing turbine blades when birds are present, and employing monitoring technology to track flight patterns. At the time of Trump's speech, a high-tech company was testing smart-camera technology that stopped the blades when an eagle approached. At a 110-turbine wind site in Wyoming, Hayes and Duke Energy were experimenting with the technology and had reduced eagle deaths by eighty-two percent.[29]

Some of these actions seemed promising, but the Trump administration was setting up a different legacy. Days before vacating the department's offices for the incoming Biden administration in January 2021, leadership at Interior eviscerated the hallowed Migratory Bird Treaty Act, at Canada's opposition. New rules let polluting industries off the hook for the so-called "incidental" killing of birds. The Biden team moved fairly rapidly to restore the old protections, yet the controversial wind-site kill permits remained in place (as did the practice of hunting with lead ammunition in national wildlife refuges).

The permits hearkened back to the old Alaskan bounties: death for industry. The ghostly cries of nuisance wildlife out where sheep grazed and people piled their refuse added yet another layer to the prospect that recalibrating the federal management of *Haliaeetus leucocephalus* lay in the bird's future: management that would be weighted not toward restoration but toward population control. Control translated into, as with wolves and bears, culling the convocation. The comeback generations were potentially facing penalties for making restoration a supreme triumph, and staring into not only the future but also the past.

❧

ANTICIPATING CLASHES BETWEEN TWO populations in which the feathered one was the likely loser, and wanting to minimize their occurrences, Doris Mager climbed on a bicycle and rode across the country. A component of her raptor rehabilitation effort at Audubon was giving educational talks at schools and to civic groups around Florida. She was the first person outside of government permitted by Fish and Wildlife to transport an eagle across state lines, and as early as 1981 she steered her outreach down interstates, taking with her an owl named El Tigre and a bald eagle named General Patton. Someone donated a Chevy van to her effort, and she and her two companions hit the blacktop, driving east and west and north and south, the van reeking of dead-rodent food and the birds' pooping and casting. (Like cats with hair balls, raptors regurgitate, or cast, undigested bones, feathers, and fur.) On hundreds of thousands of miles of roadway, people saw a silver-haired woman full of passion and purpose at the wheel of a white van with a bald eagle painted on the rear side panels. Her ride, she said, ran "like a frightened bird." The two unfrightened ones inside joined her for talks at any place that would host them.

Five years after that, Mager was pedaling. She was sixty and on a bicycle humping it from Ventura, California, to Neptune Beach, Florida, following the old Route 66 corridor for much of the first half.

Accompanied by rotating companion riders and sponsored by the discount retail giant Kmart, Mager gave "Save the Eagles" lectures in the parking lots of its stores on her route. In a support van joining the venture were a crested caracara called Cara; a horned owl that went by E.T., for "Extra Terrific"; and a bald eagle named George Washington (three years earlier, it had been hit by a car on President Washington's birthday). Near the end of her 2,850-mile tour, Mager's arms, legs, and face deeply tanned, she wheeled into the Florida state capital. On the 204th anniversary of the Great Seal of the United States, Governor Bob Graham presented her with a proclamation declaring June 20–27 Doris Mager Week. "When you try to raise money for these," she told the press while pointing to George Washington, "you have to do some strange things." In the fall, she was back in the van with El Tigre and General Patton driving six thousand miles to give lectures at schools from Florida to Pennsylvania to Indiana and back.[30]

Mager put all those miles on the road to raise awareness and instill understanding and appreciation, emphasizing the beauty and ecological significance of raptors. Without them, she said, "we'd be overrun with rats, mice," bugs, and, "as much as we like them, bunnies." What troubled her as much as anything was the senseless shooting of birds, and she knew that kids excited to use their first gun, whether a BB or pellet gun or a .22 rifle, often took aim at birds to have a target less boring than an unmoving tree, street sign, soda can, or bull's-eye on a sheet of plywood.[31]

At a school in rural Live Oak, Florida, Mager talked to some fifteen hundred kids and asked how many of them had shot a bird. Nearly half "waved their hands," she said. "They were proud of it. Nobody had told them it was wrong." *She* told them, though. She was like Atticus Finch declaring it a sin to kill a mockingbird. She turned the audience's attention to General Patton. He was sturdy and dignified, his feathers prim, his white head distinguishing him as America's bird. The kids agreed that they would never dream of shooting this resplendent creature with a mesmerizing spirit deep in his eyes, a creature with his own awareness and curiosity. But General Patton had once been shot. An unprincipled hunter took half a wing off this honorable bird. Before Mager finished her talk, she asked how many of the kids would stop shooting birds. The majority raised their hand.[32]

People like Mager were giving lectures around the country, teaching others about birds of prey. They talked mainly to a generation that was coming of age with the return of eagles and other raptors, hoping that people would learn to appreciate a wild sky with a camera or binocular and not a gun. "Billy and I felt strongly," remembered Dianne Davis (née Lefrancois), who had married Bill Davis and gave public lectures around Massachusetts with him, "that since the eagles were coming back to Massachusetts it was important to educate children about them." These were kids who one day would decide where to build roads and houses and whether it was important to make room for wildlife and to support policy that protected habitat and clean water.[33]

Mager eventually moved to North Carolina, and she continued to teach people young and old. When the frightened-bird van gave out, she got a new one. When that one gave out, she got a third, and she kept driving like Forrest Gump kept running—except *she* never stopped. She went another two decades, into the next century. She made a cross-country drive in 2019 when she was ninety-three. Just she and E.T., the great horned owl, whom she was donating to a raptor center in Connecticut.

By the early twenty-first century, every state on the continent had at least one privately funded facility, though typically more, dedicated to raptor care and rehabilitation. Many of these places combined their rehab endeavors with education programs, with the goal of reducing the number of injured birds.

Injured raptors often came into rehabilitation centers resembling death warmed over. They could be the most pathetic-looking lumps of feathers and still regain good health. Those that could be released were taken out to a clearing and triumphantly let go. Those with permanent injuries and rendered flightless were kept in captivity for the rest of their lives. They got plenty of food and the best of medical attention, and they had the companionship of other birds and humans if they desired. Many could live into their late thirties and early forties. In July 2019, cupcakes were broken out for staff and meat-eater treats for birds to celebrate the birthday of Mrs. B., a bald eagle that had lived at the Great Bend, Kansas, zoo for nearly four decades.

Under the Eagle Protection Act, the Department of the Interior was authorized to issue federal permits to zoos, museums, scientific societies, raptor centers, and qualified individuals to possess eagles. Museums, scientific institutions, and Native groups could also possess feathers and eagle body parts. Flightless eagles that interacted well with handlers and tolerated being the center of attention among a crowd of people, as had Old Abe sans Minié-ball fire, were often used for public programs. The venerable General Patton and George Washington were just a couple of the numerous eagles educating audiences.

Probably the busiest captive raptor and the most famous bald eagle in the twenty-first century was Challenger, who had been the avian star at the two delisting ceremonies in DC. Like some of the kids of the Earth Conservation Corps, Challenger had been dealt a poor hand early on. When he was a few weeks old, a storm blew him out of his nest in Louisiana. A couple of locals found him and took him home, keeping and feeding him until they gave him to the Audubon Zoo in New Orleans. The zoo arranged to have him moved to a hack site in Alabama, where he was released upon fledging.

Challenger was not the best candidate for living free. He arrived at the hack site at eight weeks of age, having been regularly exposed to humans, which is to say that he had learned to associate people with being fed. Twice in the summer after releases, he was recaptured—once because he was begging from two fishers in Alabama, the second time because after migrating north, he was found on a baseball field in Iowa, starving. Wildlife workers fattened him up and set him free again. But he quickly resorted to his old mooching ways, and he was captured a third and last time at a lake near Nashville, where a man was about to club the feathered panhandler.

Experts determined that Challenger had become too socialized to humans to survive on his own. At that point, the American Eagle Foundation (AEF) took him and named him in honor of the space shuttle crew that had been lost after takeoff two years earlier.

Not all eagles work well as educational birds. No two birds are of

the same temperament and personality, and many retain their wildness in captivity and become edgy around groups of people. Owls and
hawks tend to do better in these conditions than bald eagles. What
was special about Challenger as a permanent captive is that he liked
people and he could fly. The AEF staff glove-trained him and called
on him for special appearances like the ones at the White House. He
debuted as a free-flight performer in 1995 at a bass-fishing tournament in North Carolina, taking off from the glove of a trainer posted
in one spot, circling above participants and spectators, and landing
on the glove of another trainer in a second spot. He became the first
bald trained to fly around sports arenas, typically when the national
anthem was played, and he navigated similar flight paths at presidential inaugurations; NCAA tournaments; auto races; professional football, basketball, and baseball games; and the 9/11 memorial dedication
ceremony. Challenger completed more than 350 public appearances
before retiring in 2019, and he wasn't the only AEF flyer. Several of
the foundation's eagles together logged nearly ten thousand free-flight
performances.

Challenger wasn't the sole celebrity associated with the AEF either.
The organization was founded as a nonprofit in middle Tennessee outside Nashville in the mid-1980s, then moved to Dollywood in 1990. The
theme park's founder and namesake, Dolly Parton, offered the AEF a
new million-dollar facility that included a bird hospital, breeding area,
350-seat amphitheater, and ninety-foot-tall avian enclosure nearly the
length of two football fields. The AEF initiated a vigorous and effective
captive-breeding program using nonreleasable eagles, which included
pairs such as Crazy Horse and Juliette, Pilgrim and Mayflower, and
Eleanor and Mr. Roosevelt. Between 1992 and 2019, the organization
released 176 eagles from a hack tower in eastern Tennessee.[34]

A daughter of the region, where the rural environment informed
storytelling and songs, including her own, Parton appreciated the natural heritage of her country. Growing up, she had seen bald eagles.
When she embarked on her music career in the 1960s, they had not
bred in Tennessee since early in the previous decade. Bringing AEF to
eastern Tennessee helped restore a conspicuous part of the cultural and

natural landscape of her past and the country's organic legacy. Recognizing that the bald eagle had a unique friend in Parton, Fish and Wildlife presented her with its Partnership Award in 2003. But Parton, Mager, Dianne Davis, and anyone who handled and lectured with eagles knew that the real stars in raptor conservation and education were the birds themselves, most especially those living in the wild in their eagledoms.

Awareness and education were working. The print and broadcast media played their part in bringing people closer to eagles with both sadly tragic and uplifting stories. Nesting season generated a continuous stream of reports that put the bald eagle's unassailable domestic life in the public eye. Even in those places that had been the toughest on eagles, where atavistic ways kept symbol and species in separate spheres, attitudes had changed. Not all sheep ranchers, for example, resorted to hired guns or their own when birds of prey were about. In 2018, Tommy Moore, a fifth-generation Wyoming sheep rancher, lost some 175 lambs to eagles, primarily goldens. Sheep depredation was a growing issue in Wyoming, not because the eagle population was expanding. It wasn't. Sheep ranching was expanding and taking over habitat of the eagles' natural prey.

Moore sought and received permission from Fish and Wildlife to humanely trap and relocate the raptors and give a small number of goldens to licensed falconers. Moore set up a lottery to choose recipients, who assisted in relocation. Neither of these measures was a perfect solution. Only the few birds that, forsaking their freedom, went into falconry (Fish and Wildlife permitted two a year from Moore's ranch) were certain never to kill a lamb again. The relocated ones could find their way back to the woolly flock. Fish and Wildlife also permitted Moore to use noisemakers, as airports and landfills do, to frighten the eagles away, and, like the airports and landfills, he had limited success. "It's like shoveling snow," Kevin McGowan, an expert in bird behavior at the Cornell Lab of Ornithology, said of noisemakers: "You do it once, then you're going to have to do it again."[35]

Down in Georgia, Will Harris was using noisemakers in contending with the success of his state's restoration program of the 1980s and '90s. Not long after expanding the family cattle farm in 2010 to include organically raised free-range chickens, he invited some unexpected visitors: smart, opportunistic bald eagles. Harris's place, White Oak Pastures, was less than thirty miles from the Chattahoochee River, which traced the state line between Alabama and Georgia, but the eagles weren't hanging around the thirty-two-hundred-acre farm and thinking of fish so far away. The enterprise started with a lone eagle, which somehow caught wind that free-range chickens had come to White Oak Pastures. For the eagle, they were indeed free—free for the taking. That was fine, Harris thought. The eagle would "cull out" the weaker, unmarketable birds. Besides, he appreciated the way of eagles and nature generally.[36]

Harris was a big, infinitely likable man, who wore a boss man's Stetson and spoke with the r-dropping and r-shifting drawl of his region. In the 1990s he transformed the family farm, a possession since 1866, from the "broken system" it had become as an energy-wasting, biocide-applying, and soil-killing industrial concern into a regenerative closed-loop system. Among other changes, he stopped using chemicals, converted to grass-feeding his cattle, and raised all his livestock humanely.

That first eagle flew off with a few chickens, and Harris adjusted for the casualties. But the next year additional eagles showed up, then even more the season after that, until Harris was unintentionally hosting some eighty wintering chicken eaters, the majority of them juveniles. It must have made the young ones feel like skilled hunters when they swept in and took off with a broiler. What Harris was feeling was the loss of thousands of dollars in livestock a week.

The Farm Security Administration of the US Department of Agriculture had a depredation program that reimbursed farmers for livestock lost to federally protected predators. Harris applied for relief. But the FSA said weasels and coyotes could have just as easily been the thieves. Harris knew better. He had Great Pyrenees dogs living with the chickens full-time. They were "fantastic on mammal predation," he said, but the dogs didn't "pay a damn bit of attention to eagles." To the dogs, they were just another bird, while to the eagles, the chickens were "low-hanging fruit." Harris roofed the grazing areas with netting and monofilament line, but the eagles flew in under the low-hanging blockade and grabbed the fruit. He even tried a blow-up scarecrow. "I never heard an eagle laugh," he said, referring to a surmised reaction to the inflatable bogeyman, "but I'm sure they did." Fish and Wildlife let him bring in bird bangers and screamer guns to scare the eagles away, and they helped somewhat, but he was still losing money.[37]

White Oak was caught in a conflict similar to that of wind farms: bird conservation was undermining agricultural conservation. The paradox not surprisingly attracted attention from national publications, including *Audubon* and the *New York Times Magazine*. People started calling Harris and asking whether they could come see the eagles. Two of his daughters ran White Oak with him and suggested that, to help offset chicken losses, they give tours, host a weekend bald eagle workshop, and build cabins to rent to guests. They did, and people came. Visitors met a man who more than once had had his arm up to the shoulder in a cow to turn her breached calf, and they learned that the calving cow had not been artificially inseminated because White Oak was as natural as a skinny-dipping swimming hole on a hot summer day. He would tell visitors about his neighbors with their big conventional farms—their broken-system farms—where pesticides, fertilizers, monocrops, and meticulously tilled fields had murdered the soil, and where the runoff from these microbial graveyards wasn't doing the local streams any good. All this harm—and those neighbors were collecting generous taxpayer subsidies from the federal government. Yet the FSA wasn't cutting any checks for Will Harris.[38]

Finally, after Harris spent years fighting the confounded FSA in court, after he lost 160,000 chickens and $2.2 million, after he won one green award after another and earned humane farm-animal-care certificates, the appeals division of the FSA ruled in August 2018 that White Oak Pastures was entitled to compensation. Throughout his struggles, Harris never begrudged the eagles, understood that nature dictated their actions, and said he would never be tempted to harm them. He remained true to his word.

PEOPLE EVERYWHERE WERE NOT just learning to live with bald eagles but enjoying it. They spoke with pride when saying that eagles nested on their property, in their neighborhood, at their golf course, on the campus of the local school. It was the rare individual who resented the eagle's ubiquity. Those in the commercial fishing trade were more like those in Dutch Harbor who might grumble a bit about feathered panhandlers, but most weren't prepared to harm them as others routinely had in the past. Among sportsmen, treating bald eagles like friendly competitors and photogenic subjects had become the standard. After the hacking days, people would send MassWildlife's Bill Davis pictures they took of eagles while fishing at Quabbin Reservoir. Davis was once out there on the water with his spotting scope turned on a couple of fishers in a boat. One hooked his luck into an undersize lake trout and tossed it back. As the fish lingered on the surface recovering, a bald flew in and away with it. The two men looked at each other, stood up, and applauded.[39]

Recognizing that not everyone would have a firsthand, in-the-flesh experience with eagles in the wild, Doug Carrick, a retired accountant living on Hornby Island in British Columbia, decided to share with others his experiences with two eagles and an eagledom. At the time, he had no idea he would be bringing the world to this backyard and not only helping to advance awareness and knowledge but fostering personal relationships between people and bald eagles.

Carrick set up a closed-circuit video system in 2004 and began sitting at a monitor for hours, watching live images of a pair nesting on his wooded property. A coaxial cable ran from the monitor across the floor and through an exterior wall and outside. From there, it snaked along his property and high up to the top of a Douglas fir, where it connected to a camera in a little wooden box near an eagle's eyrie.

Carrick had advanced nest-watching technology beyond flyovers, zoom lenses, and the steel towers of the pioneer of pioneers, Francis

Herrick. At first, Carrick's images went no farther than his house. They fed to no other monitor or to the World Wide Web. Yet he did share videotapes of the nesting eagles when he gave lectures. David Hancock, a leading eagle biologist, attended one of his lectures.

Hancock was a pioneer himself. He conducted the first aerial nest surveys ever, in 1953, when planes were used to shoot eagles rather than collect scientific data on them. Hancock was fifteen at the time and accumulating solo-flight hours toward earning a pilot's license. His bird's-eye view introduced him to eagle nests around British Columbia and Washington. He saw many fewer eagles along the coast of the latter, eventually figuring out that when the bounty was still active, poachers shot eagles around Oregon and Washington and went up to Alaska to cash in the talons. Hancock kept a count of the nests and eggs he saw when flying and later incorporated the data into his graduate thesis. In the years that followed, he banded countless eaglets, once finding himself stranded in a nest for nine hours with his wife, Lyn, a wildlife photographer—an incident that Doris Mager took as inspiration for her nest-in in the next decade. David and Lyn also climbed trees to photograph and film eagles, but Carrick's 24-7 live streams mesmerized David. Hancock suggested the possibility of linking Carrick's camera to the internet to create a virtual immersive experience for a wide audience. Carrick liked the idea. With some technical assistance, they created a website and went live in 2006. People worldwide logged on to the cam view by the millions.[40]

Carrick's Hornby Island eagle cam wasn't the first to livestream online. The year before, Florida Audubon and a private development company teamed up to bring the home life of a bald eagle couple at Port St. Lucie to web users. That same season, they could log on to a site to watch a couple at Blackwater National Wildlife Refuge in Maryland hatch and raise the young. But the Hornby Island site was a blockbuster attraction, thanks in part to the old imperialistic ways of a nation with a crowned lion and white unicorn on its royal coat of arms. The Canadian Broadcasting Corporation television station near Carrick ran a story on his hobby, which had become a local marvel. Then the local marvel turned international. BBC World News picked up the story and broadcast it throughout the British Commonwealth across two hemispheres.

The Hornby eagles were overnight stars. In a single day, Hancock did seventeen international broadcast interviews. "People are blown out of their mind," he told a reporter, "to see the eagles at nest level." Viewers reached numbers that no one expected. The *Chicago Tribune* said of the eagle cam's popularity, "You'd think it was 'Desperate Housewives' or the Superbowl." This was in the days of dial-up internet service. The viewing traffic was so heavy at one point—ten to twenty million visits to the website each day—that Microsoft's servers crashed.[41]

Set up in strategic places around the globe were panda, tiger, gorilla, and alligator cams. In the US there were falcon and osprey cams. One of the latter had been online in Maryland since 2000. A new era had been born. In nest trees across the country, eagles forsook their privacy for the internet and became cast members of the most popular wildlife cams in the world. As Carrick said, if his had been a "sparrow's nest, it wouldn't have gotten the same attention."[42]

The silent-running, tree-mounted cameras went live each season typically in time to receive the eagles returning to their nests. From sunset to sundown, cam watchers saw industrious birds hauling in sticks as part of annual nest refurbishment. Every laid egg, pale white and the size of a tennis ball, weighing about four ounces, was a milestone. Viewers caught glimpses of the eggs when parents turned them over and when they traded sitting shifts. They witnessed parents' implacable devotion as mom and dad remained steadfastly atop their eggs even when they were being rained and snowed on, and when high winds were rocking their trees. Counting down the thirty-six incubation days, eagle-cam devotees anticipated an egg tooth poking through a shell, and watched another day and a half as a hatchling fought its way out. They heard tiny voices from furry-white dollops peeping for nourishment and a nest-bound adult call to its mate to bring food. They saw one parent fly in with a fish or rodent and the other rip it to shreds to feed the ravenous babies.

Ten days after a chick hatched, the lens revealed thermal down feathers coming in to replace natal ones. Around day twenty, juvenile deck feathers started growing in as loose down swirled in the air. All the while, attentive viewers watched the eaglets grow from one day to the next—from chirpy, smoky-gray muffin size to squawking football size, with wings like sandwich boards, beaks the shape of a lobster claw, eyes resembling tiny black fish eyes, and big toes like a cartoon character. Viewers saw the chicks change from being clumsy, off-kilter, and comically pathetic to coordinated and birdlike. Nearly a month along, dark juvenile flight feathers stood out more clearly on camera, and eaglets were soon bouncing and flapping their wings, tearing meat from fish and feeding themselves, and scanning the sky, pondering future prey and perhaps even the day of their first flight.

Finally, when that day came, suspense lingered at monitors until the fledged juvenile returned. Four or so weeks later, the young permanently withdrew from the camera's view, leaving separately for their first long journey, flying toward a new stage in life, across state lines, even an international divide, that had no meaning to them.

On occasion, one parent had already departed, going its own way while the other remained behind—a quiet, seemingly contemplative empty nester. Viewership tapered off. The devoted stuck around a few more days to wish the last parent a bon voyage, after which the nest finally sat empty and alone, a rattling twig or muffled wind giving the only indication that the camera was still running. Then the screen went dark and the long wait began for next year, when, camera on, the parents would find their separate ways back home, sometimes arriving at their small claim in a vast geography within hours, even minutes, of one another, and immediately setting to work refurbishing their castle.

Perhaps the one downside to the open nest life was exposure to all of nature's realities. What streamed out to the world was not edited or censored. Every moment carried the risk of revealing something brutal, gruesome, and tear-jerking. Chicks fell out of nests, got picked off by owls and falcons, succumbed to smothering mosquitoes, mites, and black flies, or got fatally caught in the twine of the nest. When food came, larger siblings muscled smaller ones out of the way; "baby bonking," viewers called it. Sometimes the smaller one starved to death. Sometimes the larger one killed the smaller one, all on camera for the world to see. Nest stewards and Fish and Wildlife officials took phone calls from frantic viewers demanding something be done, but rarely did they intervene in nature's way.

They fielded such calls in southwestern Florida in 2015 when cam fans noticed an interloper at the home of Ozzie and Harriet. The two had been a couple since 2000, as far as anyone had known, and in 2006 they built a new nest sixty feet up in a slash pine a mile from the Caloosahatchee River. There was a small pond nearby where, year after year, the two and their fledglings drank and bathed, and other ponds farther out where they sometimes fished. The tree was on the property of Dick Pritchett Real Estate, which installed a camera in 2011 and added another the next season, giving the world the first 360-degree viewing of an eagle eyrie. Then, four years later, the viewing turned ugly.

A scrappy young male, who still had some dark coloring in his white feathers, entered the couple's nesting territory. He made a play for Harriet. Pushing twenty years, Ozzie tried repeatedly to fight off the interloper, and after a series of aerial duels, Fish and Wildlife personnel found Ozzie on the ground, badly injured. Ninety-seven days he spent at a clinic before he was released. In the meantime, Harriet, seemingly neither outraged nor horrified, took up with the young paramour, who had demonstrated his superiority as a protector and, implicitly, provider. They raised a brood together. When Ozzie showed up at the nest in September, the contest for Harriet resumed. There was no mercy for the older and weaker bird. After being taken back to the clinic, Ozzie died within a few days. The turn of events devastated nest followers, and they refused to endear the new male with a name. He was forever to be identified by his Fish and Wildlife designation, M-15.

No matter what happened in front of nest cams, they were priceless educational tools. "The anticipation, the disappointment, all the emotions you can think of," said Carrick, came from being up in nest trees with eagles. Viewers developed intimate relationships with the birds they watched. They got to know their individual, sometimes idiosyncratic, behaviors, and gave them names, M-15 notwithstanding. The nest cams were not a nature film with David Attenborough's mellifluous voice nudging the viewer's emotions in one direction or the other (no offense to Sir David). No art direction or film editing manipulated a plotted narrative. Nature was editor and director. Eagles followed their own script. They turned the national symbol into a natural being—meaning not artificial, not constructed on an artist's pad or an engraved metal die. Cam watchers donated money, joined organizations, and wrote blogs. As David Hancock said, "If people are concerned about it, they will protect it."[43]

HAVING MORE EAGLES IN the wild meant brushing up sometimes against not only humans but also other wildlife. Owls and eagles commonly crossed talons, and eagles squared off against falcons and loons, to say nothing of other eagles. This was no different from when wildlife populations had been at their apex before Euramericans swept over the continent. Conflicts and displays of territoriality were common, and sometimes these kinds of events manifested unexpected and even unconventional behavior that not even scientists could explain. The mystery revealed itself most strikingly when eagles at the Upper Mississippi River National Wildlife and Fish Refuge in Illinois gave cam viewers a rousing performance in 2017 of an intereagle conflict different from Ozzie and M-15's.

Five years earlier, a couple named Valor and Hope had occupied a nest at the eighty-foot top of a silver maple. Valor was not the quintessential devoted parent. He was an unreliable provider. After eggs were laid that year, he rarely assumed his sitting duties. When chicks were in the nest and Hope called for him to bring food, he typically ignored her, forcing Hope to leave the chicks in Valor's capricious care as she herself went off to hunt. At best, he would squat at the nest's edge for a few minutes before taking flight to wherever whim took him. In the end, Hope could not sustain the brood without a fully present partner, and her two eaglets died.

When Hope returned to the nest the next year, 2013, she brought another mate with her. Valor showed up only to find he'd been ousted. He didn't fight off his rival, which seems consistent with his inertia as a parent. He didn't leave the eagledom either, and the new mate didn't chase him away. Hope and her new partner, whom the refuge's nest stewards named Valor II, remained cordial toward the original Valor.

A couple years later, Valor was part of nest life again, alongside Hope and Valor II. Having emerged from his parental torpor, he assumed the responsibilities of a proper partner. The birds formed a threesome. The refuge's visitor service manager quipped that the upper Mississippi had its "own little soap opera." In 2015, the couple and their new partner raised three eaglets.[44]

Parenting trios in the wild aren't altogether uncommon, although science isn't wise enough yet to know a lot about the domestic life of three, why threesomes happen or the purpose they might serve. What makes the Illinois family particularly stand out is the 2017 nesting season. It began with the three refurbishing the nest. Both Valors copulated with Hope; she laid two eggs, and they hatched. All the while, their domestic peculiarity picked up more cam viewers, and home life seemed tranquil.

Until March 24. Two rapacious adult eagles showed up and attacked the trio. Talons and beaks clacked and stabbed. Eagles ended up on the ground, fighting, and when feathers settled and the dust cleared, Hope had disappeared. No body was found, and no one was ever sure what happened to her.

The two Valors stood fast, though, and successfully defended their nest and eaglets. The attackers were apparently vying for the territory. Still, waggish observers maintained that the rivals were put off by the family values of the trio. Whatever the case, the attempted siege continued. Nearly every day through the month of April, the aggressors returned, and the Valors held fast to their territory. The drama went viral on the nest cam's YouTube stream, picking up viewers in eighty-nine countries, from Venezuela to Nigeria to Kuwait. The Valors continued to defend against the siege and care for their young, and to live up to their name. Their efforts paid off when the assaults finally stopped and the eaglets fledged at the end of May.

The Valors resumed their domestic partnership in the fall with a new female, who acquired the name Starr. It is unclear which Valor pursued her and locked talons with her in a courtship ritual. She laid two eggs that season; one eaglet survived and fledged. The next year, she laid three eggs, and three eaglets fledged. In 2020, the nest produced two eggs and two healthy eaglets, which fledged. The trio was still together. Unfortunately, in August a severe wind- and thunderstorm took down their nesting tree. The three didn't blink. When they came back in September, they built a new eyrie in a tree near the old one—a trio of hard workers in a three-parent household.

THE VALORS AND BALD eagles across the country brought forth lessons in acceptance and cooperative relationships, and Americans took to those lessons. Despite the challenges of coexistence for wild birds, including differences of opinion among themselves, the twenty-first century wasn't collapsing back into the abyss of gratuitous brutality. Bald eagles faithfully laid bare the truth about birds of prey and ultimately led Americans to a moral reckoning. They were reminded, as Luther Standing Bear said long ago, that animals and people drink the same water and breathe the same air; and, as Alexander Wilson said before then, that beasts and birds are "not in their ways inferior"; and, as Charles Willson Peale said before Wilson, that they "will teach thee." The bald eagle helped Americans understand more fully the world they shared with their chosen bird and realize that they were better for sharing it. It turned people dutifully toward its well-being, in turn freeing them to move beyond a burdensome past.[45]

Eagles building their durable eyries where they had not existed for decades or a century were constructing a new future. When the first chick in well over one hundred years hatched on Staten Island in 2016, *New York* magazine, playing on Donald Trump's "Make America Great Again" presidential-campaign slogan, wrote "New York City might have been made great again." Later that year, the Montezuma National Wildlife Refuge installed a thirteen-hundred-pound torch-cut-and-welded steel statue of an eagle beside the New York Thruway to commemorate the fortieth anniversary of the state's pathbreaking eagle-hacking program. New York was well on its way to hosting two hundred nests. Two years later, when the Philadelphia

Eagles won the Super Bowl in 2018, Pennsylvania had three hundred nests. When the first Super Bowl was played fifty-one years earlier, Pennsylvania had none.[46]

The New England Patriots won the Super Bowl the following year, and two months later, headlines announced that an eagle couple was tending to the first fuzzy-headed chick that Cape Cod had seen in 115 years. New Jersey had at least one nest in each of its twenty-one counties; in 1982 there had been only one in the entire state, and for six years its eggs had failed to hatch. Over in Ohio, wildlife researchers compiling reports from residents across the state counted just over seven hundred nests—a phenomenal increase from the four in 1979. The majority were congregated in the marsh region along Lake Erie, where eagles had introduced their home life to Francis Herrick a century earlier.[47]

These feathered creatures had entrenched themselves in the political and emotional life of Americans, who, while enthusiastically sharing space with the bald eagle, were intimating something exceptional about themselves. In the rest of the world, more than one-third of official national animals were threatened with extinction, and the populations of nearly half were contracting. Everywhere, the pressures on wildlife were the same: habitat loss, poison, poaching, indifference. Americans and the bald eagle managed to defy those pressures and secure the bird a place among the one-fifth of nationally emblematic animals that were either mounting comebacks or had achieved one. There was no question what America's bird had achieved.

For years after delisting, states discontinued annual census-taking, and researchers could offer only rough estimates of the number of bald eagles living in the United States. Then a census report that Fish and Wildlife released in early 2021 documented unanticipated totals. On the basis of aerial surveys from the previous two years, researchers calculated that the population in the Lower Forty-Eight had quadrupled during the previous decade. Nesting couples totaled more than seventy-one thousand, and individuals well over three hundred thousand. Alaska put its population size at over seventy thousand—likely a conservative total. Canada hosted approximately the same number.

It would not be a stretch to say that from coast to coast and from northern Canada to northern Mexico, five hundred thousand balds occupied their indigenous land—the number that some say existed when an unknown English-speaking newcomer named the species the "bald eagle."[48]

In the twenty-first century, *Haliaeetus leucocephalus* achieved the vastness across the continent that it had known before it became America's bird, and a circle closed. In the centuries in between, the species witnessed danger breaking out, the subduing of its eagledoms, and the predatory ways of the people who accused the raptor of heinous crimes. Then, in the presence of its only true predator, the bird of America came to know succor and rescue, a sovereignty restored, and freedom renewed.

EPILOGUE
Reconciliation

WHEN THE NORTHERN HEMISPHERE BANKS AWAY from the sun, and deciduous trees withdraw the last seasonal nutrients from their waning, color-changing leaves, bald eagles rise one by one from summer-to-fall convocations in Saskatchewan and fly southward. The old perch trees from which they hunted and the lakes, ponds, and oxbow rivers that were the season's emporiums begin to fall behind them. Wind buffets their faces and bodies, the air lifts and pushes, wings and tails sense changing currents, feathers ripple and, on powder-blue days, glisten. En route, they pass over rooftops and roadways, woodlands and prairie. They cross the forty-ninth parallel and the Missouri River. Some veer westward over Yellowstone and the Rocky Mountains. Some wing toward the Dakota grasslands and bison herds. Standing Rock, Northern Cheyenne, Fort Peck, Crow, Rosebud, and Pine Ridge Indian Reservations lie below, as do the Black Hills. Places named Moose Jaw, Medicine Hat, Deadwood, and Spearfish fall into the birds' wake. Except for the hushed sound of air, feathers, and breathing, their travels remain silent. No navigational calls sound. One eagle's course might duplicate another's, but each pursues its own, and most are pushing toward their respective natal territories across the western states.

Many touch down in Colorado, which in the twenty-first century came alive again with bald eagles. Only one pair in the entire state tended a nest in 1974, followed by none for the next four years. More than sixty couples raised broods in 2020, and some twelve hundred unpaired juveniles and adults wintered in the state that year. They roosted, sometimes in the hundreds, along the Colorado, Arkansas, and South Platte Rivers and other waters. What the eagles didn't leave behind in their summer places were midair brawls—the attacking pursuits and breakaways, wingovers and spirals, fish drops and catches—which had the potential of erupting with every caught fish.

Nearly one hundred migrators settled at the Rocky Mountain Arsenal National Wildlife Refuge, a sixteen-thousand-acre wedge of grassland near Denver. Birds flying in from the north negotiated airspace with planes taking off and landing at the Denver airport a few miles to the east of the refuge. The Rocky Mountains rose conspicuously to the west, and the chromium Denver skyline to the southwest. The refuge had a few lakes beside which the eagles roosted, and one couple nested in a hefty, old cottonwood with outward-bounding branches. The first eagles to winter on the site in the post-DDT era arrived in 1986. Others soon followed, and four years later the Department of the Interior turned the prairie into a national wildlife refuge. The wintering, nesting birds of America had been a principal impetus for its establishment.

The refuge inherited its unlikely name from a US Army biological-weapons facility situated on the same land during World War II. Decades of chemical manufacturing had followed, turning the prairie into a Superfund site. Cleaned up and refitted as a refuge, it supported some 330 wildlife species—bison, coyotes, white-tailed deer, kestrels, hawks, and white pelicans among them. Along with fish, the eagles preyed on prairie dogs, pocket gophers, and cottontail rabbits.

Eagles roosted and nested in parks and on lakes all around Denver, demonstrating their willingness to dwell where humans dwelled. In May 2021, storm winds battered the city and took down a cottonwood hosting a nest in one of the parks, killing an eaglet that had hatched in late March. The nest had been featured on a Facebook page with more than ten thousand friends, and spotlighted in an around-the-clock eagle cam with night-seeing technology. Viewers were devastated. "Our phones were ringing off the hook," the park naturalists told the *Denver Post*, with people calling and sharing their "many tears" over the loss.[1]

Officials took the eaglet's body to the wildlife refuge for internment in a sacred burial site, where its remains would be returned to the earth following a tribal ceremony. Cheyenne, Arapaho, Oglala Lakota, and Southern Ute, all with ancestral connections to the area, had blessed the site in a formal ceremony in 2019. It was the first of its kind on Fish and Wildlife property, and it wasn't the only sacred place at the refuge.

A car-drive away, squatting at dead center of the refuge is a twenty-two-thousand-square-foot metal-and-brick building. The nondescript facility houses the US Fish and Wildlife Service's National Eagle Repository. Here stands an end in the life cycle of eagles. Inside at any given time are the remains of hundreds, sometimes a thousand or more. The repository is a federal facility, but it exists for Native peoples. It is a place of understanding and resolution, a tangible commitment on the part of the United States to tribal spiritual beliefs and practices. America's reconciliation with the bald eagle in the twenty-first century could not be complete without the full recognition of its sacred nature in indigenous cultures.

Virtually all dead bald and golden eagles end up at the repository. If you come across an injured one, you cannot try to nurse it back to health yourself. The Bald and Golden Eagle Protection Act requires you to hand it over to the authorities. The regional director of Fish and Wildlife then makes the call on how to deal with it. The same goes for feathers. Find one on the ground, and you have to give that up too. When eagles at raptor centers molt, staffers must collect their dropped feathers and ship them to the repository. This rule even applies to the blizzards of white down that swirl around an aviary (some centers have specialized vacuum systems dedicated to the delicate fluff).

The repository was established for the purpose of distributing feathers, wings, feet, heads, and whole bodies to Indians for use in religious and cultural ceremonies. It had previously been attached to the US Fish and Wildlife Service Forensics Laboratory in Ashland, Oregon. In 1994, more than three hundred tribal members met with President Clinton, whose Secret Service code name happened to be "Eagle."

Since the wholesale slaughter of eagles in the nineteenth century and the passage of the Bald Eagle Protection Act in 1940, Indian villages had no longer quivered with feathers. Acquiring them had become difficult. Although the Eagle Protection Act enabled the winged monarch to resettle old eagledoms and establish new ones, the law failed to protect traditional Native relationships with the bird. Congress had exempted Alaska, yet not Indians, from the act's enforcement. At the very moment that the Franklin Roosevelt administration was implementing policies to reinforce Native sovereignties, the Bald Eagle Protection Act criminalized Native traditions. The Lakota, Cherokee, and others could no longer hunt eagles as had their ancestors without breaking the law. The Zuni could not take an eaglet from a nest, raise it in a stockade, and pluck its feathers. The Lenape could not send their young men up to the mountains to capture a spirit bird for one of its tail feathers.

Instead of simply venturing into eagle country to meet their religious needs, Indians had to go to bureaucrats at Fish and Wildlife with an application, even if they wanted nothing more than a single feather. The process was insulting, and the turnaround time for filling requests at the original Ashland facility was long—years long. The tedium and wait time opened up a black market that led to unnecessary and unsacred bird killings, and the arrest of tribal people involved in the market.

President Clinton's meeting prompted Fish and Wildlife to detach the repository from the Ashland forensic lab in 1995 and move it to Colorado. There it would function as a discrete entity and, if as envisioned, get feathers out the door quicker.

Another important change around the same time resulted from a successful lawsuit that challenged parts of the application process. Fish and Wildlife required applicants to justify their intended use of feathers and body parts and to have tribal governments preapprove their applications. Opponents of this gauntlet had argued that the requirements went against the country's commitment to religious freedoms. It hearkened back to the oppressive nineteenth-century Religious Crimes Code, which prohibited tribal dances and ceremonies.

No other citizens had to explain their religious activities. No brother or sister in ecclesiastical robe was obligated to disclose how they consumed their carafes of wine. Fish and Wildlife relented, dropping the intended-use and preapproval requirements. It also eliminated application fees and the cost of shipping.

Soon after the repository's relocation, Fish and Wildlife began discussions with the Zuni Pueblo, which proposed building an eagle aviary. Like existing raptor centers around the country, the native aviary would rehabilitate injured eagles and release those that could fly again. Nonreleasable birds would live at the aviary, and the Zuni would be able to keep all molted feathers and distribute them among tribal members.

By the 2010s, seven Native sovereignties were operating aviaries, six of them located in the Southwest. The one at the Comanche Nation in Oklahoma opened a second facility under the name Sia, the indigenous word for "feather." Sia was founded and headed by William Voelker, a tribal member with a doctorate from Cornell University's prestigious ornithology program (he had been classmates with Steve Sherrod of the Sutton Center, who spearheaded the Florida eagle-egg translocation program). Sia collected feathers from eagles and other migratory birds and distributed them to more than forty tribes around the country.

Even with expansion of the eagle population, which translated into more deaths and more eagles showing up in chilled boxes at the repository's loading dock, and even with changes implemented by Fish and Wildlife, demand exceeded supply. The growing popularity of pow-wows, some of which devolved into nonreligious tourist attractions and regalia contests, with costumes consuming thousands of feathers, elevated the burden of delivering orders. Adding to the problem, the federal bureaucracy initially made no budget adjustments for increasing the size of the staff after moving the repository to Colorado. The processing of requests continued to inch along. For years, only one person prepared corpses to meet the demand. He spent an average of fifteen minutes examining each bird and could complete twenty to thirty requests a day. The staff called him the "mother plucker."[2]

Only individual members of one of the 573 federally registered tribes could apply for eagle parts. Tribes themselves could not. Still, demand tested the best efforts of the mother plucker. In the fiscal year 2010, twenty-two hundred eagles came into the repository, not enough to fulfill seven thousand pending orders. The turnaround time ran two or more years. Wanting feathers to honor her son's military service, a Kiowa woman of Oklahoma waited four years for their delivery. For some, driving an hour or two to the home of a black marketeer to pick up feathers for a few hundred dollars (the black-market value in 2017 per feather was about $150) held a certain enticement.

Sensitive to the backlog and cultural traditions, Fish and Wildlife permitted select Native sovereignties to take eagles out of the wild under certain circumstances. The Hopi were able to revive an ancient ritual after obtaining authorization to remove as many as forty eaglets a year from nests, douse them in ceremonial meal, and smother them to death. The Northern Arapaho acquired permits to kill two bald eagles a year, without having to explain how the eagles would be used. Even while restrictions on obtaining religious articles from the wild had eased, the federal government, essentially a foreign government to Indians, still maintained a considerable degree of oversight in Native religious affairs.

The American relationship with Native sovereignties had always been complicated, more complicated than that with the bald eagle. As wildlife officials, policymakers, and the judiciary struggled to moderate those complications, many Americans had begun to make connections with bald eagles that went beyond the patriotic symbol and rescued species. Doris Mager said that whenever she looked into the eyes of a bald eagle, "spiritually something" took over. On his southwestern Georgia farm, where chickens had been lifted toward heaven without end, Will Harris said he could never harm a bald eagle; there "is just something spiritual about it." Such connections countered the false separation that Western society wedged between the consciousness of humans and nonhuman species, and manifested a profound awareness of the mysteries of an animal's life. Awareness could be deep and life-changing.[3]

～

IT WAS LIFE-CHANGING FOR Patrick Bradley, who said there was something "absolutely" spiritual about the bald eagle to him. Reconciliation with the once endangered species affected the former Green Beret and Vietnam combat veteran on an intimate level. Although he had no Native lineage, Bradley unexpectedly acquired a familiarity of nature that he used for his own personal betterment, and then he passed on his knowledge to others.[4]

In Vietnam, Bradley lost several friends and, with them, something within himself—a connection to reality and the living moment. After returning stateside in 1969, he, like so many other veterans, had trouble adjusting to civilian life and escaping the fight-or-flight response that perpetual violence had fused in him. While at Walter Reed National Military Medical Center, undergoing treatment for a hand injury, he slugged an officer in the face on two occasions, breaking the officer's jaw on the second. Bradley nearly went to Leavenworth, the military correctional facility. Instead, he was discharged from the service and, in May 1970, on the advice of his psychiatrist at Walter Reed, flown to Canada and the remote interior of Saskatchewan, a variegated environment of grasslands, oxbow rivers, dense woodlands, and mountain backdrops. There he experienced both transcendent beauty and earthly dangers. He was on his own, facing harsh winters and tree-toppling storms, living in shelters he built himself and on provisions occasionally air-dropped for him. Still, to survive, he had to improvise and live off the land, and avoid encounters with threatening terrain and wildlife. He would be there three years and rarely see another human being, although he was not without companionship.

Bradley was assigned to a research project counting bald eagles in an undisturbed part of Saskatchewan. In the absence of human activity, the nesting population there was healthy, and during breeding seasons migrators came up from the States, although incoming flights had diminished significantly since the introduction of DDT. At first he had doubts about his ability to contribute something useful to science. Then, after surviving his first winter and as he witnessed spring peeking through the snow, animals started to come out of their winter languor and, as he put it, let him into their world.

He soon realized that he was where he "needed to be to feel whole again." Despite the dangers of the Canadian backcountry, he experienced a "cultural shock of the most astonishing kind." In Vietnam, nature had been a "horrible thing," he said. "Nature wanted to kill me." In the North American interior, he would have felt in good company with Bartram, Jefferson, or Peale. Nature kept him alive. It elevated and saved him. He experienced something resonant and abiding, while the ceaseless pulse of "horror and the terror" that had followed him from Vietnam receded. Everything around him "melded into a spiritual feeling" that quieted nightmares and nightmarish memories and restored inner peace. In nature, he discovered a moral universe that he didn't know existed and a sensory awareness that was connected not to relentless danger but to reverence. The bald eagles he counted—fishing, nesting, and cutting across the sky—were, he said, the "logo for all this."[5]

Following his woodland reawakening, Bradley returned to the US, ate his first hamburger in three years, and went to college, where he earned a master's degree in biology. During and after college he worked odd jobs and eventually landed one as an animal trainer and handler. For a spring and summer, he was the monitor for a peregrine falcon hack site in New York. This was two years before the state's bald eagle hacking initiative, which had launched from the success of the peregrine project that Bradley worked on. He also banded birds for the Patuxent Wildlife Research Center and for the Chesapeake Bay Bald Eagle Banding Project, climbing trees like Broley had. And for five years he traveled around the country with his then wife Janeas, giving educational lectures with raptors and reaching, he estimated, a half million schoolchildren.[6]

After leaving the exhausting lecture circuit, Bradley moved to Florida and returned to animal training. In 2010 he took a position at Boyd Hill Nature Preserve in Saint Petersburg, running the raptor program there. All but three percent of the county had been built out, to the edge of Tampa Bay on the east and the Gulf on the west. Still, by the end of the decade, eagles were astir in forty nests.

As for himself, Bradley was an empty nester. His son, Skyler, was grown and in the army, fighting on the front lines in the Iraq War. When Skyler left the military after seventeen years and came home, Bradley quickly recognized PTSD symptoms. Seeing in his son what he had once known in himself, he had an idea and encouraged Skyler to come out to the nature preserve. Bradley put a red-tailed hawk on Skyler's gloved arm and had him stroll the preserve's three miles of trails winding through heavily canopied hardwoods. After that experience, Skyler returned to the preserve for more walks and volunteered to help care for the birds. He began to move beyond his combat experiences and feel whole again, as Bradley had while counting bald eagles in Canada.

Skyler's transformation sparked an idea. Bradley convinced therapists at Bay Pines VA Hospital in Saint Petersburg—where for years Charles Broley had banded eaglets on its grounds—to send PTSD patients out to the preserve. Stepping out of their passenger van, they met Bradley, a man with a broad mustache, a slightly bulging waistline, a small loop earring in his left lobe, and hair as white as a mature bald's head. Bradley exhibited a quiet enthusiasm that put the vets at ease. He never pried into their backgrounds, always listened if they had something to say, didn't push if they didn't. With the vets, he did what he'd done with his son: he put birds on their arms and told them to take a walk.

Veterans' hospitals had been employing animal-assisted therapy for some years, although using well-behaved domesticated animals. Raptors were not obvious healers. They seemed too feral and aloof. But to Bradley, they were ideal. The birds he used, all of them rescues, had experienced their own trauma, and he was matching wounded bird with wounded soldier. There was an immediate trust between the two. In each veteran, Bradley saw a latent determination beneath layers of debilitating trauma that wanted to break to the surface. Inscrutably, the trauma that owls, hawks, and eagles shared with veterans led to shared recoveries.

Bay Pines called the outings to Boyd Hill "recreational therapy," but they turned into something more for many of the veterans, including Telia Hann. Hann's maternal lineage traced back to Apache and Aztec forebears. His mother had taught him about the "old ways," which were based on the belief that nature is the embodiment of all wisdom. When Hann was a teen, she had given him a necklace with a bald eagle pendant. She told him that the winged species was a bearer of prayers between Earth and heaven. Hann's necklace had the same power as the necklaces of eagle talons that his ancestors wore: it would invoke the protection and will of the Creator. She assured him that the bald eagle will "always be there" and told him to follow it, let it be his "guide."[7]

Life had not been without its travails. Hann's father had died when he was young, and the family fell into dire straits. When he turned sixteen, Hann quit school and bounced around from one small job to another. After 9/11, he joined the army and completed two tours in Iraq, wearing not eagle feathers but the necklace his mother had given him. His main job during deployment was manning a .50-caliber machine gun atop a Humvee, and out in the haze and shellfire he felt not much safer than a fish scurrying across the surface of an eagle pond.

When he left the military, Hann suffered from PTSD, which stirred occasional suicidal thoughts, and he ended up at Bay Pines hospital. Recovery was slow. Then he rode the transport van out to Boyd Hill and met Bradley.

Hann was initially withdrawn during his first visit, but when Bradley asked whether anyone wanted to hold a red-shouldered hawk and take it for a walk in the woods, Hann didn't hesitate. Go off and "get lost in nature," Bradley told him. Hann took those words literally. He and the hawk, Rusty, were gone for two hours. The time together "flipped a switch" in his head, Hann remembered, and he went from "living in the past to being in the moment" with his feathered companion. He started to frequent the preserve for more opportunities to "flip the switch," and then transitioned into working there as a volunteer.[8]

A year after veterans started visiting Boyd Hill, Bradley left the preserve to oversee the raptor center at the George C. McGough Nature Park in Largo, a city in the same county as Saint Petersburg. He and a volunteer who had been with him at Boyd Hill, Kaleigh Hoyt, formed Avian Veteran Alliance (AVA) to give their rehabilitation program a name. There was no other raptor therapy program in the country, and AVA was making great headway with veterans from Bay Pines and in the surrounding area. The birds in residence at Largo included a bald eagle named Sarge.

Hann, who followed Bradley to volunteer at Largo, met Sarge, and while she often paid little attention to park visitors and some of the other volunteers, she exhibited an ease with Hann. Sarge was a ten-pound adult female. She had a feather deformity that rendered her flightless and that researchers had never been able to explain. Except for the absence of tail feathers, she hid her deformity by keeping herself flawlessly preened. Her colors were vibrant and clear, her eyes astute, and her taloned grip on Hann's arm sure. Of all the birds at raptor centers, bald eagles are generally the most standoffish toward people and the most hypersensitive to human moods and demeanor. With Hann, Sarge maintained a commanding yet welcoming presence. Hann, who still wore the eagle necklace his mother had given him when a teen, was in turn beguiled.

Any one of the birds at the Largo raptor center would hop on his glove without hesitation. Yet his connection with Sarge was stronger and more enduring than with the other raptors. He brought his Native American flute to the park and played it for her, and one day that switch inside him flipped and locked into place. Soon after, he enrolled in college, earned a degree, and landed a position teaching fifth graders at a local elementary school, which referred to itself as the "Home of the Eagles."

The bond that Sarge and Hann formed had reciprocal benefits. It helped Sarge grow more receptive to others. Bradley began using her as an ambassador and taking her out to schools and events, and she became known as the eagle who helped heal veterans. More than two thousand vets had walked with raptors at the park by 2020, and nearly two dozen affiliates of the AVA had opened across the country, including one in Alaska; and a rehabilitation center in the Netherlands used the AVA as a model for an animal-assisted-therapy program.

HEALING IN A BROADER sense than that with individuals was the message of Joe Biden's victory speech after the 2021 presidential election. Biden was addressing a nation divided on many lines socially, culturally, and politically. He alluded to Psalm 91, which is sometimes titled "On Eagle's Wings." "Together—on eagle's wings—we embark on the work that God and history have called upon us to do," he said. "With full hearts and steady hands, with faith in America and in each other, with a love of country—and a thirst for justice—let us be the nation that we know we can be." The eagle in his speech refers to an emissary of the divine that might deliver peace between the people of a glorious nation, a nation that affirmed its place among others with an American bird aloft on open wings.[9]

Animals and natural features of the Earth have always been a part of Western religious teachings. In the Bible they are metaphor and symbol and often at the center of allegory. Adam was made from the earth (the Hebrew word *adamah* means "land" or "soil"). Moses retreated to the mountain for forty days and nights to receive the Ten Commandments from God. The cloud and pillar of fire in the sky, and ultimately the parted and closed sea, guided the Israelites and safeguarded them from the Egyptian army. Spirit birds appear in biblical stories, too. God's children in Isaiah 40:31 "shall mount up with wings like eagles; they shall run and not be weary." In Deuteronomy 32, the eagle stands in for the figure of the Lord protecting His people, "like an eagle that stirs up its nest / and hovers over its young / that spreads its wings to catch them / and carries them aloft."

In his "on eagle's wings" speech, Biden was quoting from a Catholic version of Psalm 91 composed in 1979, when the number of nesting pairs of eagles in the Lower Forty-Eight was climbing beyond a thousand. The transcendent eagle at Biden's victory ceremony spoke evocatively to national identity, and not just from the speech. Behind and to each side of the president-elect was an American flag draped from a gilded staff topped by a bald eagle finial. For the front of his lectern, his team had fashioned an "Office of the President Elect" emblem that incorporated the founding bird from the Great Seal of the United States.

In its freer and ubiquitous state, the founding bird projected, as Francis Herrick seventy-plus years earlier had said would happen, a more honest and powerful image of national exceptionalism than before. Yet it did not thrive upon its ambassadorial calling alone. The new compact with the bald went beyond cultural considerations. Recall that when Congress passed the Bald Eagle Protection Act and then authorized each of its subsequent revisions, the looming extinction of the bird behind the national symbol gave conviction to species retention. By the time of Biden's speech, the inherent right of the species to exist had become an equally powerful rationale for retention. As bald eagles came back into their world, Americans could not imagine ever losing them again. They were a treasured signature of the natural aesthetic, which Americans embraced with invigorated fervor that recalled the days of Charles Willson Peale. Visible riding the sky, these dogged survivors were an affecting pat on America's back for doing right by the environment, and a reminder that humans and birds were of the same world.

The charismatic raptor's deification is difficult to imagine without the physical bird having reasserted its historic place in its native land. The degree to which it rebounded exceeded everyone's expectations, except perhaps those of the eagle. Living for itself rather than for humankind, it pursued the evolutionary will for self-preservation and set an example of what can be. "This is the time to heal in America," Biden said in his victory speech. He was referring to more than social and political divisions. Also on his mind was mending the human relationship with the Earth, which he depicted as the "battle to save the climate." For bald eagles, the implications of climate change and sea-level rise lay primarily in the potential diminishment of food sources and habitat. Fortunately, *Haliaeetus leucocephalus* can alter its migration patterns more easily than many other avian species can, and it is not confined to a particular climatic zone or a specific habitat, such as birds that depend on a single food source. It also has a proven track record of resilience.[10]

Americans have traveled a long distance in witnessing that resilience. On the way, they formed a palpable bond with the living species—a bond that has emerged from coming to know and learn from it and from welcoming it back into daily life. Bald eagles have become a guide, as Telia Hann's mother described to her son, for paying attention to the physical world, its nuances, its sublimity, and the history within it, to see that world as a moral universe and to form an ethical and emotional tie with it—to move beyond a purely sensory awareness to live, from deep within, in kinship with bald eagles, as Native peoples have.

During their long journey from being Benjamin Franklin and John James Audubon's morally corrupt birds to the vaunted species they ultimately became in the eyes of Americans, bald eagles showed that the improbable can be a prelude to the possible. It was not these steadfast birds that changed. They carried on as they always had, as neither immoral nor moral. What changed was American sensibilities. The story of the experience of bald eagles as America's bird is a study in awareness, transformation, and commitment. It is a story that can help us navigate unprecedented environmental challenges that have arrived decisively in the twenty-first century, alongside the bald eagles' comeback. It is a story of possibilities. If animals form a portrait of our virtues and vices, as many contend, then bald eagles have shown us, in our salutary coexistence with them, that our nature is predisposed to virtue. The living birds not only freed Americans from a burdensome past but provided a vision for the future, for the species' own organic vision of the world and life is an edifying one.

The story of *Haliaeetus leucocephalus* is also the story of itself apart from America. When bald eagles rise from the Saskatchewan interior, the Alaskan mainland, the Florida peninsula, the South Texas canyonlands, and all across their continent, in numbers that once existed and that exist once again, they are a testament to fortitude and the virtue of adaptability.

ACKNOWLEDGMENTS

WRITING A BOOK IS A TEDIOUS TASK—A TASK OFTEN described as lonely. Yet many people surround you with sizable quantities of encouragement and support.

My literary agent, Lisa Adams, is one such person. She believed in this book (and me) from the start and helped find it a home. We never doubted that home would be with Liveright and Bob Weil, my editor for this and my previous book. He has a veteran's way of gently nudging to elevate the work to your fullest potential while giving praise to the work, and he's a good one for tolerating my baseball metaphors, allegories, and similes. I'm fortunate to have him as a guide and friend, and to have Haley Bracken assisting him. And what a pleasure it is to team up again with Cordelia Calvert, publicist supreme, whose enthusiasm for my work keeps me buoyed. Nick Curley, Rebecca Homiski, Peter Miller, Jessica Murphy, and Steven Pace (from perhaps the most literary state in the Union, and truly one of my favorites) round out a stellar team at Liveright/Norton.

I don't possess the words to express how important Cynthia Barnett is to my writing career. She reads virtually everything I write, doing so with an astutely discerning eye. I am awed by her own work and forever grateful for her taking an interest in mine.

Kathy Bollerud not only read the entire book in draft but listened to me read nearly the entire book. Her cheerleading skills are surpassed by no one, and her comments are always valuable. Katie Hafner read several chapters and scrutinized each as expertly as anyone. Beate Becker (thanks for BecDowell!), Mary Santello, and Lynn Weir were selfless readers too, to say nothing of being—like Kathy and

Katie—good friends. Vickie Machado, whose friendship and company brings uplifting relief, proved such a superb reader that I wish I had tapped her abilities long before I did.

Others who read or listened to parts of the book, and in some cases did both, include Paul Armbrust (thanks for the fish and venison), Julie Bollerud, Ray Bollerud (thanks for the tequila and venison), Kirsten Colantino, Frank Corrado, John Costin, Jeannie Eastman, Isaac Eger, Katrina Farmer, Jack Feldman, Sandy Flory, Aaron Hoover, Mary Hubbard (thanks for the tequila), Lee Irby, Maggie Newman, Leslie Poole, Sonya Rudenstine, Willa Rudenstine, Justin Saarinen, Melissa Seixas, Jordan Fisher Smith, Bruce Stephenson, Lida Stinchfield, Geoff Sutton, David "Tommy T" Tomaszewski, Susan Weiler (thanks for eagle watching with me), Anne Wolf, M'Lue Zahner, and Jessica "CJ!" Zander. Thank you all.

Some of the above and many others sent newspaper and magazine articles and media releases, or contributed in other important ways. Among them were Ray Arsenault, David Auth, Joanne Auth, Julianna Barr, Patrick Bradley, Doug Brinkley, Wendy Davis, Mathias Engelmann, Mary Eno, Jackson Farmer, Anna Furr, Meg Gammage-Tucker, Chuck Gattone, Liz Gottlieb, Pat Harden, Michelle Hudson, Jackie Jones, Kena Kincaid, Gary Mormino, Lynne Mormino, Pam Murray, Andy Mele, Jean Reakes, Bubba Scales (thanks for Mark Twain), Dan Simone, Jill Vandenboogaart, William Voelker, Jim Warren, N. Adam Watson, and Jessica Wells.

I must acknowledge the archivists and librarians, who have to be the most selfless and helpful people in existence: Bridget Bihm-Manel, Jim Cusick, Jorge Gonzalez, John Nemmers, and Flo Turcotte of the University of Florida libraries. Leah Geibel of the Alaska State Archives did some important digging, as did John Worth at the University of West Florida. I also appreciate the assistance of the staffs at the American Philosophical Society Library, Carolina Raptor Center, Historical Society of Philadelphia, Florida State Archives, Smithsonian Libraries, and Hawk Mountain Sanctuary.

I am equally grateful to those who granted interviews: Patrick Bradley, Mike Collopy, Bill Davis, Dianne Davis, Tom French, David

Hancock, Telia Hann, Will Harris, Kevin Heaney, Nick Igdalsky, Doris Mager, Steve Nesbitt, Jim Ozier, Dan Reinking, Steve Sherrod, Ted Simon, Ria Warner, John Weller, Don Wolfe, and Petra Wood.

I cannot leave out Lorraine Redd, who made a vital contribution to the book as a research assistant and reader. She rummaged deep and found information I would likely have missed. Her insight made the book a much better one, and her sense of the natural world has long influenced me to an extent she may not imagine. Her friendship means the world to me.

From the start, Preston Cook graciously made his vast archive of bald eagle documents and artifacts available. And he hosted me on a research trip, when in the evenings we sat beside the Mississippi River, watching eagles and sipping tequila. His *American Eagle: A Visual History of Our National Emblem* is an inspiration.

Mark Techler, Dave McCally, and Rae Nell Hunter passed away during the course of my writing this book. In addition to being longtime friends, they were always important intellectual lights to me. My sisters and I lost our mother, Becky Davis. She fed all the critters—avian and terrestrial—in her backyard, setting an example for us to appreciate wildlife.

I was fortunate to have Stephanie Hiebert as the copy editor for my previous book, and I was grateful when she agreed to lend her considerable expertise on this one. She's a master at tightening leggy sentences and preventing me from going where I shouldn't go with my writing.

This book benefited immensely from a generous fellowship granted by the Carnegie Corporation, which funded research and covered a year's salary so I could write. The University of Florida tacked on a one-year sabbatical, and the Rothman Family Endowed Chair in the Humanities provided additional and significant support. My colleagues in the history department, led skillfully by our department chair, Elizabeth Dale, during the time I wrote this book, set the bar high as scholars and encouraging colleagues, as do Dean Dave Richardson, Associate Dean Chris McCarty, and Associate Dean David Pharies of the College of Liberal Arts and Sciences. UF President Kent Fuchs nominated me for the Carnegie Fellowship and goes out of his way to express his appreciation for my work.

I fear I might have forgotten someone to thank, and hope I have not. The one person I will never forget is Marina Powdermaker, to whom this book is dedicated. She read it in draft through chapter 7, until she could no longer read. She then insisted on having those seven chapters and the newly written eighth read to her when she was bedridden, starting over from the beginning, until she could no longer concentrate. Well before she learned that the time had come to leave this life, we had many invigorating conversations about her art and my writing and the creative process behind each. Her art hangs on my wall, and I wish that she could have seen this book on her shelf and that I could thank her in person for her contributions to it. She understood and could articulate the spirit of the bald eagle as well as anyone.

Notes

Prologue

1. Passages in Mark V. Barrow Jr.'s excellent book on species extinction—*Nature's Ghosts: Confronting Extinction from the Age of Jefferson to the Age of Ecology* (University of Chicago Press, 2015), 234–45, 249–50—offer exceptions.

Introduction: Haliaeetus Leucocephalus: *The Species*

1. William R. Muchlberger, Patricia W. Dickerson, J. Russel Dyer, and David V. LeMone, *Structure and Stratigraphy of Trans-Pecos Texas: El Paso to Guadalupe Mountains and Big Bend July 20–29, 1989*, Field Trip Guidebook, vol. 317 (American Geophysical Union, 1989), xi.
2. John Burroughs, *Birds and Bees: Essays* (Houghton, Mifflin, 1887), 50.
3. Burroughs, *Birds and Bees*.
4. Arthur Barlowe, "The First Voyage to Roanoke. 1584. The First Voyage Made to the Coasts of America, with Two Barks Wherein Were Captains M. Philip Amadas and M. Arthur Barlowe, Who Discovered Part of the Country Now Called Virginia, Anno 1584. Written by One of the Said Captains, and Sent to Sir Walter Raleigh, Knight, at Whose Charge and Direction, the Said Voyage Was Set Forth," Old South Leaflets no. 92, North Carolina Collection, University of North Carolina at Chapel Hill, electronic edition, https://docsouth.unc.edu/nc/barlowe/barlowe.html. For full studies of the "Lost Colony," as Roanoke came to be known, see Andrew Lawler, *The Secret Token: Myth, Obsession, and the Search for the Lost Colony of Roanoke* (Anchor Books, 2018); James Horn, *A Kingdom Strange: The Brief and Tragic History of the Lost Colony of Roanoke* (Basic Books, 2010).
5. Thomas Hariot, *A Briefe and True Report of the New Found Land of Virginia* (1588; repr., Dodd, Mead, 1903), x. The seventeenth-century New England chronicler William Wood wrote that America's white-headed, white-tailed eagles were "commonly called Gripes." William Wood, *Wood's New-England's Prospect* (John Wilson and Son, 1865), 30.

Chapter 1: *Two Myths / Searching for a Seal*

1. Benjamin Franklin to Sarah Bache, January 26, 1784, in Benjamin Franklin, *The Works of Benjamin Franklin: Containing Several Political and Historical Tracts Not Included in Any Former Edition, and Many Letters Official and Private Not Hitherto Published*, vol. 10 (Benjamin Franklin Stevens, 1882), 58.

2. Paul Leicester Ford, ed., *The Writings of Thomas Jefferson* (G. P. Putnam's Sons, 1899), 312–13.

3. Franklin to Bache, January 26, 1784.

4. Benjamin Franklin, *Poor Richard's Almanac* (1757; repr., H. M. Caldwell, 1900), 130; Jill Lepore, *The Story of America: Essays on Origins* (Princeton University Press, 2010), 46.

5. Benjamin Franklin, *Experiments and Observations on Electricity* (1769; repr., Cambridge University Press, 2019), 153.

6. William Temple Franklin, *The Private Correspondence of Benjamin Franklin* (A. J. Valpy, 1817).

7. US Continental Congress, *Journals of the American Congress from 1774–1788: In Four Volumes*, vol. 4, *April 1, 1782 to November 1, 1788, Inclusive* (Way and Gideon, 1823), 79.

8. Howard Corning, ed., *Journal of John James Audubon: Made during His Trip to New Orleans in 1820–1821* (Business Historical Society, 1929), 61–62; Franklin to Bache, January 26, 1784.

9. Pauline Maier, *American Scripture: Making the Declaration of Independence* (Vintage Books, 1997), 144, 147, 149; Jill Lepore, *These Truths: A History of the United States* (W. W. Norton, 2018), xv.

10. US Continental Congress, *Journals of the Continental Congress 1774–1789*, vol. 5 (Government Printing Office, 1906), 518; Edmund Cody Burnett, ed., *Letters of Members of the Continental Congress*, vol. 1, *August 29, 1774 to July 4, 1776* (Carnegie Institution of Washington, 1921), 530.

11. Janine Rogers, *Eagle* (Reaktion Books, 2015), 85; Ernest Ingersoll, *Birds in Legend, Fable and Folklore* (Longmans, Green, 1923), 28–29.

12. Moncure D. Conway, "The English Ancestry of Washington," *Harper's New Monthly Magazine* 82 (December 1890–May 1891): 877; Gordon S. Wood, *Friends Divided: John Adams and Thomas Jefferson* (Penguin, 2017), 58.

13. Charles A. L. Totten, *The Seal of History: Our Inheritance in the Great Seal of "Manasseh," the United States of America* (Our Race Publishing, 1897), 8.

14. Richard S. Patterson and Richardson Dougall, *The Eagle and the Shield: A History of the Great Seal of the United States* (US Department of State, 1978), 18.

15. *Pierre du Simitière: His American Museum 200 Years After* (exhibition pamphlet) (Library Company of Philadelphia, 1985).

16. Benjamin Franklin, *The Autobiography of Benjamin Franklin* (Applewood Books, 2008), 85; Thomas S. Kidd, "How Benjamin Franklin, a Deist, Became the Founding Father of a Unique Kind of American Faith," *Washington Post*, June 28, 2017; Thomas S. Kidd, *Benjamin Franklin: The Religious Life of a Founding Father* (Yale University Press, 2017), 7.

17. Patterson and Dougall, *Eagle and the Shield*, 20.

18. Philip M. Isaacson, *The American Eagle* (New York Graphic Society, 1975), 22; US Continental Congress, *Journals of the Continental Congress, 1774–1789*, 5:689–91.

19. Polly Redford, *Raccoons and Eagles: Two Views of American Wildlife* (E. P. Dutton, 1963), 135.

20. John James Audubon, *Ornithological Biography, or an Account of the Habits of the Birds of the United States of America*, vol. 1 (Adam Black, 1831), 168.

21. Frank Abial Flower, *Old Abe, the Eighth Wisconsin War Eagle: A Full Account and Enlistment, Exploits in War and Honorable as Well as Useful Career in Peace* (Curran and Bowen, 1885), 13.

22. David R. Brigham, "'Ask the Beasts, and They Shall Teach Thee': The Human Lessons of Charles Willson Peale's Natural History Displays," *Huntington Library Quarterly* 59 (1996): 193; Robert Shlaer, "An Eagle's Eye: Quality of the Retinal Image," *Science* 176 (May 26, 1972): 920–22.

23. Redford, *Raccoons and Eagles*, 127.

24. Patterson and Dougall, *Eagle and the Shield*, 38.

25. William Barton to George Washington, August 28, 1788; and George Washington to

William Barton, September 7, 1788, both in *The Papers of George Washington Digital Edition* (University of Virginia Press, Rotunda, 2008) http://rotunda.upress.virginia.edu/founders/GEWN-04-06-0-0446.

26. Patterson and Dougall, *Eagle and the Shield*, 61–62.
27. Eugene Zieber, *Heraldry in America* (Haskell House, 1969), 99.
28. Patterson and Dougall, *Eagle and the Shield*, 62.
29. "John Adams Diary 21, 15 August–3 September 1774," *Adams Family Papers: An Electronic Archive*, Massachusetts Historical Society, https://www.masshist.org/digitaladams/archive/doc?id=D21&bc=%2Fdigitaladams%2Farchive%2Fbrowse%2Fdiaries_by_date.php.
30. "Samuel Holten's Diary, Letters of Delegates to Congress, vol. 14, October 1, 1779–March 31, 1780," *American Memory*, http://memory.loc.gov/ammem/amlaw/lwdg.html.
31. Patterson and Dougall, *Eagle and the Shield*, 75; US Continental Congress, *Journals of the American Congress from 1774–1788*, 4:39.
32. Patterson and Dougall, *Eagle and the Shield*, 84–89.
33. Patterson and Dougall, *Eagle and the Shield*, 97–99; Joachim Camerarius, *Symbolorum et emblematum ex re herbaria desumtorum centuria una collecta*, vol. 3 (Johannis Ammonii, 1654), 3, 15.
34. Douglas V. Campana and Pam J. Crabtree, "Soldiers' Diet at Valley Forge: An Analysis of the Faunal Remains from the 2000 Excavation Season," *Bulletin of the Florida Museum of Natural History* 44 (2003): 199–204.
35. For more on the natural and cultural history of the American shad, see John McPhee, *The Founding Fish* (Farrar, Straus and Giroux, 2003). The "Shad-e-o!" quote comes from page 249.
36. McPhee, *Founding Fish*, 156.
37. Wallace Stevens, "Some Friends from Pascagoula," in *The Collected Poems of Wallace Stevens*, ed. John N. Serio and Chris Beyers (Vintage Books, 1982), 134–35.
38. Yves Prévost, "Osprey-Bald Eagle Interactions at a Common Foraging Site," *Auk* 96 (April 1979): 413–14; Charles P. Schaadt and Larry M. Ryan, "Innate Fishing Behavior of Ospreys," *Raptor Research* 16 (Summer 1982): 61–62; Richard O. Bierregaard, Alan F. Poole, and Brian Washburn, "Ospreys (Pandion Haliaetus) in the 21st Century: Populations, Migration, Management, and Research Priorities," *Journal of Raptor Research* 48 (December 2014): 301–8.
39. David Hackett Fischer, *Liberty and Freedom: A Visual History of America's Founding Ideas* (Oxford University Press, 2005), 148.
40. François Jean Chastellux, *Travels in North-America, in the Years 1780-81-82* (G. G. J. and J. Robinson, 1787), 387.
41. Greg Breining, *Return of the Eagle: How America Saved Its National Symbol* (Lyon Press, 2008), 26; Fischer, *Liberty and Freedom*, 149.

CHAPTER 2: *Buttons and Coins / National Identity, National Expansion, and Everywhere the "Monarch of the Air"*

1. *Pennsylvania Packet* (Philadelphia), August 1, 1782.
2. Richard S. Patterson and Richardson Dougall, *The Eagle and the Shield: A History of the Great Seal of the United States* (US Department of State, 1978), 86–87; Philip M. Isaacson, *The American Eagle* (New York Graphic Society, 1975), 6.
3. *Pennsylvania Packet* (Philadelphia), March 12, 1789; Isaacson, *American Eagle*, 45.
4. Donald Jackson and Dorothy Twohig, eds., *The Diaries of George Washington*, vol. 5, *July 1786–December 1789* (University Press of Virginia, 1979), 474–75; Isaacson, *American Eagle*, 46–47.
5. "National Standards and Emblems," *Harper's New Monthly Magazine* 47 (June–November 1873): 178; Edmund Cody Burnett, ed., *Letters of Members of the Continental Congress*, vol. 1, *August 29, 1774 to July 4, 1776* (Carnegie Institution of Washington, 1921), 253.

6.   "Eagle on New Quarter Defended by Expert as Bald, Not Golden," *New York Times*, August 21, 1932.

7.   Maude M. Grant, "The Eagle, Our National Emblem," in *The Junior Instructor*, ed. Walter J. Beecher and Grace B. Faxon (F. A. Owen, 1921), 162.

8.   Elliott Coues, *The History of the Lewis and Clark Expedition*, vol. 3 (Francis P. Harper, 1893; repr., Dover, 1979), 879.

9.   C. A. Hartnagel and Sherman C. Bishop, "The Mastodons, Mammoths, and Other Pleistocene Mammals of New York State," *New York State Bulletin* 241 (January–February 1921): 21; Richard Conniff, "Mammoths and Mastodons: All American Monsters," *Smithsonian*, April 2010, https://www.smithsonianmag.com/science-nature/mammoths-and-mastodons-all-american-monsters-8898672; Donald E. Stanford, "The Giant Bones of Claverack, New York, 1705," *New York History* 40 (January 1959): 47.

10. Abigail Adams to Mrs. Shaw, November 21, 1786, in *Letters of Mrs. Adams, the Wife of John Adams*, ed. Charles Francis Adams (Freeman and Bolles, 1848), 310.

11. Barbara Arneil, *John Locke and America: The Defence of English Colonialism* (Clarendon, 1996), 42.

12. Nathaniel Parker Willis, *American Scenery; or, Land, Lake, and River Illustrations of Transatlantic Nature* (George Virtue, 1840), iiv; Adams to Shaw, November 21, 1786, 309.

13. Hartnagel and Bishop, "Mastodons, Mammoths and Other Pleistocene Mammals," 62; Oliver Jensen, "The Peales," *American Heritage* 6 (April 1955), https://www.americanheritage.com/peales.

14. Mark V. Barrow Jr., *Nature's Ghosts: Confronting Extinction from the Age of Jefferson to the Age of Ecology* (University of Chicago Press, 2015), 19; Martin J. S. Rudwick, *Georges Cuvier, Fossil Bones, and Geological Catastrophes: New Translations and Interpretations of the Primary Texts* (University of Chicago Press, 1997), 89, 96; Conniff, "Mammoths and Mastodons."

15. Cornelius Vermeule, *Numismatic Art in America: Aesthetics of the United States Coinage* (Whitman, 2007), 22, 40, 132.

16. Mical Elizabeth Tawney, "Charles Willson Peale's Philadelphia Museum: A Space of Amusement, Education, and American Citizenship" (master's thesis, University of Virginia School of Architecture, 2015), 36, 140, 153.

17. David C. Ward, *Charles Willson Peale: Art and Selfhood in the Early American Republic* (University of California Press, 2004), 170.

18. Edward P. Alexander, *Museum Masters: Their Museums and Their Influence* (Altamira Press, 1995), 64.

19. Jensen, "Peales"; Lillian B. Miller, ed., *The Selected Papers of Charles Willson Peale and His Family*, vol. 5, *The Autobiography of Charles Willson Peale* (Smithsonian Institution, 1983), 271–72.

20. Ward, *Charles Willson Peale*, 222; Alexander, *Museum Masters*, 222; Joseph Kastner, *A World of Naturalists* (John Murray, 1978), 153.

21. Charles A. Miller, *Jefferson and Nature: An Interpretation* (Johns Hopkins University Press, 1988), 7.

22. John Adams to Benjamin Waterhouse, June 11, 1811, in *The Papers of Thomas Jefferson Digital Edition*, ed. James P. McClure and J. Jefferson Looney (University of Virginia Press, Rotunda, 2008–2021), https://rotunda.upress.virginia.edu/founders/TSJN-02-02-02-0021.

23. Miller, *Jefferson and Nature*; Thomas Jefferson, *Notes on the State of Virginia* (Prichard and Hall, 1788; electronic version, University of North Carolina at Chapel Hill Libraries, 2006), 18, 56, 59; Thomas Jefferson to Archibald Cary, January 7, 1786, *Founders Online*, National Archives, https://founders.archives.gov/documents/Jefferson/01-09-02-0143.

24. John P. Kaminski, ed., *Jefferson in Love: The Love Letters between Thomas Jefferson and Maria Cosway* (Rowan & Littlefield, 1999), 50.

25. *Proceedings of the American Philosophical Society* 50, no. 198 (January–April 1911): 1;

Wilson Jeremiah Moses, *Thomas Jefferson: A Modern Prometheus* (Cambridge University Press, 2019), 267.

26. Andrea Wulf, *The Brother Gardeners: A Generation of Gentlemen Naturalist and the Birth of an Obsession* (Vintage Books, 2010), 99, 107.

27. Edward J. Cashin, *William Bartram and the American Revolution on the Southern Frontier* (University of South Carolina Press, 2000), 196; William Bartram, *Travels through North & South Carolina, Georgia, East & West Florida, the Cherokee Country, the Extensive Territories of the Muscogulges, or Creek Confederacy, and the Country of the Chactaws; Containing an Account of the Soil and Natural Productions of Those Regions, Together with Observations on the Manners of the Indians* (James and Johnson, 1793; electronic ed., University of North Carolina at Chapel Hill, 2001), 73; William Bartram, *The Travels of William Bartram*, ed. Francis Harper, Naturalist's Edition (University of Georgia Press, 1998), xxxv.

28. Bartram, *Travels of William Bartram* (1998), 99.

29. Bartram, *Travels through North & South Carolina* (1793/2001), 8, 199, 352.

30. Bartram, *Travels of William Bartram* (1998), 66–67.

31. Bartram, *Travels of William Bartram* (1998), xxiv.

32. Bartram, *Travels of William Bartram* (1998), 2.

33. Keith Thomson, *A Passion for Nature: Thomas Jefferson and Natural History* (Thomas Jefferson Foundation, 2008), 111, 113; Thomas Jefferson to Benjamin Smith Barton, February 27, 1803, *Founders Online*, National Archives, https://founders.archives.gov/documents/Jefferson/01-39-02-0499.

34. Meriwether Lewis, *History of the Expedition under the Command of Captains Lewis and Clark, to the Sources of the Missouri, Thence across the Rocky Mountains and down the River Columbia to the Pacific Ocean Performed during the Years 1804-5-6* (Bradford and Inskeep; Abm. H. Inskeep, 1814), 188, 310.

35. François-René de Chateaubriand, *Recollections of Italy, England and America, with Essays on Various Subjects, in Morals and Literature* (Henry Colburn, 1815), 206.

36. François Jean Chastellux, *Travels in North-America, in the Years 1780-81-82*, vol. 1 (1828; Applewood Books, 2009), 233, 250, 339, 357.

37. Alexis de Tocqueville, *Memoirs, Letters, and Remains of Alexis de Tocqueville* (Ticknor and Fields, 1862), 185; Alexis de Tocqueville, *The Republic of the United Sates of America, and Its Political Institutions, Reviewed and Examined*, trans. Henry Reeves (A. S. Barnes, 1851), 20, 25, 349.

38. Zebulon Pike and Elliot Coues, *The Expeditions of Zebulon Montgomery Pike, to the Headwaters of the Mississippi River, through Louisiana Territory, and in New Spain, during the Years 1805-6-7* (1810; repr., Francis P. Harper, 1895), 497.

39. F. G. Young, ed., *The Correspondence and Journals of Captain Nathaniel J. Wyeth, 1831-6* (University of Oregon Press, 1899), 119, 212.

40. Andrea Wulf, *Founding Gardeners: The Revolutionary Generation, Nature, and the Shaping of the American Nation* (Vintage Books, 2011), 171.

41. Isaac Thomas, ed., *Selections from Washington Irving*, The Students' Series of English Classics (Leach, Shewell, & Sanborn, 1894), 244, 245.

42. Washington Irving, *A Tour on the Prairies* (John Murray, 1835), 56, 212.

43. Roderick Nash, *Wilderness and the American Mind* (Yale University Press, 1982), 70; *Sacramento (CA) Daily Union* 43 (September 14, 1872).

44. James A. Craig, *Fitz H. Lane: An Artist's Voyage through Nineteenth-Century America* (History Press, 2006), 119; Philip Kennicott, "Philadelphia Might Have Been a Sylvan Athens. Instead, We Got Washington, Our Muddy Rome," *Washington Post*, August 1, 2019.

45. Gordon S. Wood, *Empire of Liberty: A History of the Early Republic, 1789–1815* (Oxford University Press, 2009), 723.

46. Louis Legrand Noble, *The Life Works of Thomas Cole* (Sheldon, Blakeman, 1856), 277; Thomas Cole, "Essay on American Scenery," *American Monthly Magazine* 1 (January 1836), 12; Nash, *Wilderness and the American Mind*, 81.

47. Isaacson, *American Eagle*, 47.
48. George Catlin, *Letters and Notes on the Manners, Customs, and Condition of the North American Indians*, vol. 1 (Willis P. Hazard, 1857), 19, 168, 257.
49. Catlin, *Letters and Notes*, 397.
50. Wulf, *Founding Gardeners*, 171; William Cullen Bryant, ed., *Picturesque America, or, the Land We Live In* (D. Appleton, 1872), 3, 4; Nash, *Wilderness*, 71, 75.
51. William Cullen Bryant, *The Complete Poetical Works of William Cullen Bryant* (Knight and Son, 1854), 93.
52. George Ticknor, *Life, Letters, and Journals of George Ticknor* (James R. Osgood, 1876), 34; Benson J. Lossing, "Monticello," *Harper's New Monthly Magazine* 38 (July 1853), 148.
53. Lossing, "Monticello," 148.

Chapter 3: *Twice-Baked Turkey / The Bald Eagle in Early Science, or Ornithologists with Guns*

1. Edward H. Burtt Jr. and William E. Davis Jr., *Alexander Wilson: The Scot Who Founded American Ornithology* (Belknap Press of Harvard University Press, 2013).
2. William Souder, *Under a Wild Sky: John James Audubon and the Making of* The Birds of America (North Point, 2004); Richard Rhodes, *John James Audubon: The Making of an American* (Alfred A. Knopf, 2004); Gregory Nobles, *John James Audubon: The Nature of the American Woodsman* (University of Pennsylvania Press, 2017).
3. Audubon quotes that follow come from his *Ornithological Biography, or an Account of the Habits of the Birds of the United States of America*, vol. 1 (Adam Black, 1831), 160–69.
4. Benjamin Franklin to Sarah Bache, January 26, 1784, in Benjamin Franklin, *The Works of Benjamin Franklin: Containing Several Political and Historical Tracts Not Included in Any Former Edition, and Many Letters Official and Private Not Hitherto Published* (Benjamin Franklin Stevens, 1882), 10:58.
5. New-York Historical Society, "Talking Turkey and John James Audubon," Behind the Scenes, November 2, 2017, http://behindthescenes.nyhistory.org/talking-turkey-and-john-james-audubon; Rhodes, *John James Audubon*, 237. Thanks to Preston Cook for measuring the text length of these essays.
6. New-York Historical Society, "Talking Turkey"; Carolyn E. DeLatte, *Lucy Audubon: A Biography* (Louisiana State University Press, 1982), 77; D. T. A. Tyler, "Audubon's Seal," *Bird-Lore* 1 (October 1899): 173.
7. John James Audubon, *Journal of John James Audubon: Made during His Trip to New Orleans in 1820–21* (Business Historical Society, 1929), 61.
8. Audubon, *Journal*, 63.
9. Audubon, *Journal*, 62, 63.
10. Alexander Wilson, *Memoir and Remains of Alexander Wilson, the American Ornithologist*, ed. Alexander B. Grosart, vol. 2 (A. Gardner, 1876), 129, 389; Homer, *Iliad*, trans. Rodney Merrill (University of Michigan Press, 2007), 68.
11. John Alexander Moore, *Science as a Way of Knowing: Foundations of Modern Biology* (Harvard University Press, 1993), 173.
12. Mark V. Stalmaster, *The Bald Eagle* (Universe Books, 1987), 5.
13. Scott Weidensaul, *A World on the Wing: The Global Odyssey of Migratory Birds* (W. W. Norton, 2021), 55.
14. Jessica Metcalfe, Kim Schmidt, Wayne Kerr, Christopher Guglielmo, and Scott MacDougal-Shackleton, "White-Throated Sparrows Adjust Behavior in Response to Manipulations of Barometric Pressure and Temperature," *Animal Behavior* 86 (December 2013): 1285–90.
15. Tim Birkhead, *The Wonderful Mr. Willughby* (Bloomsbury Publishing, 2018).
16. Francis Willughby and John Ray, *Ornithologiae libri tres* (John Martyn, 1676).
17. Polly Redford, *Raccoons and Eagles: Two Views of American Wildlife* (E. P. Dutton, 1963), 139.

18. Redford, *Raccoons and Eagles*, 138.

19. John Brickell, *The Natural History of North Carolina: With an Account of the Trade, Manners, and Customs of the Christian and Indian Inhabitants* (James Carson, 1737), 171.

20. William Bartram, *The Travels of William Bartram*, ed. Francis Harper, Naturalist's Edition (University of Georgia Press, 1998), 5.

21. Mark Catesby, *The Natural History of Carolina, Florida, and the Bahama Islands*, vol. 1 (C. Marsh, 1754), 1.

22. "On the Hibernation of Swallows, by the Late Colonel Antes. Communicated by Dr. Barton," *Transactions of the American Philosophical Society* 6 (1809): 59–60.

23. Burtt and Davis, *Alexander Wilson*, 302.

24. Catesby, *Natural History of Carolina*, 1:1, 2.

25. Robert Henry Welker, *Birds and Men: American Birds in Science, Art, Literature, and Conservation, 1800–1900* (Atheneum, 1966), 12.

26. Alexander Wilson, *The Poems and Literary Prose of Alexander Wilson: The American Ornithologist*, ed. Alexander B. Grosart, vol. 1 (A. Gardner, 1876), 118; Alexander Wilson, *The Life and Letters of Alexander Wilson*, ed. Clark Hunt (American Philosophical Society, 1983), 191.

27. Wilson, *Memoir and Remains*, 359–61; Alexander Wilson, *Wilson's American Ornithology* (T. L. Magagnos, 1854), 129, 154, 189, 498, 499, 526, 564; Welker, *Birds and Men*, 18.

28. Alexander Wilson to William Bartram, May 22, 1807, in Witmer Stone, "Some Unpublished Letters of Alexander Wilson and John Abbot," *Auk* 23 (October 1906): 362.

29. Wilson, *Memoir and Remains*, 363; Wilson, *Life and Letters*, 86.

30. Walter Faxon, "John Abbot's Drawings of the Birds of Georgia," *Auk* 13 (July 1896): 212; C. Lucy Brightwell, *Difficulties Overcome: Scenes in the Life of Alexander Wilson, the Ornithologist* (Sampson, Low, Sona, 1861), 139; Alexander Wilson, *American Ornithology, or the Natural History of the Birds of the United States*, 9 vols. (Bradford and Inskeep, 1808–14), 1:4.

31. Wilson, *American Ornithology*, 1:iv.

32. Wilson, *Memoir and Remains*, 1:xl.

33. "Wilson's American Ornithology," *Blackwood's Edinburgh Magazine* 30 (July–December 1831): 279.

34. Wilson, *American Ornithology*, 4:89.

35. Wilson, *American Ornithology*, 4:332.

36. Welker, *Birds and Men*, 20; Wilson, *Life and Letters*, 378.

37. Wilson, *American Ornithology*, 4:95.

38. Wilson, *American Ornithology*, 4:95; John Lawson, *History of North Carolina* (1831 manuscript; facsimile, Observer Printing House, 1903), 80.

39. Wilson, *American Ornithology*, 4:90, 99.

40. Wilson, *American Ornithology*, 4:89.

41. John James Audubon, *Audubon and His Journals*, ed. Maria R. Audubon (John C. Nimmo, 1898), 169.

42. Souder, *Under a Wild Sky*, 286.

43. John James Audubon, *The Audubon Reader*, ed. Richard Rhodes (Everyman's Library, 2006), 78; Audubon, *Ornithological Biography*, 321; the Cooper quote comes from Nobles, *John James Audubon*, 45, 186–87.

44. Nobles, *John James Audubon*, 45; Wilson, *Life and Letters*, 93.

45. Audubon, *Ornithological Biography*, 167.

46. Bruce E. Beans, *Eagle's Plume: The Struggle to Preserve the Life and Haunts of America's Bald Eagle* (Scribner, 1996), 83.

47. Wilson, *American Ornithology*, 6:62; John James Audubon, *The Birds of America*, vol. 1 (J. B. Chevalier, 1840), 238–39; Souder, *Under a Wild Sky*, 110; Rhodes, *John James Audubon*, 201.

48. Audubon, *Ornithological Biography*, 1:166.

49. Audubon, *Ornithological Biography*, 1:161.

50. Audubon, *Ornithological Biography*, 1:161–62.
51. Audubon, *Ornithological Biography*, 1:61.
52. Audubon, *Ornithological Biography*, 1:60–61. Ornithologists in the twentieth century would name the larger northern bald eagle *Haliaeetus leucocephalus washingtoniensis*.
53. Wilson, *American Ornithology*, 2:145 and 4:89, 91.
54. Wilson, *American Ornithology*, 4:100.

Chapter 4: *Perches / Bird of Paradox*

1. B. H. Latrobe to Charles Willson Peale, April 18, 1806; and Charles Willson Peale to Benjamin Henry Latrobe, April 21, 1806, in *The Selected Papers of Charles Willson Peale and His Family*, ed. Lillian B. Miller, vol. 2, *Charles Willson Peale, the Artist as Museum Keeper, 1791–1810* (Smithsonian Institution, 1983), 958–59.
2. Catherine Frances Cavanaugh, "Jefferson Davis' Imprint on the National Capitol," *New Age* (January 1908): 512–14; David Hackett Fischer, *Liberty and Freedom: A Visual History of America's Founding Ideas* (Oxford University Press, 2005), 299–300.
3. William C. Allen, *History of the United States Capitol: A Chronicle of Design, Construction, and Politics* (University Press of the Pacific, 2005), 326.
4. Allen, *History of the United States Capitol*, 327.
5. Allen, *History of the United States Capitol*, 327.
6. Neltje Blanchan, *Game Birds: Life Histories of One Hundred and Seventy Birds of Prey, Game Birds and Water-Fowls* (Doubleday, Page, 1922), 327.
7. "He Caught 'Old Abe,'" *Indian Leader* 13 (April 1915): 15.
8. "He Caught 'Old Abe,'" *Indian Leader*, 15.
9. David McLain, "That 'War Eagle' Had Quite a Record," *Leader-Telegram* (Eau Claire, WI), July 1, 1976; Frank Abial Flower, *Old Abe, the Eighth Wisconsin War Eagle: A Full Account and Enlistment, Exploits in War and Honorable as Well as Useful Career in Peace* (Curran and Bowen, 1885), 31.
10. Flower, *Old Abe*, 41.
11. Flower, *Old Abe*, 33; McLain, "That 'War Eagle' Had Quite a Record."
12. Flower, *Old Abe*, 52; Joseph O. Barrett, *The Soldier Bird: "Old Abe": The Live War Eagle of Wisconsin, That Served a Three Years' Campaign in the Great Rebellion* (Atwood & Culver, 1876), 58–59.
13. H. H. Bennett, *Old Abe at the Wisconsin State Capitol*, ca. 1875, photograph, Wisconsin Historical Society, https://www.wisconsinhistory.org/Records/Image/IM7534; Emily Thacher Bennett, "Bird Histories," *Audubon Magazine* 2 (February 1888–January 1889): 106.
14. Flower, *Old Abe*, 56.
15. Flower, *Old Abe*, 56.
16. "Old Abe,'" *Inter Ocean* (Chicago), May 13, 1875.
17. "The Past Week," *Weekly Herald* (Chippewa Falls, WI), May 21, 1875; "'Old Abe,'" *Inter Ocean*.
18. Bennett, "Bird Histories," 106.
19. "Shooting a Bald Eagle," *New York Times*, March 10, 1888; "A Saloon Keeper's Hobby," *New York Times*, July 31, 1886.
20. National Eagle Roost Registry (website), Center for Conservation Biology, accessed December 2020, https://ccbbirds.org/what-we-do/research/species-of-concern/species -of-concern-projects/national-eagle-roost-registry.
21. Mark V. Stalmaster and James R. Newman, "Behavioral Responses of Wintering Bald Eagles to Human Activity," *Journal of Wildlife Management* 42 (January 1979): 506–13.
22. "Shooting a Bald Eagle," *New York Times*.
23. "Shooting a Bald Eagle," *New York Times*.
24. Noel Burch, *Life to Those Shadows* (University of California Press, 1990), 131.

25. Ira H. Gallen, *D. W. Griffith: Master of Cinema* (Friesen Press, 2015), 120.

26. Alexander Wilson, *American Ornithology, or the Natural History of the Birds of the United States*, 9 vols. (Bradford and Inskeep, 1808–14), 1:37; Thomas Nuttall, *A Manual of the Ornithology of the United States and of Canada* (Hilliard and Brown, 1832), 75.

27. Blanchan, *Game Birds*, 327; *American Eagle Protection: Hearing before the Committee on Agriculture, House of Representatives, Seventy-First Congress, Second Session, January 31, 1930*, serial D (Government Printing Office, 1930), 17.

28. Ken Ross, *Pioneering Conservation in Alaska* (University of Colorado Press, 2006), 283, 284.

29. John McPhee, *The Control of Nature* (Farrar, Straus and Giroux, 1990), 6.

30. John L. O'Sullivan, "Annexation," *United States Magazine and Democratic Review* 17 (July–August 1845): 6; Julius W. Pratt, "The Origin of 'Manifest Destiny,'" *American Historical Review* 32 (July 1927): 795.

31. Georges-Louis Leclerc, Comte de Buffon, *Natural History, General and Particular*, English ed., vol. 5 (William Smellie, 1781), 122.

32. A pathbreaking history of the feather trade is Robin W. Doughty, *Feather Fashions and Bird Preservation: A Study in Nature Protection* (University of California Press, 1975). See also William T. Hornaday, *Our Vanishing Wild Life: Its Extermination and Preservation* (Charles Scribner's Sons, 1913), 124.

33. Thomas Dunlap, "Values for Varmints: Predator Control and Environmental Ideas, 1920–1939," *Pacific Historical Review* 53 (May 1984): 144.

34. Mark Catesby, *The Natural History of Carolina, Florida, and the Bahama Islands*, vol. 1 (C. Marsh, 1754), 1.

35. Polly Redford, *Raccoons and Eagles: Two Views of American Wildlife* (E. P. Dutton, 1963), 172–73; Barton W. Evermann, "Birds of Carroll County, Indiana," *Auk* 5 (1888): 350.

36. "An Eagle Killed by a Lady," *Baltimore Sun*, February 27, 1857; "Eagle Shot," *Orleans Independent Standard* (Irasburgh, VT), November 18, 1864; "Old Times," *Ironton County (MO) Register*, February 16, 1899; "Bald Eagle Shot," *Philadelphia Inquirer*, September 28, 1861; "Bird of Freedom Winged," *Boston Globe*, February 5, 1892.

37. "A Bald Eagle Story," *Star Democrat* (Easton, MD), May 25, 1858.

38. "The Bald Eagle's Bad Habits," *Sun* (New York), July 23, 1905; "A Bald Eagle Shot," *Trenton (NJ) State Gazette*, May 5, 1853; "Battle with an Eagle," *Roanoke (VA) Times*, January 30, 1890; "Large Gray Eagle Killed," *Muncie (IN) Evening Press*, November 1896.

39. "Eagle's Fight for Its Mate," *Boston Globe*, April 13, 1904.

40. "Miscellaneous Items," *Brooklyn Daily Eagle*, August 12, 1872; "Bald Eagle Shot," *Detroit Free Press*, April 28, 1900; "Eagle Escapes and Carries Away Pig," *Pittsburgh Press*, February 7, 1916; William Cullen Bryant, ed., *Picturesque America, or, the Land We Live In* (D. Appleton, 1872), 87.

41. Albert K. Fisher, *The Hawks and Owls of the United States in the Their Relation to Agriculture*, US Department of Agriculture Bulletin, no. 3 (Government Printing Office, 1893), 99; T. Gilbert Pearson, "The Bald Eagle," *Bird-Lore* 17 (1915): 404.

42. Redford, *Raccoons and Eagles*, 172.

43. Wallace Stegner, *The American West as a Living Space* (University of Michigan Press, 1987), 41.

44. Dunlap, "Values for Varmints," 144; Frederick H. Dale, "Eagle 'Control' in Northern California," *Condor* 38 (September–October 1936): 208–20.

45. Frederick H. Dale, "Eagle 'Control' in Northern California," *Condor* 38 (September–October 1936): 208–20.

46. "Birdman Bags Predatory Pests," *Santa Ynez Valley (CA) News*, June 12, 1936; "Aviator Uses Plane to Kill Coyotes, Eagles," *Sacramento (CA) Bee*, February 13, 1937.

47. "Eye Witness Tells of Red Bluff Crash," *San Francisco Examiner*, July 22, 1938.

48. Richard S. Patterson and Richardson Dougall, *The Eagle and the Shield: A History of the Great Seal of the United States* (US Department of State, 1978), 201, 203, 252, 255, 265, 270.

49.   Homer Saint-Gaudens, ed., "Roosevelt and Our Coin Designs," *Century Magazine* 99 (November 1919–April 1920): 721; Stephen L. Levine, "'A Serious Art and Literature of Our Own': Exploring Theodore Roosevelt's Art World," in *A Companion to Theodore Roosevelt*, ed. Serge Ricard (Wiley-Blackwell, 2011), 165.

50.   "First Eagle Scout," *Brooklyn Daily Eagle*, September 22, 1912; "Rockville Centre Scout Wins High Honor," *Brooklyn Daily Eagle*, April 14, 1912.

51.   "Protest," *San Francisco Chronicle*, May 7, 1871; "Eagle Baseball Team Is Again Organized," *Green Bay (WI) Semi-weekly Gazette*, March 27, 1912; "Colored Football Team Wants Games," *Times Dispatch* (Richmond, VA), September 7, 1913.

52.   "The Eagle," *American Sentinel* (Middleton, CT), September 7, 1831.

CHAPTER 5: *Feather Straight Up / Native Peoples and the Spirit Bird*

1.   Luther Standing Bear, *My Indian Boyhood* (1931; repr., University of Nebraska Press, 2006), 78, 85.

2.   Luther Standing Bear, *My People the Sioux* (1928; repr., University of Nebraska Press, 2006), 5–6.

3.   Luther Standing Bear, *Land of the Spotted Eagle* (Houghton Mifflin, 1933; repr., University of Nebraska Press, 2006), 38.

4.   Standing Bear, *Land of the Spotted Eagle*, xiii.

5.   Standing Bear, *My Indian Boyhood*, 29, 68; Standing Bear, *Land of the Spotted Eagle*, 193.

6.   John Alden Mason, *The Ethnology of the Salinan Indians* (University of California Publications, 1912), 191–92.

7.   Robert Gordon Latham, *The Natural History of the Varieties of Man* (J. Van Voorst, 1850); John Wesley Powell, *Indian Linguistic Families of America, North of Mexico* (Government Printing Office, 1891), 101–2.

8.   Bernard Romans, *A Concise Natural History of East and West Florida*, ed. Kathryn Holland Braund (University of Alabama Press, 1999), 142.

9.   Duane H. King, ed., *The Memoir of Lt. Henry Timberlake: The Story of a Soldier, Adventurer, and Emissary to the Cherokees, 1756–1765* (Museum of the Cherokee Indian Press, 2007), 22.

10.   Frank G. Speck, *The Creek Indians of Taskigi Town*, Memoirs of the American Anthropological Association, vol. 2, pt. 2 (New Era Printing, 1907), 145–46.

11.   Standing Bear, *My Indian Boyhood*, 78; John Fire (Lame Deer) and Richard Erdoes, *Lame Deer, Seeker of Visions* (Pocket Books, 1972), 137–38.

12.   Standing Bear, *Land of the Spotted Eagle*, 45, 192; Michael Edmonds, "Flights of Fancy: Birds and People in the Old Northwest," *Wisconsin Magazine of History* 83 (Spring 2000): 165; Elsie Clews Parsons, ed., *Hopi Journal of Alexander M. Stephen* (Columbia University Press, 1936), 706.

13.   David James Duncan, *My Story Told by Water: Confessions, Druidic Rants, Reflections, Bird-Watchings, Fish-Stalkings, Visions, Songs and Prayers Refracting Light, from Living Rivers, in the Age of the Industrial Park* (Sierra Club Books, 2001), 105.

14.   Zitkala-Sa, *Dreams and Thunder: Stories, Poems, and* The Sun Dance Opera, ed. P. Jane Hafen (University of Nebraska Press, 2001), 145.

15.   Love Miller, "Bird Remains from an Oregon Midden," *Condor* 59 (January 1957): 60–63; Mark V. Stalmaster, *The Bald Eagle* (Universe Books, 1987), 150.

16.   Richard C. Adams, *Legends of the Delaware Indians and Picture Writing* (Washington, DC, 1905), 56–57.

17.   Standing Bear, *Land of the Spotted Eagle*, 158; Peter Matthiessen, *Birds of Heaven: Travels with Cranes* (North Point, 2001), 141.

18.   Frank G. Speck, *A Study of the Delaware Indian Big House: In Native Text Dictated by Witapanóxwe* (Pennsylvania Historical Commission, 1931), 23.

19. Standing Bear, *Land of the Spotted Eagle*, 27; Virgil Wyaco, *A Zuni Life: A Pueblo Indian in Two Worlds* (University of New Mexico Press, 1998), 2, 52.

20. John Witthoft, "Will West Long, Cherokee Informant," *American Anthropologist* 50 (1948): 359.

21. Witthoft, "Will West Long," 359.

22. Frank G. Speck and Leonard Broom, *Cherokee Dance and Drama* (University of Oklahoma Press, 1993), 4.

23. King, *Memoir of Lt. Henry Timberlake*, 19–20.

24. Speck and Broom, *Cherokee Dance and Drama*, 40, 42.

25. King, *Memoir of Lt. Henry Timberlake*, 18, 21.

26. Reuben Gold Thwaites, ed., *Travels and Explorations of the Jesuit Missionaries in New France, 1610–1791*, vol. 59, *Lower Canada, Illinois, Ottawas, 1667–1669* (Burrows Brothers, 1899), 129; Edmonds, "Flights of Fancy," 177.

27. Speck and Broom, *Cherokee Dance and Drama*, 44.

28. Speck and Broom, *Cherokee Dance and Drama*, 41.

29. Standing Bear, *My Indian Boyhood*, 79–80.

30. Julio Blas, Sonia Cabezas, Jordi Figuerola, and Lidia Lopez-Jimenez, "Carotenoids and Skin Coloration in a Social Raptor," *Journal of Raptor Research* 47 (June 2013): 174–84; George A. Lozano, "Carotenoids, Parasites, and Sexual Selection," *Oikos* 70 (June 1994): 309–11.

31. Standing Bear, *My Indian Boyhood*, 78.

32. Wyaco, *Zuni Life*, 42–43.

33. Virginia More Roediger, *The Ceremonial Costumes of the Pueblo Indians: Their Evolution, Fabrication, and Significance in Prayer Drama* (University of California Press, 1991), 72.

34. George Parker Winship, *The Coronado Expedition, 1540–1542* (Government Printing Office, 1896), 516.

35. J. Walter Fewkes, "Property-Right in Eagles among the Hopi," *American Anthropologist* 2, no. 4 (1900): 692.

36. Fewkes, "Property-Right in Eagles," 700.

37. Standing Bear, *My Indian Boyhood*, 79–84.

38. Standing Bear, *My Indian Boyhood*, 83; Robert L. Hall, *An Archaeology of the Soul: North American Indian Belief and Ritual* (University of Illinois Press, 1997), 167.

39. William Bartram, *The Travels of William Bartram*, ed. Francis Harper, Naturalist's Edition (University of Georgia Press, 1998), 287.

40. Gregory A. Waselkov and Kathryn E. Holland Braund, eds., *William Bartram on the Southeastern Indians* (University of Nebraska Press, 1995), 150; George H. Pepper and Gilbert L. Wilson, *An Hidatsa Shrine and the Beliefs Respecting It*, Memoirs of the American Anthropological Association, vol. 2, pt. 4 (New Era Printing, 1908), 315; Hall, *Archaeology of the Soul*, 98.

41. Standing Bear, *Land of the Spotted Eagle*, 192.

CHAPTER 6: *Eagledom / Pulling Back from Extinction—the First Time*

1. "Eagle and John," *News-Journal* (Mansfield, OH), December 19, 1924.

2. Francis Hobart Herrick, *The American Eagle: A Study in Natural and Civil History* (D. Appleton-Century, 1934), 4.

3. "Prof. Herrick Retires from Teaching" (newspaper clipping), *Alumnae Folio*, December 1, 1928, Francis Hobart Biographical Information, Case Western Reserve University Archives.

4. Francis H. Herrick, *The Home Life of Wild Birds: A New Method of the Study and Photography of Birds* (G. P. Putnam's Sons, 1902).

5. Herrick, *American Eagle*, 45.

6. "Hobbies" (newspaper clipping, unidentified source), n.d., Francis Hobart Biographical Information, Case Western Reserve University Archives; Herrick, *American Eagle*, 45.

7. Herrick, *American Eagle*, 1, 28–29; Ralph H. Ilmer and E. R. Kalmbach, *The Bald Eagle and Its Economic Status* (US Department of the Interior, Bureau of Sport Fisheries and Wildlife, 1964), 13.

8. "Eagle and John," *News-Journal*; "Eagles May Pay Death Penalty for Raids on Lambs, Fowl," *Battle Creek (MI) Inquirer*, August 2, 1925.

9. Herrick, *American Eagle*, 10–12.

10. Herrick, *American Eagle*, 10–12.

11. Herrick, *American Eagle*, 5; "Eagle and John," *News-Journal*; "Audubon Law," *Nature Magazine* 15 (January 1930): 140.

12. Francis H. Herrick, "Can We Save Our National Bird?" in *Save the Bald Eagle! Fighting the Good Fight*, Emergency Conservation Committee Circular 26, January 1935, https://books.google.com/books?newbks=1&newbks_redir=0&id=239WAAAAMAAJ&q=c an+we+save+our+national+bird#v=onepage&q=bald%20eagle&f=false.

13. John Burroughs, "Wild Life about My Cabin," in *The Writings of John Burroughs*, vol. 13 (Houghton, Mifflin, 1905), 154–56.

14. Burroughs, "Wild Life about My Cabin," 156.

15. William Finley and Irene Finley, "A War against American Eagles," *Nature Magazine* 2 (November 1923): 269.

16. "Monster Bald Eagle Is Shot near the City," *Gazette* (York, PA), September 12, 1910; "Black Eagle Stuffed," *Fort Wayne (IN) Sentinel*, February 25, 1903; "Bald Eagle Shot," *Fall River (MA) Daily Evening News*, January 24, 1902.

17. Edward Howe Forbush, *Useful Birds and Their Protection* (Massachusetts State Board of Agriculture, 1913), 366.

18. "Monster Bald Eagle Shot across Narrows from Park," *Daily Province* (Vancouver, BC), November 13, 1911; "Bald-Headed Eagle," *Ottawa Citizen*, February 8, 1907.

19. William T. Hornaday, *Taxidermy and Zoological Collecting: A Complete Handbook for the Amateur Taxidermist, Collector, Osteologist, Museum-Builder, Sportsman, and Traveller* (Charles Scribner's Sons, 1891), 2; William T. Hornaday, *Our Vanishing Wild Life: Its Extermination and Preservation* (Charles Scribner's Sons, 1913), 8; "The American Eagle Dying Out," *American Review of Reviews* 62 (August 1920): 208.

20. Spencer Trotter, "Notes on the Ornithological Observations of Peter Kalm," *Auk* 20 (July 1903): 251.

21. Mark V. Barrow, *A Passion for Birds: American Ornithology after Audubon* (Princeton University Press, 1998), 156–57.

22. Paul J. Baicich, Margaret A. Barker, and Carroll L. Henderson, *Feeding Wild Birds in America: Culture, Commerce, and Conservation* (Texas A&M University Press, 2015), 8.

23. Margaret Porter Griffin, *The Amazing Bird Collection of Young Mr. Roosevelt: The Determined Independent Study of a Boy Who Became America's 26th President* (Xlibris, 2014), 10; Barrow, *Passion for Birds*, 141.

24. John Taliaferro, *Grinnell: America's Environmental Pioneer and His Restless Drive to Save the West* (W. W. Norton, 2019), 30, 149–51, 154.

25. Burroughs, "Wild Life about My Cabin," 155–56.

26. Harvey Frommer, *Old-Time Baseball: America's Pastime in the Gilded Age* (Rowman & Littlefield, 2017), 134.

27. Thomas S. Roberts, "The Camera as an Aid in the Study of Birds," *Bird-Lore* 1 (April 1899): 35.

28. "1900," *Bird-Lore* 2 (December 1900): 199.

29. Henry Ford, *My Life and Work* (Garden City Publishing, 1922), 237.

30. Florence Merriam Bailey, *Birds of Village and Field: A Bird Book for Beginners* (Houghton, Mifflin, 1898), 282, 283.

31. John Muir, *Travels in Alaska* (Houghton Mifflin, 1915), 251.

32. Charles Keeler, "Days among Alaska Birds," in William H. Dall, Charles Keeler,

Henry Gannett, William H. Brewer, C. Hart Merriam, George Bird Grinnell, and M. L. Washburn, *Alaska*, vol. 2, *History, Geography, Resources* (Doubleday, Page, 1901), 207; Muir, *Travels in Alaska*, 256.

33.   "Preposterous Anti-eagle Tales from Alaska," Emergency Conservation Committee Circular, January 1935, https://books.google.com/books?newbks=1&newbks_redir =0&id=239WAAAAMAAJ&q=can+we+save+our+national+bird#v=onepage&q=b ald%20eagle&f=false.

34.   Finley and Finley, "War against American Eagles," 269.

35.   Albert K. Fisher, *The Hawks and Owls of the United States in Their Relation to Agriculture*, US Department of Agriculture Bulletin no. 3 (Government Printing Office, 1893), 3, 100; Mark V. Barrow Jr., *Nature's Ghosts: Confronting Extinction from the Age of Jefferson to the Age of Ecology* (University of Chicago Press, 2015), 238–39.

36.   T. Gilbert Pearson, "The Bald Eagle," *Bird-Lore* 17 (1915): 404; Barrow, *Nature's Ghosts*, 241–42; Stephen Fox, *The American Conservation Movement: John Muir and His Legacy* (University of Wisconsin Press, 1981), 155–56, 165.

37.   "Crow 'Raid' Protest Flood Shows Nation-wide Censure," *Christian Science Monitor*, April 11, 1924; Barrow, *Passion for Birds*, 148.

38.   "Eagle Is Condemned without a Hearing," *News-Herald* (Franklin, PA), November 27, 1926.

39.   Barrow, *Nature's Ghosts*, 246; "The Bald Eagle, Our National Emblem," *Fighting the Good Fight* 6 (1928): 6.

40.   Irving Brant, *Adventures in Conservation with Franklin D. Roosevelt* (Northland, 1989), 15.

41.   Brant, *Adventures in Conservation*, 15; Barrow, *Nature's Ghosts*, 246; Willard Van Name, "Bird Collecting and Ornithology," *Science* 41, no. 1066 (June 4, 1915): 824.

42.   Waldron DeWitt Miller, Willard G. Van Name, and Davis Quinn, *A Crisis in Conservation: Serious Danger of Extinction of Many North American Birds*, in Pamphlets on Biology: Kofold Collection, vol. 1328, June 1929, https://books.google.com/books?id =fFYXAQAAIAAJ&printsec=frontcover#v=onepage&q&f=false, 10, 12; *American Eagle Protection, Hearing before the Committee on Agriculture, House of Representatives, Seventy-First Congress, Second Session, January 31, 1930*, serial D (Government Printing Office, 1930), 9.

43.   Dyana Z. Furmansky, *Rosalie Edge, Hawk of Mercy: The Activist Who Saved Nature from the Conservationists* (University of Georgia Press, 2009), 108; Van Name, "Bird Collecting and Ornithology," 823–25.

44.   Rosalie Edge, "Questions Truth of News Story That Golden Eagle Attacked Child," *Brooklyn Daily Eagle*, April 20, 1935; Jim Stevens, "Out of the Woods," *Lebanon (OR) Express*, May 6, 1948.

45.   Furmansky, *Rosalie Edge*, 108.

46.   Furmansky, *Rosalie Edge*, 111.

47.   "Important Offer Refused by the Audubon Society," *The Bald Eagle, Our National Emblem*, Emergency Conservation Committee Circular, April 1930, https://books .google.com/books?newbks=1&newbks_redir=0&id=239WAAAAMAAJ&q=can+we +save+our+national+bird#v=onepage&q=bald%20eagle&f=false.

48.   Furmansky, *Rosalie Edge*, 113–14; "The United States Biological Survey Must Bear the Responsibility for Allowing the Establishment and Continuance (Now for Nearly Eighteen Years) of This Obnoxious and Unjustified Bounty," in *Save the Bald Eagle! Shall We Allow Our National Emblem to Become Extinct?* (Emergency Conservation Committee, 1935), 25.

49.   "American Bald Eagle Is Near Extinction," *Popular Science* (March 1930): 62.

50.   "American Eagle Endangered as Press Debates Its Morals," *Evening Star* (Washington, DC), January 24, 1930.

51.   "Mississippi Federation of Women's Clubs Passes Very Important Resolutions of General Interest," *Clarion-Ledger* (Jackson, MS), May 4, 1930; "Facts about Tennessee Eagles," *Nashville (TN) Banner*, December 21, 1930.

52. *American Eagle Protection: Hearing,* 7.

53. *American Eagle Protection: Hearing,* 16.

54. Ellsworth D. Lumley, *The Two Eagles of North America* (Emergency Conservation Committee, 1939), 2 (digital copy provided by the Western History and Genealogy Department, Denver Public Library).

55. *American Eagle Protection: Hearing,* 10, 11, 16; "The Bald Eagle," *Bird-Lore* 32 (1930): 164.

56. Herrick, "Can We Save Our National Bird?"; Herrick, *American Eagle*; "The American Eagle," *Nashville (TN) Banner,* December 9, 1934; "A Noted Authority Writes of Eagles," *Hartford (CT) Courant,* December 9, 1934; "America's Eagle," *New York Times,* December 2, 1934.

57. Barrow, *Nature's Ghosts,* 255; Bald Eagle Protection, H.R. 5590, March 11, 1940, Proquest Congressional Collection, https://congressional.proquest.com/legisinsight?id=HRG-1940-HAG-0018&type=HEARING.

58. "Blue Cross Not to Relax Effort to Win Protection of Bald Eagle," *Springfield (MA) Republican,* January 14, 1940.

59. "Measures to Protect Bald Eagle Have Gained National Attention," *Springfield (MA) Republican,* February 26, 1939; "Fish Will Support Bald Eagle Bill," *Springfield (MA) Republican,* February 2, 1940.

60. Bald Eagle Protection, H.R. 5590.

61. Bald Eagle Protection, H.R. 5590.

62. Theodore Roosevelt to Ellsworth D. Lumley, June 30, 1939, frontispiece in Lumley, *Two Eagles of North America.*

63. "'Old Abe' Recalled as Most Famous Eagle," *Casper Star-Tribune* (Casper, WY), May 29, 1930; "The Bald Eagle," *Daily Courier* (Connellsville, PA), April 5, 1940; 86 Cong. Rec. 6447 (May 20, 1940).

64. 86 Cong. Rec. 6446 (May 20, 1940); "Misunderstood Eagle," *Miami (FL) News,* June 11, 1939.

65. Committee on Agriculture and Forestry, "Preserving from Extinction the American Eagle Emblem of the Sovereignty of the United States of America," S. Doc. No. 75-1589 (May 14, 1940).

66. On Roosevelt's conservation commitment, see Douglas Brinkley, *Rightful Heritage: Franklin D. Roosevelt and the Land of America* (Harper, 2016).

CHAPTER 7: *Birds in a Band / Poison Rain, Blessed Relief*

1. Fred Bodsworth, "How to Catch an Eagle," *MacLean's Magazine* 65 (February 1, 1963): 23.

2. "Eagle Bander," *New Yorker,* June 19, 1954, 20.

3. Bodsworth, "How to Catch an Eagle," 22, 23; Polly Redford, *Raccoons and Eagles: Two Views of American Wildlife* (E. P. Dutton, 1963), 197.

4. Myrtle Jeanne Broley, *Eagle Man: Charles Broley's Adventures with American Eagles* (Pellegrinni & Cudahy, 1952), 70.

5. Broley, *Eagle Man,* 70.

6. Gary Schmidgall, ed., *Walt Whitman: Selected Poems, 1855–1892* (St. Martin's Griffin, 1999), 371.

7. G. R. West, "How It Got Its Name," *Quad-City Times* (Davenport, IA), February 23, 1880; Francis Hobart Herrick, *The American Eagle: A Study in Natural and Civil History* (D. Appleton-Century, 1934), 69–70; George E. Goodrich, *The Centennial History of the Town of Dryden, 1797–1897* (J. Giles Ford, 1898), 38–39.

8. Matthew R. Halley, "Audubon's Famous Banding Experiment: Fact or Fiction?" *Archives of Natural History* 45 (2018): 118–21.

9. J. MacCracken, *The Canadian Bird Bander's Training Manual* (Public Works and Government Services Canada, 1999), 6.

10. "Charles Broley, Famed 'Eagle Man,' Dies at 79," *Tampa (FL) Tribune*, May 5, 1959; Bruce E. Beans, *Eagle's Plume: The Struggle to Preserve the Life and Haunts of America's Bald Eagle* (Scribner, 1996), 82–83; Charles L. Broley, "The Plight of the American Bald Eagle," *Audubon*, July–August 1958, 162; Charles L. Broley, "The Bald Eagle in Florida," *Atlantic Naturalist* 12 (July 1957): 230–31.

11. Jackson M. Abbot, "Bald Eagle Survey Report," *Atlantic Naturalist* 14 (October–December 1959): 252–58.

12. Marjory Kinnan Rawlings, *Cross Creek* (Charles Scribner's Sons, 1942), 271.

13. Chandler S. Robbins, *The Status of the Bald Eagle, Summer of 1959*, Wildlife Leaflet 418 (US Department of the Interior, Fish and Wildlife Service, Bureau of Sport Fisheries and Wildlife, 1960), 3–4.

14. Broley, "Plight of the American Bald Eagle," 163.

15. Beans, *Eagle's Plume*, 109.

16. Rachel Carson, "Road of the Hawks," in Rachel Carson, *Lost Woods: The Discovered Writing of Rachel Carson*, ed. Linda Lear (Beacon, 1998), 31.

17. Elena Conis, "Beyond Silent Spring: An Alternative History of DDT," *Distillations*, February 14, 2017, https://www.sciencehistory.org/distillations/beyond-silent-spring -an-alternate-history-of-ddt#:~:text=Environment-,Beyond%20Silent%20Spring %3A%20An%20Alternate%20History%20of%20DDT,often%20conflict%20with %20the%20facts.&text=A%20Belgian%20advertisement%20for%20the%20DDT %2Dpowered%20insecticide%2C%20Insectoline.

18. Elena Conis, "DDT Disbelievers: Health and the New Economic Poisons in Georgia after World War II," *Southern Spaces*, October 28, 2016, https://southernspaces.org/2016/ ddt-disbelievers-health-and-new-economic-poisons-georgia-after-world-war-ii.

19. Beans, *Eagle's Plume*, 91.

20. Channing Cope, "DDT Experiment Proves Successful," *Atlanta Constitution*, August 29, 1945.

21. Cope, "DDT Experiment Proves Successful"; Conis, "Beyond Silent Spring."

22. Cope, "DDT Experiment Proves Successful"; Conis, "DDT Disbelievers."

23. Conis, "Beyond Silent Spring."

24. Beans, *Eagle's Plume*, 91, 93.

25. "DDT Warning," *Time*, August 7, 1944, 66; "DDT—Deadly New Bug Killer," *Science Digest* 16 (1944): 15; "DDT—Handle with Care," *Railroad Workers Journal* 6 (1945): 28; Conis, "Beyond Silent Spring."

26. Conis, "DDT Disbelievers."

27. Conis, "DDT Disbelievers."

28. Beans, *Eagle's Plume*, 90–91.

29. Jason R. Richardson, Ananya Roy, Stuart L. Shalat, Richard T. von Stein, Muhammad M. Hossain, Brian Buckley, Marla Gearing, Allan I. Levey, and Dwight C. German, "Elevated Serum Pesticide Levels and Risks for Alzheimer Disease," *JAMA Neurology* 71 (March 2014): 284–90, https://jamanetwork.com/journals/jamaneurology/fullarticle /1816015; Columbia University's Mailman School of Public Health, "DDT Exposure Tied to Breast Cancer Risk for All Women through Age 54: Six-Decade-Long Study Finds 40-Year Induction Period between Time of Exposure and Diagnosis," *ScienceDaily*, February 13, 2019, https://www.sciencedaily.com/releases/2019/02/190 213124347.htm; Rosanna Xia, "DDT's Toxic Legacy Can Harm Granddaughters of Women Exposed, Study Says," *Los Angeles Times*, April 24, 2021.

30. Beans, *Eagle's Plume*, 99; W. C. Krantz, B. M. Mullern, George E. Bagley, A. Sprunt IV, F. J. Ligas, and W. B. Robertson Jr., "Organochlorine and Heavy Metals in Bald Eagle Eggs," *Pesticides Monitoring Journal* 4 (December 1970): 136–40; "Labeling DDT," *Pests and Their Control* 13 (1945): 12.

31. David B. Peakall, J. L. Lincer, R. W. Risebrough, J. B. Pritchard, and W. B. Kinter, "DDE-Induced Egg-Shell Thinning: Structural and Physiological Effects on Three Species," *Comparative and General Pharmacology* 4 (September 1973): 305–13.

32. Ken Ross, *Pioneering Conservation in Alaska* (University of Colorado Press, 2006), 311.

33. Walter J. Larson, "Eagle Hunting in Alaska," *Harding's Magazine* 49 (February 1929), 17.

34. Ross, *Pioneering Conservation in Alaska*, 311.

35. "Concerning the Alaska Eagle" (news article), *Seward Gateway*, December 25, 1940, Office of the District and Territorial Governor, General Correspondence, 1934–1953, RG101, SR 727 (hereafter cited as "ODTG"), Alaska State Archies, Juneau, Alaska (hereafter cited as "ASA").

36. Ross, *Pioneering Conservation in Alaska*, 313.

37. "Summary Analysis of the Bald Eagles from Alaska, Collected by Hosea R. Sarber, 1940," ODTG, ASA; Ernest Gruening to Morris Llewellyn Cooke, December 16, 1949, ODTG, ASA; Ralph H. Imler and E. R. Kalmbach, *The Bald Eagle and Its Economic Status*, US Department of the Interior, Fish and Wildlife Service, Circulars 30 (Government Printing Office, 1955), 39.

38. Willard G. Van Name, "Saving the Bald Eagle: Extinction of the Species Feared If Slaughter Continues," *New York Times*, December 4, 1949; Gruening to Cooke, December 16, 1949; R. N. DeArmond, "Shoot the Damn Things! Alaska's War against the American Bald Eagle," in *Bald Eagles in Alaska*, ed. Bruce A. Wright and Phil Schempf (Bald Eagle Research Institute, 2008), 273.

39. "US Eagle May Take Offense," *Austin American-Statesman*, March 17, 1949.

40. National Audubon Society, "Alaska Eagle Bounty Nullified by Federal Regulation" (news release), May 1952, ODTG , ASA; "Alaskan Eagles Now Protected; Local Group Acted for Ruling," *Palladium-Item* (Richmond, IN), July 15, 1952.

41. "Bald Eagle Killed," *Daily Chronicle* (Centralla, WA), May 7, 1953.

42. George E. Sokolsky, "Congress Complicates Its Duties," *Orlando (FL) Sentinel*, November 24, 1962.

43. *Protection for the Golden Eagle. Hearing before a Subcommittee of the Committee on Commerce, United States Senate, Eighty-Seventh Congress, Second Session, on S.J. Res. 105 and H.J. Res. 489, Joint Resolutions to Provide Protection for the Golden Eagle, June 26, 1962* (US Government Printing Office, 1962), 115; "Joint Resolution to Provide Protection for the Golden Eagle," P. L. No. 87-884, Stat. 76, October 24, 1962, govinfo.gov https://www.govinfo.gov/content/pkg/STATUTE-76/pdf/STATUTE-76-Pg1246.pdf#page=1.

44. Dyana Z. Furmansky, *Rosalie Edge, Hawk of Mercy: The Activist Who Saved Nature from the Conservationists* (University of Georgia Press, 2009), 241, 283n24.

45. Rachel Carson, *Silent Spring* (Houghton Mifflin, 1962), 17.

46. E. B. White, "The Deserted Nation," *New Yorker*, October 6, 1966, 53.

47. Carson, *Silent Spring*, 118.

48. "Myrtle Jeanne Broley," *Washington Post*, March 13, 1958; "Wife of 'Eagle Man' Dies Here," *Tampa (FL) Tribune*, March 12, 1958.

49. *Santa Barbara Oil Spill, Hearings before the Subcommittee on Minerals, Materials, and Fuels of the Committee on Interior and Insular Affairs, United States Senate, 91st Congress, 1st Session, S. 1219, May 19, 20, 1969* (US Government Printing Office, 1969), 147; Meir Rinde, "Richard Nixon and the Rise of American Environmentalism," *Distillations*, June 2, 2017, https://www.sciencehistory.org/distillations/richard-nixon-and-the-rise-of-american-environmentalism.

50. J. Brooks Flippen, *Nixon and the Environment* (University of New Mexico Press, 2000), 25.

51. Basanta Kumara Behera and Ram Prasad, *Environmental Technology and Sustainability: Physical, Chemical and Biological Technologies for Clean Environmental Management* (Elsevier, 2020), 6; Kate Wheeling, "When Michigan Students Put the Car on Trial," *Smithsonian Magazine*, April 2020, https://www.smithsonianmag.com/history/when-michigan-students-put-car-trial-180974374.

52. *Permit the Use of DDT: Hearings before the Subcommittee on Forests of the Committee on Agriculture, House of Representatives, 93rd Congress, 1st Session, on H.R. 10796, October 23, 24, and 25, 1973* (US Government Printing Office, 1974), 22.

53. Beans, *Eagle's Plume*, 104, 105.

CHAPTER 8: *Eagle Lady / Persistence and Restoration*

1. Jeff Klinkenberg, "Outdoors," *St. Petersburg (FL) Times*, June 15, 1979; "Move Over, You Birds, Here's the Eagle Lady," *Capital News Journal* (Salem, OR), June 15, 1979; Jeff Klinkenberg, email to the author, August 11, 2020.
2. Klinkenberg, "Outdoors."
3. Klinkenberg, "Outdoors."
4. Klinkenberg, "Outdoors."
5. Doris Mager, interview by the author, August 17, 2020.
6. Klinkenberg, "Outdoors"; Mager, interview, August 17, 2020.
7. Nancy Burns-Fusaro, "'The Eagle Lady' Says Goodbye to an Old Friend," *Westerly Sun* (Pawcatuck, RI), April 26, 2019; Kristine Liao, "Meet the Woman Who Lived in a Bald Eagle's Nest to Save Raptors," *Audubon*, October 29, 2019, https://www.audubon .org/news/meet-woman-who-lived-bald-eagles-nest-save-raptors#:~:text=Conservatio n,Meet%20The%20Woman%20Who%20Lived%20in%20a%20Bald%20Eagle's%20 Nest,in%20restoring%20Florida's%20eagle%20population.
8. Burns-Fusaro, "'Eagle Lady.'"
9. George Korda, "Eagle Lady Comes Home from the Roost," *Florida Today* (Cocoa, FL), June 21, 1979.
10. "She's up a Tree Raising Money to Save Eagles," *Life* 2 (November 1979): 56.
11. Korda, "Eagle Lady Comes Home."
12. "The Bald Bird Lives! America's Endangered Eagle Is on the Increase," *Life* 2 (November 1979): 53; Klinkenberg, "Outdoors."
13. National Environmental Policy Act of 1969, 16 CFR § 1.82 Declaration of Policy.
14. Craig Collins, *Toxic Loopholes: Failures and Future Prospects for Environmental Law* (Cambridge University Press, 2010), 56.
15. Bruce Chadwick, "From Beautiful People to Beautiful Animals," *Daily News* (New York), April 15, 1983; Leslie R. Myers, "Rare Animals Are Losing Their Grip," *Clarion-Ledger* (Jackson, MS), December 11, 1983.
16. James W. Grier, James B. Elder, Francis J. Gramlich, Nancy F. Green, Joel V. Kussman, John E. Mathison, and James P. Mattsson, *Northern States Bald Eagle Recovery Plan* (n.p., 1983), 3; Myers, "Rare Animals Are Losing Their Grip."
17. US Department of the Interior, "Secretary Hickel Halts Issuance of 'Blanket' Permits to Kill Golden Eagles" (news release), March 6, 1970, https://www.fws.gov/news/ Historic/NewsReleases/1970/19700306.pdf.
18. Nathaniel Pryor Reed, *Travels on the Green Highway: An Environmentalist's Journey* (Reed Publishing, 2016), 18.
19. Frank Graham Jr., "Will the Bald Eagle Survive to 2076?" *Audubon* 78 (March 1976): 99.
20. *Protection for the Golden Eagle. Hearing before a Subcommittee of the Committee on Commerce, United States Senate, Eighty-Seventh Congress, Second Session, on S.J. Res. 105 and H.J. Res. 489, Joint Resolutions to Provide Protection for the Golden Eagle, June 26, 1962* (US Government Printing Office, 1962), 38–39.
21. Bill Davis, interview by the author, September 28, 2020.
22. William J. Davis, "The Eagles Have Landed," *Massachusetts Wildlife* 34 (1983): 4.
23. Bill Davis, "Bringing Back Magnificence: 30 Years of Bald Eagle Restoration," *Massachusetts Wildlife* 61 (2013): 19; Bev Kaufman, "Bald Eaglets Survive Trip," *Fort Lauderdale (FL) News*, June 13, 1982.
24. Davis, "Eagles Have Landed," 7.
25. Bill Davis, interview, September 28, 2020.
26. Dianne Benson Davis, *Eagle One: Raising Bald Eagles—A Wildlife Memoir* (Chandler House, 2013), 140; Lefrancois used the word "woman" instead of "girl" and an exclamation point instead of a period. Erma J. Fisk, *The Peacocks of Baboquivari* (W. W. Norton, 1983), 34.

27. Davis, *Eagle One*, 182; Dianne Benson Davis, interview by the author, October 16, 2020.
28. Dianne Benson Davis, interview, October 16, 2020.
29. Susan Weiler, conversation with the author, November 29, 2020.
30. Tom French, interview by the author, September 29, 2020.
31. Christopher Maurer, *Fortune's Favorite Child: The Uneasy Life of Walter Anderson* (University Press of Mississippi, 2003), 256.
32. Redding S. Sugg Jr., *The Horn Island Logs of Walter Inglis Anderson* (University Press of Mississippi, 1985), 202.
33. "Bald Eagle Population Exceeds 11,000 Pairs in 2007," Center for Biological Diversity, June 2007, https://www.biologicaldiversity.org/species/birds/bald_eagle/report/#WI.
34. Steve Sherrod, interview by the author, October 29, 2020; Don Wolfe, interview by the author, October 29, 2020.
35. Steve Nesbitt, interview by the author, August 18, 2020.
36. Petra Wood, interview by the author, October 29, 2020; Marjory Kinnan Rawlings, *Cross Creek* (Charles Scribner's Sons, 1942), 7; Petra Bohall Wood and Michael W. Collopy, "Effects of Egg Removal on Bald Eagle Productivity in Northern Florida," *Journal of Wildlife Management* 57 (January 1993): 1–9.
37. Michael Wallis, "Into the Air, Little Baldies," *Life* 11 (May 1988): 55; Mike Collopy, interview by the author, October 29, 2020; Wood, interview, October 29, 2020.
38. Sherrod, interview, October 29, 2020; Wolfe, interview, October 29, 2020.
39. Bobby Cleveland, "Coast Receives More Bald Eagles," *Clarion-Ledger* (Jackson, MS), April 2, 1988.
40. Ted Simon, interview by the author, October 30, 2020.
41. John Weller, interview by the author, November 1, 2020.
42. Weller, interview, November 1, 2020.
43. Dan Reinking, interview by the author, October 30, 2020.
44. Davis, "Eagles Have Landed," 7; "Eagle 'Betsy' Flies Northwest from Quabbin," *Berkshire Eagle* (Pittsfield, MA), August 10, 1982.
45. Davis, "Bringing Back Magnificence," 21.
46. Weller, interview, November 1, 2020.

Chapter 9: *Bird on Top / New Century, New Age*

1. Bill Gifford, "How a Hollywood Producer and a Bunch of Teenagers Brought Bald Eagles Back to DC," *Washingtonian*, June 5, 2016; "Challenger Invited to White House to Celebrate Resurgence of America's National Symbol," July 2, 1999, American Eagle Foundation, https://www.eagles.org/challenger-meets-president-clinton-at-the-white-house-1999.
2. "Challenger Invited to White House."
3. D'Vera Cohn, "Eagle-Eyed Discovery: Surprise Hatchling Marks First Sight of Species in District since 1940s," *Washington Post*, August 10, 2000; "Challenger Invited to White House."
4. "Challenger Invited to White House."
5. "The Eagle Has Landed," *Hartford (CT) Courant*, July 4, 1999.
6. "Challenger Invited to White House."
7. Tim Post, "Deadline for Delisting Bald Eagles Extended," *MPR News*, Minnesota Public Radio, February 8, 2007, https://www.mprnews.org/story/2007/02/06/eagledevelopment; Kirsti Marohn, "Man May Sue over Bald Eagles Again, *St. Cloud (MN) Times*, January 19, 2007.
8. The initial deadline imposed by the court was February 16, 2007; it was then extended to June 29. Edmund Contoski v. P. Lynn Scarlett, United States Fish and Wildlife Service, United States District Court, Minnesota, Civil no. 05-2528, August 10, 2006.
9. World Wildlife Fund, "Statement on Removal of Bald Eagle from Endangered Species List" (press release), June 28, 2007, https://www.worldwildlife.org/press-releases

/statement-on-removal-of-bald-eagle-from-endangered-species-list; "Environmental Defense Declares Bald Eagle 'Back from the Brink,' Launches New Effort to Restore Other Species," Environmental Defense Fund, May 11, 2004, https://www.edf.org /news/environmental-defense-declares-bald-eagle-back-brink-launches-new-effort -restore-other-species.

10. Noelle Walker, "Hill Country Rancher Faces Indictments for Allegedly Killing Bald Eagles," NBC 5 Dallas-Fort Worth, October 30, 2017, https://www.nbcdfw.com/news /local/hill-country-rancher-faces-indictments-for-allegedly-killing-bald-eagles/44878.

11. "Pair Guilty of Killing Federally Protected Birds: Bald Eagle and Hundreds of Great Blue Herons and Ospreys Killed by Owner of Trout Farm" (press release), Department of Justice, US Attorney Michael J. Sullivan, District of Massachusetts, April 10, 2007, www.fws.gov/home/feature/2007/conviction.pdf; Tom French, interview by the author, September 29, 2020.

12. D. W. Stinson, J. W. Watson, and K. R. McAllister, *Washington State Status Report for the Bald Eagle* (Washington Department of Fish and Wildlife, Wildlife Program, October 2007), 18; Stephen Lee, "ND Buffalo Rancher Admits Killing 6 SD Bald Eagles with Prairie Dog Poison," *Capital Journal* (Pierre, SD), February 2, 2020.

13. Leah K. Manning, Arno Wünschmann, Anibal G. Armién, Michelle Willette, Kathleen MacAulay, Jeff. B. Bender, John P. Buchweitz, and Patrick Redig, "Lead Intoxication in Free-Ranging Bald Eagle (*Haliaeetus leucocephalus*)," *Veterinary Pathology* 56 (March 1, 2019): 289–99.

14. Alan Yuhas, "Bald Eagles: Scientists Decry Overturn of Ban That Would Save American Symbol," *Guardian*, March 16, 2017.

15. Marla Cone, "Chemical Firms Settle DDT Suit," *Los Angeles Times*, December 20, 2000.

16. Veronika Kiverson, Karin L. Lemkau, Oscar Pizarro, Danna R. Yoerger, Carl Kaiser, Robert K. Nelson, Catherine Carmichael, Blair G. Paul, Christopher M. Reddy, and David Valentin, "Ocean Dumping of Containerized DDT Waste Was a Sloppy Process," *Environmental Science Technology* 53 (March 2019): 2971–80; Rosanna Xia, "A Toxic Secret Lurks in Deep Sea," *Los Angeles Times*, October 25, 2020.

17. Francis Hobart Herrick, *The American Eagle: A Study in Natural and Civil History* (D. Appleton-Century, 1934), 40–42.

18. Myrtle Jeanne Broley, *Eagle Man: Charles Broley's Adventures with American Eagles* (Pellegrinni & Cudahy, 1952), 25.

19. Mark V. Stalmaster and James R. Newman, "Behavioral Responses of Wintering Bald Eagles to Human Activity," *Journal of Wildlife Management* 42 (January 1979): 506–13; James R. Newman, William H. Brennan, and Lenore M. Smith, "Twelve-Year Changes in Nesting Patterns of Bald Eagles (*Haliaeetus leucocephalus*) on San Juan Island, Washington," *Murrelet* 58 (Summer 1977): 37–39.

20. Jeff Klinkenberg, "Bald Eagles: Fighting the Fall toward Extinction," *St. Petersburg (FL) Times*, October 26, 1980; William Booth, "How to Protect Habitat? Bald Eagle's Removal from Endangered Species List Delayed, *Washington Post*, July 4, 2000. A superb book on the dusky seaside sparrow is Mark J. Walters, *A Shadow and a Song: The Struggle to Save an Endangered Species* (Chelsea Green, 1992).

21. Eric Staats, "Former Development Company Employee Charged with Destroying Eagle Nest," *Daily News* (Naples, FL), January 5, 2006.

22. Brian A. Millsap, Emily R. Bjerre, Mark C. Otto, Guthrie S. Zimmerman, and Nathan L. Zimpfer, *Bald and Golden Eagles: Demographics and Estimation of Sustainable Take in the United States, 2016 Update* (US Fish and Wildlife, Division of Migratory Bird Management, 2016), 97.

23. Suncoast Waterkeeper to USFWS Staff, November 6, 2018, "New Developments: Eagle Nest Taking Permit MB47140C-0, Effective March 12, 2018" (report provided to author by Andre Mele); Dennis Maley, "Bald Eagle Nest Disappears and Reappears near Long Pointe Project," *Bradenton (FL) Times*, November 1, 2016.

24. Richard A. Oppel Jr., "Bald Eagles, Symbol of America, Are Dumping Trash on the

Seattle Suburbs," *New York Times*, April 2, 2019; Richard A. Oppel Jr., "Majestic Symbol of Strength Is Also a Revolting Nuisance," *New York Times*, April 3, 2019.

25. Mike Cherney, "Bald Eagles: Angry Birds Are Ripping Drones Out of the Sky," *Wall Street Journal*, September 30, 2017.

26. Kyle Hopkins, "Bald Eagles Pose an Increasing Risk at U.S. Airports. Here's How Officials Are Trying to Protect Them," *Los Angeles Times*, June 7, 2016.

27. "Wildlife Strike Database," Federal Aviation Administration, accessed November 2020, https://wildlife.faa.gov/search; Brian E. Washburn, Michael J. Begier, and Sandra E. Wright, "Collisions between Eagles and Aircraft: An Increasing Problem in the Airport Environment," *Journal of Raptor Research* 49 (June 2015): 192–200.

28. Conrad Duncan, "Trump Is Ranting Again about Wind Turbines Killing Bald Eagles—Here's Why He's Wrong," *Indy100*, May 15, 2019, https://www.indy100.com/news/trump-bald-eagles-dead-wind-turbines-8914861; David Edwards, "Trump Claims He's Seen Piles of Dead Bald Eagles underneath Windmills," Raw Story, May 14, 2019, https://www.rawstory.com/2019/05/trump-claims-hes-seen-piles-of-dead-bald-eagles-underneath-windmills-you-see-them-all-over-the-place.

29. "Wind Power and Birds," National Audubon Society, July 21, 2020, https://www.audubon.org/news/wind-power-and-birds; Michael Hutchins, "Top 10 Myths about Wind Energy and Birds," *Birdcalls: News and Perspectives on Bird Conservation*, American Bird Conservancy, December 6, 2017, https://abcbirds.org/top-10-myths-wind-energy-birds/?gclid=Cj0KCQiA0MD_BRCTARIsADXoopbAH7zii6aA7TS966N6MRo2fAdiixNV9e9whDSDIBUKw-1BzO4Or3QaAv-PEALw_wcB; "Wind Turbines against Nature," Institute for Energy Research, July 19, 2019, https://www.instituteforenergyresearch.org/renewable/wind/wind-turbines-against-nature/?gclid=Cj0KCQiA0MD_BRCTARIsADXoopaV-QCa3NH5AsXaSmofUtHckl4Yfqa3hC2527We2Rac_VFqS4BD9uYaAqHVEALw_wcB; Jessica Wells, "How Duke Energy Is Using Technology to Save Eagles at a Wyoming Wind Site," *Illumination*, Duke Energy, June 22, 2021, https://illumination.duke-energy.com/articles/how-duke-energy-is-using-technology-to-save-eagles-at-a-wyoming-wind-site; Melissa Seixas, email to the author, June 30, 2021.

30. UPI, "'Eagle Lady' Nears End of Trip," *Tampa (FL) Tribune*, June 21, 1986; "Graham Honors 'Eagle Lady,'" *Florida Today* (Cocoa, FL), June 21, 1986.

31. Dick Bothwell, "Doris Mager May be Patton's Buddy, but to El Tigre She's Mom," *St. Petersburg (FL) Times*, November 23, 1979.

32. Nancy Burns-Fusaro, "'The Eagle Lady' Says Goodbye to an Old Friend," *Westerly Sun* (Pawcatuck, RI), April 26, 2019.

33. Dianne Benson Davis, *Eagle One: Raising Bald Eagles—A Wildlife Memoir* (Chandler House, 2013), 304–5.

34. Richard Powelson, "National Zoo Gets New Wings," *Citizen's Voice* (Wilkes-Barre, PA), July 5, 2003.

35. Oppel, "Bald Eagles, Symbol of America."

36. Will Harris, interview by the author, March 7, 2018.

37. Harris, interview, March 7, 2018.

38. Susan Matthews, "Let Them Eat Chicken," *Audubon* 75 (Fall 2016): 36–42; Wyatt Williams, "National Burden," *New York Times Magazine*, January 22, 2017, 2–29, 65.

39. Bill Davis, email to the author, November 29, 2020; Bill Davis, interview by the author, September 28, 2020.

40. "Eagle-Eye Hancock Flies High in Search of Nests," *Vancouver (BC) Sun*, December 15, 1965; David Hancock, interview by the author, October 26, 2020; Doris Mager, interview by the author, August 17, 2020.

41. Heather Travis, "Internet Users Tune in to B.C. Eagles," *Vancouver (BC) Sun*, April 3, 2006; Hartford Courant, "Bird's-Eye View on Wildlife," *Chicago Tribune*, May 17, 2006.

42. Hartford Courant, "Bird's-Eye View on Wildlife."

43. Travis, "Internet Users Tune in to B.C. Eagles"; Canadian Press, "New Eagle Nest Cam Takes Over after Hornby Island Failure," *Edmonton Journal*, May 9, 2006.

44. Nara Schoenberg, "Eaglets Have Two Daddies—and a Mom," *Chicago Tribune*, March 4, 2020.

45. Luther Standing Bear, *Land of the Spotted Eagle* (Houghton Mifflin, 1933; repr., University of Nebraska Press, 2006), 45; Alexander Wilson, *American Ornithology, or the Natural History of the Birds of the United States*, 9 vols. (Bradford and Inskeep, 1808–14), 4:100; Edward P. Alexander, *Museum Masters: Their Museums and Their Influence* (Altamira Press, 1995), 64.

46. Jen Kirby, "Staten Island Bird-Watchers Spot What Could Be the First Bald Eagle Born in NYC in More than 100 Years," *New York*, Intelligencer, September 6, 2016, https://nymag .com/intelligencer/2016/09/bird-watchers-spot-a-baby-bald-eagle-on-staten-island .html.

47. David Figura, "What's with the Huge Bald Eagle Statue off the NYS Thruway at Montezuma?" Syracuse.com, November 1, 2016, https://www.syracuse.com/outdoors /2016/11/whats_with_the_huge_bald_eagle_statue_off_the_nys_thruway_at_ montezuma.html#:~:text=The%20eye%2Dopening%20statue%2C%20dedicated,from %20wing%20tip%20to%20tip.

48. Millsap et al., *Bald and Golden Eagles*, 7–8; *U.S. Fish and Wildlife Service Final Report: Bald Eagle Population Size: 2020 Update* (US Fish and Wildlife Service, Division of Migratory Bird Management, 2020).

Epilogue: *Reconciliation*

1. John Mayer, "Tragedy on Standley Lake: Famous Baby Eagle Dies After Its Nest Collapses," *Denver Post*, May 14, 2021.

2. Steve Hendrix and Dana Hedgpeth, "Eagle Makes Journey from Metro Tracks to U.S. Repository," *Washington Post*, March 17, 2012; Preston Cook, "National Eagle Repository Visit, September 22, 2010," personal notes provided to the author; Jennifer Nalewicki, "Inside a Remarkable Repository That Supplies Eagle Parts to Native Americans and Science," *Smithsonian Magazine*, September 6, 2016, https://www.smithsonianmag .com/science-nature/national-eagle-repository-eagles-go-to-native-american-tribes -and-science-$20960306.

3. Kristine Liao, "Meet the Woman Who Lived in a Bald Eagle's Nest to Save Raptors," *Audubon*, October 29, 2019, https://www.audubon.org/news/meet-woman-who -lived-bald-eagles-nest-save-raptors#:~:text=Conservation,Meet%20The%20Woman %20Who%20Lived%20in%20a%20Bald%20Eagle's%20Nest,in%20restoring %20Florida's%20eagle%20population; Doris Mager, interview by the author, August 17, 2020; Will Harris, interview by the author, March 7, 2018.

4. Patrick Bradley, interview by the author, January 8, 2021.

5. Bradley, interview, January 8, 2021.

6. Raptor Information Center, *Chesapeake Bald Eagle Banding Project, 1981 Report and Five-Year Summary* (National Wildlife Federation Raptor Information Center, 1981), 31.

7. Telia Hann, interview by the author, January 9, 2021.

8. Dava Guerin and Terry Bivens, *The Eagle on My Arm: How the Wilderness and Birds of Prey Saved a Veteran's Life* (University Press of Kentucky, 2020), 110.

9. "Transcript of President-Elect Joe Biden's Victory Speech," Associated Press, November 7, 2020, https://apnews.com/article/election-2020-joe-biden-religion-technology-race -and-ethnicity-2b961c70bc72c2516046bffd378e95de.

10. "Transcript of President-Elect Joe Biden's Victory Speech."

INDEX

MORE FROM
JACK E. DAVIS

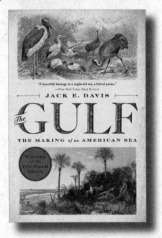

Winner of the Pulitzer Prize for History
Winner of the Kirkus Prize for Nonfiction
Finalist for the National Book Critics Circle Award for Nonfiction
A *New York Times* Notable Book of the Year
Named one of the Best Books of the Year by the *Washington Post*, NPR,
Library Journal*, and *gCaptain
A *Booklist* Editors' Choice Selection
Longlisted for the Andrew Carnegie Medal for Excellence

In this "cri de coeur about the Gulf's environmental ruin" (*New York Times*), "Davis has written a beautiful homage to a neglected sea" (*New York Times Book Review*, front page).

Hailed as a "nonfiction epic . . . in the tradition of Jared Diamond's bestseller *Collapse*, and Simon Winchester's *Atlantic*" (*Dallas Morning News*), Jack E. Davis's *The Gulf* is "by turns informative, lyrical, inspiring and chilling for anyone who cares about the future of 'America's Sea'" (*Wall Street Journal*). Illuminating America's political and economic relationship with the environment from the age of the conquistadors to the present, Davis demonstrates how the Gulf's fruitful ecosystems and exceptional beauty empowered a growing nation. Filled with vivid, untold stories from the sportfish that launched Gulfside vacationing to Hollywood's role in the country's first offshore oil wells, this "vast and well-told story shows how we made the Gulf . . . [into] a 'national sacrifice zone'" (Bill McKibben). The first and only study of its kind, *The Gulf* offers "a unique and illuminating history of the American Southern coast and sea as it should be written" (Edward O. Wilson).

Liveright Publishing Corporation

A Division of W. W. Norton & Company
Celebrating a Century of Independent Publishing